Hands-On Simulation Modeling with Python

Develop simulation models for improved efficiency and precision in the decision-making process

Giuseppe Ciaburro

BIRMINGHAM—MUMBAI

Hands-On Simulation Modeling with Python

Copyright © 2022 Packt Publishing

All rights reserved. No part of this book may be reproduced, stored in a retrieval system, or transmitted in any form or by any means, without the prior written permission of the publisher, except in the case of brief quotations embedded in critical articles or reviews.

Every effort has been made in the preparation of this book to ensure the accuracy of the information presented. However, the information contained in this book is sold without warranty, either express or implied. Neither the author nor Packt Publishing or its dealers and distributors, will be held liable for any damages caused or alleged to have been caused directly or indirectly by this book.

Packt Publishing has endeavored to provide trademark information about all of the companies and products mentioned in this book by the appropriate use of capitals. However, Packt Publishing cannot guarantee the accuracy of this information.

Publishing Product Manager: Ali Abidi
Content Development Editor: Joseph Sunil
Technical Editor: Rahul Limbachiya
Copy Editor: Safis Editing
Project Coordinator: Farheen Fathima
Proofreader: Safis Editing
Indexer: Tejal Soni
Production Designer: Prashant Ghare
Marketing Coordinator: Shifa Ansari

First published: July 2020

Second edition: November 2022

Production reference: 1251122

Published by Packt Publishing Ltd.
Livery Place
35 Livery Street
Birmingham
B3 2PB, UK.

ISBN 978-1-80461-688-8

www.packt.com

Dedicated to my family, to my sons Luigi and Simone, to my wife Tiziana and to the memory of my parents.

Contributors

About the author

Giuseppe Ciaburro holds a Ph.D. in environmental technical physics and two master's degrees in chemical engineering and acoustic and noise control. He works at the Built Environment Control Laboratory – Università degli Studi della Campania "Luigi Vanvitelli." He has over 20 years of work experience in programming, first in the field of combustion and then in acoustics and noise control. His core programming knowledge is in Python and R, and he has extensive experience working with MATLAB. An expert in acoustics and noise control, Giuseppe has wide experience in teaching professional ITC courses (about 20 years), dealing with e-learning as an author. He has several publications to his credit: monographs, scientific journals, and thematic conferences. He is currently researching machine learning applications in acoustics and noise control. He was recently included in the world's top 2% scientists list by Stanford University.

I would like to thank the Packt publishing staff who have been very helpful throughout the preparation of the book.

About the reviewer

Srikanth Sivaramakrishnan is a simulation engineer specializing in the automotive industry with more than a decade of experience in dynamic simulations, control systems, data analysis, statistical techniques, and machine learning for use in vehicle dynamics, tire modeling, and control systems. He has published and has been a technical reviewer for research papers in multiple automotive journals and for multiple conferences. He currently works as a lead motorsports simulation engineer for General Motors. He holds an MSc in mechanical engineering from Virginia Polytechnic Institute and State University, Blacksburg, and a bachelor's degree in technology from the National Institute of Technology, Tiruchirappalli, India.

Table of Contents

Preface · xv

Part 1: Getting Started with Numerical Simulation

1

Introducing Simulation Models · 3

Technical requirements	3
Introducing simulation models	4
Decision-making workflow	4
Comparing modeling and simulation	5
Pros and cons of simulation modeling	5
Simulation modeling terminology	6
Classifying simulation models	8
Comparing static and dynamic models	8
Comparing deterministic and stochastic models	8
Comparing continuous and discrete models	9
Approaching a simulation-based problem	9
Problem analysis	10
Data collection	10
Setting up the simulation model	10
Simulation software selection	11
Verification of the software solution	12
Validation of the simulation model	13
Simulation and analysis of results	13
Exploring Discrete Event Simulation (DES)	14
Finite-state machine (FSM)	16
State transition table (STT)	17
State transition graph (STG)	17
Dynamic systems modeling	17
Managing workshop machinery	17
Simple harmonic oscillator	19
The predator-prey model	21
How to run efficient simulations to analyze real-world systems	22
Summary	24

2

Understanding Randomness and Random Numbers 25

Technical requirements	25	Exploring generic methods for random distributions	51
Stochastic processes	26	The inverse transform sampling method	52
Types of stochastic processes	27	The acceptance-rejection method	53
Examples of stochastic processes	27		
The Bernoulli process	28	Random number generation using Python	54
Random walk	28	Introducing the random module	54
The Poisson process	30	Generating real-value distributions	58
Random number simulation	31	Randomness requirements for security	59
Probability distribution	31	Password-based authentication systems	59
Properties of random numbers	33	Random password generator	60
The pseudorandom number generator	34		
The pros and cons of a random number generator	34	Cryptographic random number generator	62
Random number generation algorithms	34	Introducing cryptography	62
Linear congruential generator	35	Randomness and cryptography	63
Random numbers with uniform distribution	37	Encrypted/decrypted message generator	64
Lagged Fibonacci generator	39		
Testing uniform distribution	42	Summary	68
Chi-squared test	43		
Uniformity test	46		

3

Probability and Data Generation Processes 69

Technical requirements	69	Exploring probability distributions	76
Explaining probability concepts	70	The probability density function	77
Types of events	70	Mean and variance	78
Calculating probability	71	Uniform distribution	79
Probability definition with an example	71	Binomial distribution	83
		Normal distribution	86
Understanding Bayes' theorem	73	Generating synthetic data	90
Compound probability	74	Real data versus artificial data	90
Bayes' theorem	75		

Synthetic data generation methods	91	Simulation of power analysis	98
Data generation with Keras	93	The power of a statistical test	98
Data augmentation	93	Power analysis	99
		Summary	103

Part 2: Simulation Modeling Algorithms and Techniques

4

Exploring Monte Carlo Simulations — 107

Technical requirements	108	Performing numerical integration using Monte Carlo	122
Introducing the Monte Carlo simulation	108	Defining the problem	122
Monte Carlo components	108	Numerical solution	124
First Monte Carlo application	109	Min-max detection	126
Monte Carlo applications	110	The Monte Carlo method	127
Applying the Monte Carlo method for Pi estimation	110	Visual representation	129
Understanding the central limit theorem	115	Exploring sensitivity analysis concepts	131
		Local and global approaches	132
Law of large numbers	115	Sensitivity analysis methods	133
The central limit theorem	116	Sensitivity analysis in action	133
Applying the Monte Carlo simulation	119	Explaining the cross-entropy method	136
Generating probability distributions	120	Introducing cross-entropy	137
Numerical optimization	120	Cross-entropy in Python	138
Project management	121	Binary cross-entropy as a loss function	140
		Summary	142

5

Simulation-Based Markov Decision Processes — 143

Technical requirements	144	Overview of Markov processes	146
Introducing agent-based models	144	The agent-environment interface	146

Exploring MDPs	148	Simulating a 1D random walk	157
Understanding the discounted cumulative reward	151	Simulating a weather forecast	160
Comparing exploration and exploitation concepts	152	**Bellman equation explained**	**165**
		Dynamic programming concepts	165
Introducing Markov chains	**153**	Principle of optimality	166
Transition matrix	154	Bellman equation	166
Transition diagram	154	**Multi-agent simulation**	**167**
Markov chain applications	**155**	**Schelling's model of segregation**	**168**
Introducing random walks	155	Python Schelling model	169
One-dimensional random walk	156	**Summary**	**174**

6

Resampling Methods 175

Technical requirements	**176**	Bootstrap resampling using Python	188
Introducing resampling methods	**176**	Comparing Jackknife and bootstrap	192
Sampling concepts overview	177	**Applying bootstrapping regression**	**192**
Reasoning about sampling	178	**Explaining permutation tests**	**201**
Pros and cons of sampling	178	**Performing a permutation test**	**202**
Probability sampling	179	**Approaching cross-validation techniques**	**207**
How sampling works	179		
Exploring the Jackknife technique	**179**	Validation set approach	208
Defining the Jackknife method	179	Leave-one-out cross-validation	208
Estimating the coefficient of variation	181	k-fold cross-validation	209
Applying Jackknife resampling using Python	182	Cross-validation using Python	210
Demystifying bootstrapping	**187**	**Summary**	**212**
Introducing bootstrapping	187		
Bootstrap definition problem	188		

7

Using Simulation to Improve and Optimize Systems — 213

Technical requirements — 214
Introducing numerical optimization techniques — 214
Defining an optimization problem — 214
Explaining local optimality — 216

Exploring the gradient descent technique — 217
Defining descent methods — 217
Approaching the gradient descent algorithm — 217
Understanding the learning rate — 220
Explaining the trial and error method — 220
Implementing gradient descent in Python — 221

Understanding the Newton-Raphson method — 225
Using the Newton-Raphson algorithm for root finding — 225
Approaching Newton-Raphson for numerical optimization — 226
Applying the Newton-Raphson technique — 227
The secant method — 231

Deepening our knowledge of stochastic gradient descent — 231
Approaching the EM algorithm — 232
EM algorithm for Gaussian mixture — 234

Understanding Simulated Annealing (SA) — 240
Iterative improvement algorithms — 240
SA in action — 241

Discovering multivariate optimization methods in Python — 247
The Nelder-Mead method — 248
Powell's conjugate direction algorithm — 251
Summarizing other optimization methodologies — 253

Summary — 253

8

Introducing Evolutionary Systems — 255

Technical requirements — 255
Introducing SC — 256
Fuzzy logic (FL) — 257
Artificial neural network (ANN) — 257
Evolutionary computation — 257

Understanding genetic programming — 258
Introducing the genetic algorithm (GA) — 258
The basics of GA — 259
Genetic operators — 260

Applying a GA for search and optimization — 263
Performing symbolic regression (SR) — 268
Exploring the CA model — 273
Game-of-life — 275
Wolfram code for CA — 276

Summary — 280

Part 3: Simulation Applications to Solve Real-World Problems

9

Using Simulation Models for Financial Engineering — 285

Technical requirements	285
Understanding the geometric Brownian motion model	286
Defining a standard Brownian motion	286
Addressing the Wiener process as random walk	287
Implementing a standard Brownian motion	288
Using Monte Carlo methods for stock price prediction	290
Exploring the Amazon stock price trend	290
Handling the stock price trend as a time series	295
Introducing the Black-Scholes model	297
Applying the Monte Carlo simulation	298
Studying risk models for portfolio management	302
Using variance as a risk measure	302
Introducing the Value-at-Risk metric	303
Estimating VaR for some NASDAQ assets	305
Summary	311

10

Simulating Physical Phenomena Using Neural Networks — 313

Technical requirements	314
Introducing the basics of neural networks	314
Understanding biological neural networks	314
Exploring ANNs	316
Understanding feedforward neural networks	321
Exploring neural network training	322
Simulating airfoil self-noise using ANNs	324
Importing data using pandas	325
Scaling the data using sklearn	328
Viewing the data using Matplotlib	330
Splitting the data	333
Explaining multiple linear regression	335
Understanding a multilayer perceptron regressor model	337
Approaching deep neural networks	340
Getting familiar with convolutional neural networks	341
Examining recurrent neural networks	342
Analyzing long short-term memory networks	343
Exploring GNNs	343
Introducing graph theory	343
Adjacency matrix	345
GNNs	345

Simulation modeling using neural network techniques	346	Summary	353
Concrete quality prediction model	346		

11

Modeling and Simulation for Project Management — 355

Technical requirements	355	Changing the probability of a fire starting	366
Introducing project management	356	Scheduling project time using the Monte Carlo simulation	369
Understanding what-if analysis	356		
Managing a tiny forest problem	357	Defining the scheduling grid	369
Summarizing the Markov decision process	357	Estimating the task's time	370
Exploring the optimization process	358	Developing an algorithm for project scheduling	371
Introducing MDPtoolbox	359	Exploring triangular distribution	372
Defining the tiny forest management example	360	Summary	377
Addressing management problems using MDPtoolbox	363		

12

Simulating Models for Fault Diagnosis in Dynamic Systems — 379

Technical requirements	379	Fault diagnosis model for a motor gearbox	385
Introducing fault diagnosis	380		
Understanding fault diagnosis methods	380	Fault diagnosis system for an unmanned aerial vehicle	396
The machine-learning-based approach	382		
		Summary	404

13

What's Next? — 405

Summarizing simulation modeling concepts	405	Addressing the Markov decision process	408
		Analyzing resampling methods	410
Generating random numbers	406	Exploring numerical optimization techniques	411
Applying Monte Carlo methods	407	Using artificial neural networks for simulation	412

Applying simulation models to real life — 413
Modeling in healthcare — 413
Modeling in financial applications — 414
Modeling physical phenomenon — 414
Modeling fault diagnosis system — 415
Modeling public transportation — 416

Modeling human behavior — 417

Next steps for simulation modeling — 417
Increasing the computational power — 418
Machine-learning-based models — 418
Automated generation of simulation models — 420

Summary — 420

Index — 423

Other Books You May Enjoy — 436

Preface

This book is a comprehensive guide to understanding various computational statistical simulations using Python.

This book starts with the foundation required to understand various methods and techniques before delving into complex topics. Developers working with simulation models will be able to put their knowledge to work with this practical guide. The book takes a hands-on approach to implementation and associated methodologies that will have you up and running and productive in no time.

Complete with step-by-step explanations of essential concepts, practical examples, and self-assessment questions, you will begin by exploring the numerical simulation algorithms, including an overview of relevant applications. You'll learn how to use Python to develop simulation models and understand how to use several Python packages. You will then explore various numerical simulation algorithms and concepts, such as a Markov Decision Process, Monte Carlo methods, and bootstrapping techniques.

By the end of this book, you will be able to construct simulation models.

Who this book is for

This book is for data scientists, simulation engineers, or anyone who is already familiar with the basic computational methods but now wants to implement various simulation techniques such as Monte-Carlo methods and statistical simulation using Python.

What this book covers

Chapter 1, Introducing Simulation Models, introduces simulation models, their various types, and examples of simulation model applications in the world.

Chapter 2, Understanding Randomness and Random Numbers, explores the stochastic process and various random number simulation concepts. You will learn how to distinguish between pseudo- and non-uniform numbers, and explore various methods for random distribution evaluation. Finally, you will explore some cryptography techniques.

Chapter 3, Probability and Data Generating Processes, introduces the concept of probability theory, and how to calculate the probability of a cause that triggers certain events. You will also learn how to work with discrete and continuous distributions, and finally, you will learn various techniques and tools for data generation.

Chapter 4, *Exploring Monte Carlo Simulations*, explores the Monte Carlo simulation and some of its applications. You will discover how to generate a sequence of numbers randomly distributed according to a Gaussian distribution. You will also learn how to implement the Monte Carlo method practically, and finally, you will learn about the basics of sensitivity analysis and cross-entropy.

Chapter 5, *Simulation-Based Markov Decision Processes*, will see you get to grips with the Markov process and understanding the whole interaction between agent and environment. You will learn to use Bellman equations. You will also discover what Markov chains are, learn how to use them, and simulate random walks using them.

Chapter 6, *Resampling Methods*, tells you how to obtain robust estimates of confidence intervals and standard errors of population parameters. You will learn how to estimate the distortion and standard error of a statistic, perform a test for statistical significance, and finally, validate a forecast model.

Chapter 7, *Using Simulation to Improve and Optimize Systems*, shows the basic concepts of optimization techniques, and how to implement them. You will understand the difference between numerical and stochastic optimization techniques. You'll learn how to implement the Stochastic Gradient Descent and estimate missing or latent variables and optimize model parameters. Finally, you will discover how to use optimization methods in real-life applications.

Chapter 8, *Introducing Evolutionary Systems*, explores the basic concepts of soft computing, genetic programming, and various evolutionary systems. You will learn how to apply genetic algorithms for search and optimization, and you will also explore a cellular automation model.

Chapter 9, *Using Simulation Models for Financial Engineering*, shows some practical use cases for using simulation methods in a financial context.

Chapter 10, *Simulating Physical Phenomena Using Neural Networks*, explores artificial neural networks and how to implement them, and also how neural network algorithms work. You will learn about deep neural networks. Finally, you will explore graph neural networks, and learn how to use neural networks to simulate physical phenomena, such as particles and point objects, and Lattice Boltzmann modeling of fluid flows.

Chapter 11, *Modeling and Simulation for Project Management*, covers various project management concepts and explores a few practical use cases of simulation modeling for project management.

Chapter 12, *Simulation Models for Fault Diagnosis in Dynamic Systems*, covers fault diagnosis in various systems and explores a few use cases of simulation modeling for fault detection and diagnosis in various systems such as UAVs.

Chapter 13, *What's Next?*, summarizes simulation modeling and explores the main applications in real life that make use of it. You will also learn about the future of simulation modeling.

To get the most out of this book

Software/hardware covered in the book	Operating system requirements
Python 3.x	Windows, macOS, or Linux

If you are using the digital version of this book, we advise you to type the code yourself or access the code from the book's GitHub repository (a link is available in the next section). Doing so will help you avoid any potential errors related to the copying and pasting of code.

Download the example code files

You can download the example code files for this book from GitHub at https://github.com/PacktPublishing/Hands-On-Simulation-Modeling-with-Python-Second-Edition. If there's an update to the code, it will be updated in the GitHub repository.

We also have other code bundles from our rich catalog of books and videos available at https://github.com/PacktPublishing/. Check them out!

Conventions used

There are a number of text conventions used throughout this book.

`Code in text`: Indicates code words in text, database table names, folder names, filenames, file extensions, pathnames, dummy URLs, user input, and Twitter handles. Here is an example: "The numpy library is a Python library containing numerous functions that can help us in the management of multidimensional matrices."

A block of code is set as follows:

```
np.random.seed(4)
n = 1000
sqn = 1/np.math.sqrt(n)
z_values = np.random.randn(n)
```

When we wish to draw your attention to a particular part of a code block, the relevant lines or items are set in bold:

```
AmznData = pd.read_csv('AMZN.csv',header=0,
            usecols = ['Date','Close'],parse_dates=True,
            index_col='Date')
```

Bold: Indicates a new term, an important word, or words that you see onscreen. For instance, words in menus or dialog boxes appear in **bold**. Here is an example: "Select **System info** from the **Administration** panel."

> **Tips or important notes**
> Appear like this.

Get in touch

Feedback from our readers is always welcome.

General feedback: If you have questions about any aspect of this book, email us at `customercare@packtpub.com` and mention the book title in the subject of your message.

Errata: Although we have taken every care to ensure the accuracy of our content, mistakes do happen. If you have found a mistake in this book, we would be grateful if you would report this to us. Please visit `www.packtpub.com/support/errata` and fill in the form.

Piracy: If you come across any illegal copies of our works in any form on the internet, we would be grateful if you would provide us with the location address or website name. Please contact us at `copyright@packt.com` with a link to the material.

If you are interested in becoming an author: If there is a topic that you have expertise in and you are interested in either writing or contributing to a book, please visit `authors.packtpub.com`.

Share Your Thoughts

Once you've read *Hands-On Simulation Modeling with Python, Second Edition*, we'd love to hear your thoughts! Scan the QR code below to go straight to the Amazon review page for this book and share your feedback.

`https://packt.link/r/1-804-61688-5`

Your review is important to us and the tech community and will help us make sure we're delivering excellent quality content.

Download a free PDF copy of this book

Thanks for purchasing this book!

Do you like to read on the go but are unable to carry your print books everywhere? Is your eBook purchase not compatible with the device of your choice?

Don't worry, now with every Packt book you get a DRM-free PDF version of that book at no cost.

Read anywhere, any place, on any device. Search, copy, and paste code from your favorite technical books directly into your application.

The perks don't stop there, you can get exclusive access to discounts, newsletters, and great free content in your inbox daily

Follow these simple steps to get the benefits:

1. Scan the QR code or visit the link below

```
https://packt.link/free-ebook/9781804616888
```

2. Submit your proof of purchase
3. That's it! We'll send your free PDF and other benefits to your email directly

Part 1: Getting Started with Numerical Simulation

In this section, the basic concepts of simulation modeling are addressed. This section will help you understand the fundamental concepts and elements of numerical simulation.

This part covers the following chapters:

- Chapter 1, *Introducing Simulation Models*
- Chapter 2, *Understanding Randomness and Random Numbers*
- Chapter 3, *Probability and Data-Generating Processes*

1
Introducing Simulation Models

A simulation model is a tool capable of processing information and data and predicting the responses of a real system to certain inputs, providing effective support for analysis, performance evaluation, and decision-making processes. The term **simulation** refers to reproducing the behavior of a system. In a simulation, we can use a real model, perhaps in scale, or a numerical model, which reproduces reality using a computer. An example of a real model is a scale model of an airplane, which is then placed in a wind tunnel to run simulated tests to estimate suitable performance measures.

Although, over the years, physicists have developed theoretical laws that we can use to obtain information on the performance of dynamic systems, often, the application of these laws to a real case takes too long. In these cases, it is convenient to construct a numerical simulation model that allows us to simulate the behavior of the system under certain conditions. This elaborated model will allow us to test the functionality of the system in a simple and immediate way, saving considerable resources in terms of time and money.

In this chapter, we're going to cover the following main topics:

- Introducing simulation models
- Classifying simulation models
- Approaching a simulation-based problem
- Exploring discrete event simulation
- Dynamical systems modeling
- How to run efficient simulations to analyze real-world systems

Technical requirements

In this chapter, an introduction to simulation techniques will be discussed. To deal with the topics at hand, it is necessary that you have a basic knowledge of algebra and mathematical modeling.

Introducing simulation models

Simulation uses digital models developed to replicate the characteristics of a system. The system functioning is simulated using probability distributions to randomly generate system events, and statistical observations are obtained from the simulated system. It plays a very important role, especially in the design of a stochastic system and in the definition of its operating procedures.

By not working directly on the real system, many scenarios can be simulated simply by changing the input parameters, thus limiting the costs and reducing time. In this way, it is possible to quickly try alternative policies and design choices and model systems of great complexity by studying their behavior and evolution over time.

> **Important note**
> Simulation is used when working on real systems is not convenient due to high costs, technical impossibility, and the non-existence of a real system. The simulation predicts what will happen to the real system if certain inputs are used: changing these input parameters reproduces different scenarios by identifying the most convenient one from various points of view.

Decision-making workflow

In a decision-making process, the starting point is identifying the problem that requires a change and, therefore, a decision. The identified problem is then analyzed to highlight what needs to be studied for the decisions to be made; that is, the relevant elements are chosen, the relationships that bind them are highlighted, and the objectives to be achieved are defined. At this point, a formal model is built, which allows the simulation of the system to understand its behavior and identify the decisions to be made. The following diagram depicts the workflow that allows us to make a decision, starting from the observation of the problem scenario:

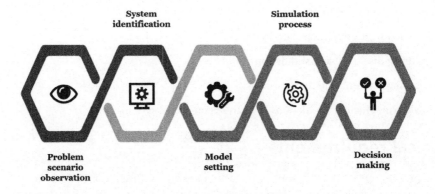

Figure 1.1: Decision-making workflow

The model building is a two-way process:

- Definition of conceptual models
- Continuous interaction between the model and reality by comparison

In addition, learning also has a participatory characteristic: it proceeds through the involvement of different actors. The models also allow you to analyze and propose organized actions so that you can modify the current situation and produce the desired solution.

Comparing modeling and simulation

To start, we will clarify the differences between modeling and simulation. A model is a representation of a physical system, while simulation is the process of seeing how a model-based system would work under certain conditions.

Modeling is a design methodology based on producing a model implementation for a system and representing its functionality. In this way, it is possible to predict the system's behavior and the effects of variations or modifications that are made to it. Even if the model is a simplified system representation, it must still be close enough to the functional nature of the real system, but without becoming too complex and difficult to handle.

> **Important note**
> Simulation is the process that puts the model into operation and allows you to evaluate its behavior under certain conditions. Simulation is a fundamental tool for modeling because, without necessarily resorting to physical prototyping, the developer can verify the functionality of the modeled system with the project specifications.

In the simulation, the system is studied by reproducing all possible operating conditions, even the impossible ones, without worrying about the costs involved in such experiments.

The simulation is the transposition in logical-mathematical-procedural terms of a conceptual model of the real system. This conceptual model can be defined as the set of processes that take place in the evaluated system and whose whole allows us to understand the operating logic of the system itself.

Pros and cons of simulation modeling

Simulation is a tool that's widely used in a variety of fields, from operational research to the application industry. This technique can be made successful by it overcoming the difficulties that each complex procedure contains. The following are the pros and cons of simulation modeling. Let's start with the concrete advantages that can be obtained from the use of simulation models (**pros**):

- It reproduces the behavior of a system in reference to situations that cannot be directly experienced

- It represents real systems, even complex ones, while also considering the sources of uncertainty
- It requires limited resources in terms of data
- It allows experimentation in a limited time
- The models that are obtained are easily interpretable

As anticipated, since it is a technique capable of reproducing complex scenarios, it has some limitations (**cons**):

- The simulation provides indications of the behavior of the system but not exact results
- The analysis of the output of a simulation could be complex, and it could be difficult to identify which may be the best configuration
- The implementation of a simulation model could be laborious and, moreover, it may take a long time to carry out a significant simulation
- The results that are returned by the simulation depend on the quality of the input data: it cannot provide accurate results in the case of inaccurate input data
- The complexity of the simulation model depends on the complexity of the system it intends to reproduce

Nevertheless, simulation models represent the best solution for the analysis of complex scenarios.

Simulation modeling terminology

In this section, we will analyze the elements that make up a model and those that characterize a simulation process. We will give a brief description of each so that you understand their meaning and the role they play in the numerical simulation process.

System

The context of an investigation is represented through a system, that is, the set of elements that interact with each other. The main problem linked to this element concerns the system boundaries, that is, which elements of reality must be inserted into the system that it represents and which are left out, and the relationships that exist between them.

State variables

A system is described in each instant of time by a set of variables. These are called state variables. For example, in the case of a weather system, the temperature is a state variable. In discrete systems, the variables change instantly at precise moments of time that are finite. In continuous systems, the variables vary in terms of continuity with respect to time.

Events

An event is defined as any instantaneous event that causes the value of at least one of the status variables to change. The arrival of a blizzard for a weather system is an event, as it causes the temperature to drop suddenly. There are both external events and internal events.

Parameters

Parameters represent essential terms when building a model. They are adjusted during the model simulation process to ensure that the results are brought into the necessary convergence margins. They can be modified iteratively through sensitivity analysis or in the model calibration phase.

Calibration

Calibration represents the process by which the parameters of the model are adjusted to adapt the results to the data observed in the best possible way. When calibrating the model, we try to obtain the best possible accuracy. A good calibration requires eliminating, or minimizing, errors in data collection and choosing a theoretical model that is the best possible description of reality. The choice of model parameters is decisive and must be done in such a way as to minimize the deviation of its results when applied to historical data.

Accuracy

Accuracy is the degree of correspondence of the simulation result that can be inferred from a series of calculated values with the actual data, that is, the difference between the average modeled value and the true or reference value. Accuracy, when calculated, provides a quantitative estimate of the quality expected from a forecast. Several indicators are available to measure accuracy. The most widely used ones are **mean absolute error** (**MAE**), **mean absolute percentage error** (**MAPE**), and **mean square error** (**MSE**).

Sensitivity

The sensitivity of a model indicates the degree to which the model's outputs are affected by changes in the selected input parameters. Sensitivity analysis identifies the sensitive parameters for the output of the model. It allows us to determine which parameters require further investigation so that we have a more realistic evaluation of the model's output values. Furthermore, it allows us to identify which parameters are not significant for the generation of a certain output and, therefore, can possibly be eliminated from the model. Finally, it tells us which parameters should be considered in a possible and subsequent analysis of the uncertainty of the output values provided by the model.

Validation

This is the process that verifies the accuracy of the proposed model. The model must be validated to be used as a tool to support decisions. It aims to verify whether the model that's being analyzed corresponds conceptually to our intentions. The validation of a model is based on the various techniques of multivariate analysis, which, from time to time, study the variability and interdependence of attributes within a class of objects.

Now that we have understood what simulation models are and the various pros, cons, and features they have, we will learn how to classify them in the next section.

Classifying simulation models

Simulation models can be classified according to different criteria. The first distinction is between static and dynamic systems. So, let's see what differentiates them.

Comparing static and dynamic models

A system is an object with a finite number of degrees of freedom that evolves over time according to a deterministic law. A system can be represented as a black box that can be stimulated by a stress (input) *x (t)* and that produces an effect (output) *y (t)*. The behavior of the system is fully described by the following equation:

$$y(t) = f(x(t))$$

Static models are the representation of a system in an instant of time or representative models of a system in which the time variable plays no role. An example of a static simulation is a Monte Carlo model.

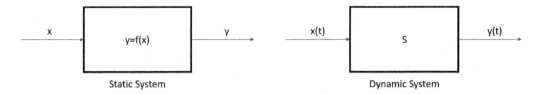

Figure 1.2: Static versus dynamic system representation

Dynamic models, on the other hand, describe the evolution of the system over time. In the simplest case, the state of the system at time t is described by a function x (t). For example, in population dynamics, x (t) represents the population present at time t. The equation that regulates the system is dynamic: it describes the instantaneous variation of the population or the variation in fixed time intervals.

Comparing deterministic and stochastic models

A model is deterministic when its evolution, over time, is uniquely determined by its initial conditions and characteristics. These models do not consider random elements and lend themselves to be solved with exact methods that are derived from mathematical analysis. In deterministic models, the output is well determined once the input data and the relationships that make up the model have been specified, despite the time required for data processing being particularly long. For these systems, the transformation rules univocally determine the change of state of the system. Examples of deterministic systems can be observed in some production and automation systems.

Stochastic models, on the other hand, can be evolved by inserting random elements into the evolution. These are obtained by extracting them from statistical distributions. Among the operational characteristics of these models, there is not just one relationship that fits all. There are also probability density functions, which means there is no one-to-one correspondence between the data and system history.

A final distinction is based on how the system evolves over time: therefore, we distinguish between continuous and discrete simulation models.

Comparing continuous and discrete models

Continuous models represent systems in which the state of the variables changes continuously as a function of time. For example, a car moving on a road represents a continuous system since the variables that identify it, such as position and speed, can change continuously with respect to time.

In discrete models, the system is described by an overlapping sequence of physical operations interspersed with inactivity pauses. These operations begin and end in well-defined instances (events). The system undergoes a change of state when each event occurs, remaining in the same state throughout the interval between the two subsequent events. This type of operation is easy to treat with the simulation approach.

> **Important note**
> The stochastic, deterministic, continuous, or discrete nature of a model is not its absolute property and depends on the observer's vision of the system itself. This is determined by the objectives and the method of study, as well as by the experience of the observer.

Now that we've analyzed the different types of models in detail, we will learn how to develop a numerical simulation model.

Approaching a simulation-based problem

To tackle a numerical simulation process that returns accurate results, it is crucial to rigorously follow a series of procedures that partly precede and partly follow the actual modeling of the system. We can separate the simulation process workflow into the following individual steps:

1. Problem analysis
2. Data collection
3. Setting up the simulation model

4. Simulation software selection
5. Verification of the software solution
6. Validation of the simulation model
7. Simulation and analysis of results

To fully understand the whole simulation process, it is essential to analyze in depth the various phases that characterize a study based on simulation.

Problem analysis

In this initial step, the goal is to understand the problem by trying to identify the aims of the study and the essential components, as well as the performance measures that interest them. Simulation is not simply an optimization technique, and therefore, there is no parameter that needs to be maximized or minimized. However, there is a series of performance indices whose dependence on the input variables must be verified. If an operational version of the system is already available, the work is simplified as it is enough to observe this system to deduce its fundamental characteristics.

Data collection

This represents a crucial step in the whole process since the quality of the simulation model depends on the quality of the input data. This step is closely related to the previous one. In fact, once the objective of the study has been identified, data is collected and subsequently processed. Processing the collected data is necessary to transform it into a format that can be used by the model. The origin of the data can be different: sometimes, the data is retrieved from company databases, but often, direct measurements in the field must be made through a series of sensors that, in recent years, have become increasingly smart. These operations weigh down the entire study process, thus lengthening their execution times.

Setting up the simulation model

This is the most crucial step of the whole simulation process; therefore, it is necessary to pay close attention to it. To set up a simulation model, it is necessary to know the probability distributions of the variables of interest. In fact, to generate various representative scenarios of how a system works, it is essential that a simulation generates random observations from these distributions.

For example, when managing stocks, the distribution of the product being requested and the distribution of time between an order and the receipt of the goods is necessary. On the other hand, when managing production systems with machines that can occasionally fail, it will be necessary to know the distribution of time until a machine fails and the distribution of repair times.

If the system is not already available, it is only possible to estimate these distributions by deriving them, for example, from the observation of similar, already existing systems. If, from an analysis of the data, it is seen that this form of distribution approximates a standard type distribution, the standard theoretical distribution can be used by carrying out a statistical test to verify whether the data can be well represented by that probability distribution. If there are no similar systems from which observable data can be obtained, other sources of information must be used, such as machine specifications, instruction manuals for the same, and experimental studies.

As we've already mentioned, constructing a simulation model is a complex procedure. Referring to simulating discrete events, constructing a model involves the following steps:

1. Defining the state variables
2. Identifying the values that can be taken by the state variables
3. Identifying the possible events that change the state of the system
4. Realizing a simulated time measurement, that is, a simulation clock, that records the flow of simulated time
5. Implementing a method for randomly generating events
6. Identifying the state transitions generated by events

After following these steps, we will have the simulation model ready for use. At this point, it will be necessary to implement this model in a dedicated software platform; let's see how.

Simulation software selection

The choice of the software platform that you will perform the numerical simulation on is fundamental for the success of the project. In this regard, we have several solutions that we can adopt. This choice will be made based on our knowledge of programming. Let's see what solutions are available:

- **Simulators**: These are application-oriented packages for simulation. There are numerous interactive software packages for simulation, including MATLAB, COMSOL Multiphysics, Ansys, SolidWorks, Simulink, Arena, AnyLogic, and SimScale. These pieces of software represent excellent simulation platforms whose performance differs based on the application solutions provided. These simulators allow us to elaborate on a simulation environment using graphic menus without the need to program.

    They are easy to use and many of them have excellent modeling capabilities, even if you just use their standard features. Some of them provide animations that show the simulation in action, which allows you to easily illustrate the simulation to non-experts. The limitations presented by this software solution are the high costs of the licenses, which can only be faced by large companies, and the difficulty in modeling solutions that have not been foreseen by the standards.

- **Simulation languages**: A more versatile solution is offered by the different simulation languages available. There are solutions that facilitate the task of the programmer who, with these languages, can develop entire models or sub-models with a few lines of code that would otherwise require much longer drafting times, with a consequent increase in the probability of error. An example of a simulation language is the **general-purpose simulation system** (**GPSS**). This is a generic programming language that was developed by IBM in 1965. In it, a simulation clock advances in discrete steps, modeling a system as transactions enter the system and are passed from one service to another. It is mainly used as a process flow-oriented simulation language and is particularly suitable for application problems. Another example of a simulation language is SimScript, which was developed in 1963 as an extension of Fortran. SimScript is an event-based scripting language, so different parts of the script are triggered by different events.

- **GPSS**: A general-purpose programming language is designed to be able to create software in numerous areas of application. They are particularly suitable for the development of system software such as drivers, kernels, and anything that communicates directly with the hardware of a computer. Since these languages are not specifically dedicated to a simulation activity, they require the programmer to work harder to implement all the mechanisms and data structures necessary in a simulator. On the other hand, by offering all the potential of a high-level programming language, they offer the programmer a more versatile programming environment. In this way, you can develop a numerical simulation model perfectly suited to the needs of the researcher. In this book, we will use this solution by devoting ourselves to programming with Python. This software platform offers a series of tools that have been created by researchers from all over the world that make the elaboration of a numerical modeling system particularly easy. In addition, the open source nature of the projects written in Python makes this solution particularly inexpensive.

Now that we've chosen the software platform we're going to use and have elaborated on the numerical model, we need to verify the software solution.

Verification of the software solution

In this phase, a check is carried out on the numerical code. This is known as debugging, which consists of ensuring that the code correctly follows the desired logical flow without unexpected blocks or interruptions. The verification must be provided in real time during the creation phase because correcting any concept or syntax errors becomes more difficult as the complexity of the model increases.

Although verification is simple in theory, debugging large-scale simulation code is a difficult task due to virtual competition. The correctness or otherwise of executions depends on time, as well as on the large number of potential logical paths. When developing a simulation model, you should divide the code into modules or subroutines in order to facilitate debugging. It is also advisable to have more than one person review the code, as a single programmer may not be a good critic. In addition, it can be helpful to perform the simulation when considering a large variety of input parameters and checking that the output is reasonable.

> **Important note**
> One of the best techniques that can be used to verify a discrete-event simulation program is one based on tracking. The status of the system, the content of the list of events, the simulated time, the status variables, and the statistical counters are shown after the occurrence of each event and then compared with hand-made calculations to check the operation of the code.

A track often produces a large volume of output that needs to be checked event by event for errors. Possible problems may arise, including the following:

- There may be information that hasn't been requested by the analyst
- Other useful information may be missing, or a certain type of error may not be detectable during a limited debugging run

After the verification process, it is necessary to validate the simulation model.

Validation of the simulation model

In this step, it is necessary to check whether the model that has been created provides valid results for the system in question. We must check whether the performance measurements of the real system are well approximated by the measurements generated by the simulation model. A simulation model of a complex system can only approximate it. A simulation model is always developed for a set of objectives. A model that's valid for one purpose may not be valid for another.

> **Important note**
> Validation checks that the level of accuracy between the model and the system is respected. It is necessary to establish whether the model adequately represents the behavior of the system. The value of a model can only be defined in relation to its use. Therefore, validation is a process that aims to determine whether a simulation model accurately represents the system for the set objectives.

In this step, the ability of the model to reproduce the real functionality of the system is ascertained; that is, it is ensured that the calibrated parameters, relative to the calibration scenario, can be used to correctly simulate other system situations. Once the validation phase is over, the model can be considered transferable and, therefore, usable for the simulation of any new control strategies and new intervention alternatives. As widely discussed in the literature on this subject, it is important to validate the model parameters that were previously calibrated based on data other than that used to calibrate the model, always with reference to the phenomenon specific to the scenario being analyzed.

Simulation and analysis of results

A simulation is a process that evolves during its realization, and where the initial results help lead the simulation toward more complex configurations. Attention should be paid to some details. For example, it is necessary to determine the transient length of the system before reaching stationary conditions if you want performance measures of the system at full capacity. It is also necessary to determine the duration of the simulation after the system has reached equilibrium. In fact, it must always be kept in mind that a simulation does not produce the exact values of the performance measures of a system since each simulation is a statistical experiment that generates statistical observations regarding the performance of the system. These observations are then used to produce estimates of performance measures. Increasing the duration of the simulation can increase the accuracy of these estimates.

The simulation results return statistical estimates of a system's performance measures. The fundamental point is that each measurement is accompanied by a confidence interval, within which it can vary. These results could immediately highlight a better system configuration than the others, but more often, more than one candidate configuration will be identified. In this case, further investigations may be needed to compare these configurations.

Now that we've analyzed a generalized simulation-based problem, we will learn how to approach a Discrete Event Simulation.

Exploring Discrete Event Simulation (DES)

A DES is a dynamic system whose states can assume logical or symbolic values rather than numerical ones. The behavior of the system is characterized by the occurrence of instantaneous events that arise with an irregular timing not necessarily known a priori. The behavior of such systems is described in terms of states and events.

A DES considers the system as a sequence of discrete events over time: each event occurs at a particular instant *t* and implements a change of state in the system. Therefore, between two consecutive events, it is assumed that there is no change of state, and therefore once the event is concluded at a certain instant *t1*, it is possible to go directly to the instant of the start of the next event *t2*.

In DES, the evolutions of the system over time are represented with variables that instantly change value in defined instants of time, included in a well-defined numerical interval, which determines the occurrence of events. The objects in a simulation system of this type are distinct elements, each having its own characteristics that determine its behavior within the model, which are called tokens. Discrete event simulation models are characterized by the following points:

- State variables that take on discrete values
- Transitions from one state to another that occur in discrete moments

To simulate a discrete event model, it is necessary to identify the state variables of the system and the classes of events that give rise to state transitions. Generally, in a discrete event simulation system, the state variables between one event and the next remain constant.

An example of a discrete event simulation is given by a queue of customers arriving at the checkout and needing to be served by one or more cashiers. In this case, the system entities are the customers in the queue and the cashiers. The events are only `customer-approach` and `customer-leaves` because we can include the `cashier-serves-customer` event in the logic of the other two. They implement a change in the states of the system; in this case, we can consider the following as states:

- The number of customers in the queue, represented by an integer from *0* to *n*

- The state of the cashier, represented by a Boolean variable that indicates whether it is free (F) or busy (B) (*Figure 1.3*)

The simulation also needs random variables: the first represents how often a new customer approaches, and the second is the time it takes the cashier to serve a customer.

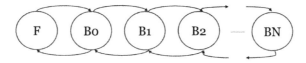

Figure 1.3: Transition diagram of cash desk queue

All events of a discrete event simulation system can be classified into the following categories:

- External events, independent of the behavior of the modeled system, are always present in the list of active events
- Internal events, which are a function of the system status, may or may not be present in the list of active events

The processes of arrival and departure of customers at a cash desk are completely characterized by probability distributions. It is essential to be able to evaluate, based on the data available, the probability distributions relating to the inter-arrival times of the tokens in the activities and the service of activities.

Another distinctive element of DES models is that the use of resources and activity times are specific to every single element of the system and are sampled using probability distributions. The rules governing the order in which the activities occur in the model are defined during the implementation phase of the model and are dependent on both how the workflow is structured and the characteristics of the individual entities.

The fundamental elements of a DES system are, therefore, the following:

- **State Variables**: Variables that describe the state of a system at any instant in time. In a DES model, the state variables take on discrete values, and the transitions from one state to another take place in discrete instants.
- **Events**: Any instantaneous occurrence that causes the value of at least one of the state variables to change.
- **Activity**: Temporal actions between two events, the duration of which is known a priori at the beginning of the execution of the activity.
- **Entities and logical relationships**: Entities are the tangible elements present in the real world, while logical relationships connect the different objects together, defining the general behavior

of the model. Entities can be dynamic if they flow within the system or static entities. Entities can also be characterized by attributes that provide a data value assigned to the entity.

- **Resources**: Elements of the system that provide a service to entities.
- **Simulation time**: Keeps track of the logical relationships between entities in the time range to be simulated.

There are two ways to define the progress of the simulation time:

- **Advancement of time to the next event**: Time flows according to the events; it passes from the previous moment to the next one only according to the events associated with these moments
- **Advancement of time in predetermined increments**: Time flows regardless of the events; it passes from one instant before to the next regardless of the events associated with those instants

Finite-state machine (FSM)

To describe a DES in a simple way, we can adapt the FSM model. It is the first abstraction of a machine equipped with memory that executes algorithms. It introduces the fundamental concept of the state, which can be defined as a particular condition of the machine. As a result, the machine reacts with a certain output to a specific input. Since the output also depends on the state, FSM is an automaton intrinsically equipped with an internal memory that can therefore influence the answers given by the automaton. A system of this nature can be formally described with a quintuple of variables:

$$FSM = f(S, I, O, \delta, s_0)$$

Here, we have the following variables:

- S is the finite set of possible states
- I is the possible set of input values
- O is the set of possible output values
- δ is the function that links inputs and the current state with the neighboring state
- s_0 is the initial state

The state contains the information necessary to calculate the system output, and in general, the inference of the sequence of inputs that brought the system to a certain state is not possible.

A system in which the output depends only on the current state is called Moore's Machine. If the output depends on both the current state and the inputs, then it is called the Mealy Machine. In Moore's machine, the outputs are connected to the current state of the system, whereas in Mealy's machine, they are connected to the transitions from one state to another. The two automata are equivalent, that is, a system that can be created with one of the two models can always be created with the other model as well. In general, Mealy automata have fewer states than the corresponding Moore automata, even if they are more complicated to synthesize.

State transition table (STT)

The state of a state machine evolves in the presence of an event on the input or on the state itself. A sequential machine can be described by means of the state table. The column indices are the symbols of the input values, while the row indices are the state symbols that indicate, in the case of a Moore automaton, each possible current state, and for each combination of inputs, the next state reached by the automaton is indicated. Finally, a further column of the table expresses the relationship between the system status and the corresponding output.

State transition graph (STG)

The STG allows you to completely describe a finite state automaton through an oriented and labeled graph. It is a very useful graphic representation in the initial phases of the project, in which we pass from a formal description of the machine to its behavioral model. A node of the graph corresponds to each state. If a transition from state A to state B is possible in relation to input i, then on the graph, there will exist an arc oriented from the node corresponding to A to the node corresponding to B, labeled with i (*Figure 1.2*). In the case of Moore automata, in which the output is a function of the current state only, the indication of the output u corresponding to a state A is shown inside the node that identifies state A.

After exploring DES, we will now look at how to model dynamic systems.

Dynamic systems modeling

In this section, we will analyze a real case of modeling a production process. In this way, we will learn how to deal with the elements of the system and how to translate the production instances into the elements of the model. A model is created to study the behavior of a system over time. It consists of a set of assumptions about the behavior of the system being expressed using mathematical-logical-symbolic relationships. These relationships are between the entities that make up the system. Recall that a model is used to simulate changes in the system and predict the effects of these changes on the real system. Simple models are resolved analytically using mathematical methods, while complex models are numerically simulated on the computer, where the model results are treated like data from a real system.

Managing workshop machinery

In this section, we will look at a simple example of how a discrete event simulation of a dynamic system is created. A discrete event system is a dynamic system whose states can assume logical or symbolic values, rather than numerical ones, and whose behavior is characterized by the occurrence of instantaneous events within an irregular timing sequence, not necessarily known a priori. The behavior of these systems is described in terms of states and events.

In a workshop, there are two machines that we will call A_1 and A_2. At the beginning of the day, five jobs need to be carried out: W_1, W_2, W_3, W_4, and W_5. The following table shows the time (minutes) we need to work on the machines:

	A1	A2
W_1	30	50
W_2	0	40
W_3	20	70
W_4	30	0
W_5	50	20

Table 1.1: Table showing work time on the machines

A zero indicates that a job does not require that machine. Jobs that require two machines must pass through A_1 and then through A_2. Suppose that we decide to carry out the jobs by assigning them to each machine so that when they become available, the first executable job is started first, in order from 1 to 5. If at the same time, more jobs can be executed on the same machine, we will execute the one with a minor index first.

The purpose of modeling is to determine the minimum time needed to complete all the work. The events in which state changes can occur in the system are as follows:

- A job becomes available for a machine
- A machine starts a job
- A machine ends a job

Based on these rules and the evaluation times indicated in the previous table, we can insert the sequence of the jobs, along with the events scheduled according to the execution times, into a table:

Time (minutes)	A1		A2	
	Start	End	Start	End
0	W1		W2	
30	W3	W1		
40			W1	W2
50	W4	W3		
80	W5	W4		
90			W3	W1
130		W5		
160			W5	W3
180				W5

Table 1.2: Table of job sequences

This table shows the times of the events in sequence, indicating the start and end of the work associated with the two machines available in the workshop. At the end of each job, a new job is sent to each machine according to the rules set previously. In this way, the deadlines for the work and the subsequent start of another job are clearly indicated. This is just as easy as it is to identify the time interval in which each machine is used and when it becomes available again. The table solution we have proposed represents a simple and immediate way of simulating a simple dynamic discrete system.

The example we just discussed is a typical case of a dynamic system in which time proceeds in steps in a discrete way. However, many dynamic systems are best described by assuming that time passes continuously. In the next section, we will analyze the case of a continuous dynamic system.

Simple harmonic oscillator

Consider a mass m, resting on a horizontal plane, without friction, and attached to a wall by an ideal spring, of elastic constant k. Suppose that when the horizontal coordinate x is zero, the spring is at rest. The following diagram shows the scheme of a simple harmonic oscillator:

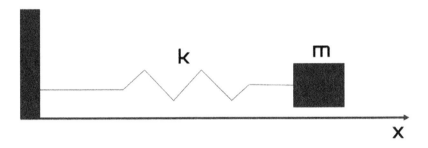

Figure 1.4: The scheme of a harmonic oscillator

If the block of mass m is moved to the right with respect to its equilibrium position ($x > 0$), the spring being elongated pulls it to the left. Conversely, if the block is placed to the left of its equilibrium position ($x < 0$), then the spring is compressed and pushes the block to the right. In both cases, we can express the component along the x axis of the force due to the spring according to the following formula:

$F_x = -k * x$

Here, we have the following:

- F_x is the force
- k is the elastic constant
- x is the horizontal coordinate that indicates the position of the mass m

From the second law of dynamics, we can derive the component of acceleration along *x* as follows:

$$a_x = -\frac{k}{m} * x$$

Here, we have the following:

- a_x is the acceleration
- k is the elastic constant
- m is the mass of the block
- x is the horizontal coordinate that indicates the position of the mass m

If we indicate with $\frac{dv}{dt} = a_x$ the rate of change of the speed and with $\frac{dx}{dt} = v$ the speed, we can obtain the evolution equations of the dynamic system, as follows:

$$\begin{cases} \dfrac{dx}{dt} = v \\ \dfrac{dv}{dt} = -\omega^2 * x \end{cases}$$

Here, we have the following:

$$\omega^2 = -\frac{k}{m}$$

For these equations, we must associate the initial conditions of the system, which we can write in the following way:

$$\begin{cases} x(0) = x_0 \\ v(0) = v_0 \end{cases}$$

The solutions to the previous differential equations are as follows:

$$\begin{cases} x(t) = x_0 \cos(\omega t) + \dfrac{v_0}{\omega} \sin(\omega t) \\ v(t) = v_0 \cos(\omega t) - x_0 \omega \sin(\omega t) \end{cases}$$

In this way, we obtained the mathematical model of the analyzed system. To study the evolution of the oscillation phenomenon of the mass block m over time, it is enough to vary the time and calculate the position of the mass at that instant and its speed.

In decision-making processes characterized by high levels of complexity, the use of analytical models is not possible. In these cases, it is necessary to resort to models that differ from those of an analytical type for the use of the calculator as a tool not only for calculation, such as in mathematical programming models but also for representing the elements that make up reality while studying the relationships between them.

The predator-prey model

In the field of simulations, simulating the functionality of production and logistic processes is considerably important. These systems are, in fact, characterized by high complexity, numerous interrelationships between the different processes that pass through them, segment failures, unavailability, and the stochasticity of the system parameters.

To understand how complex the analytical modeling of some phenomena is, let's analyze a universally widespread biological model. This is the predator-prey model, which was developed independently by the Italian researcher Vito Volterra and the American biophysicist Alfred Lotka.

On an island, there are two populations of animals: prey and predators. The vegetation of the island provides the prey with nourishment in quantities that we can consider as unlimited, while the prey is the only food available for predators. We can consider the birth rate of the prey constant over time; this means that in the absence of predators, the prey would grow by exponential law. Their mortality rate, on the other hand, depends on the probability they have of falling prey to a predator and, therefore, on the number of predators present per unit area.

As for predators, the mortality rate is constant, while their growth rate depends on the availability of food and, therefore, on the number of prey per unit area present on the island. We want to study the trend of the size of the two populations over time, starting from a known initial situation (number of prey and predators).

To carry out a simulation of this biological system, we can model it by means of the following system of finite difference equations, where $x(t)$ and $y(t)$ are the numbers of prey and predators at time t, respectively:

$$\begin{cases} \dfrac{dx}{dt} = \alpha * x(t) - \beta * x(t) * y(t) \\ \dfrac{dy}{dt} = \gamma * y(t) - \delta * x(t) * y(t) \end{cases}$$

Here, we have the following:

- α, β, γ, δ are positive real parameters related to the interaction of the two species
- $\frac{dx}{dt}$ is the instantaneous growth rate of the prey
- $\frac{dy}{dt}$ is the instantaneous growth rate of the predators

The following hypotheses are underlined in this model:

- In the absence of predators, the number of prey increases according to an exponential law, that is, at a constant rate
- Similarly, in the absence of prey, the number of predators decreases at a constant rate

This is a deterministic and continuous simulation model. In fact, the state of the system, represented by the size of the two populations in each instance of time, is univocally determined by the initial state and the parameters of the model. Furthermore, at least in principle, the variables – that is, the size of the populations – vary continuously over time.

The differential system is in normal form and can be solved with respect to the derivatives of maximum order but cannot be separated into variables. It is not possible to solve this in an analytical form, but with numerical methods, it is solved immediately (Runge-Kutta method). The solution obviously depends on the values of the four constants and the initial values.

After analyzing how to model dynamic systems, let's see step by step how to perform a simulation of a real system.

How to run efficient simulations to analyze real-world systems

To perform a correct simulation, the simulation software must be properly structured. The program structure must consist of three software levels:

- The simulation executive software controls the execution of the model program. It sequences operations and manages modularity problems by separating generic control from model details. Two techniques are possible: the synchronous execution technique and the event-scanning technique.
- The model program implements the model of the system to be simulated and is executed by the simulator.
- The routine tools generate random numbers by deriving them from typical probability distributions to obtain statistics and to manage problems related to modularity by separating generic functions from those specific to the model.

The main cycle of a simulation generally consists of three phases:

- **Start phase**: The termination condition is initialized to false; the status variables are initialized, the clock is set to the simulation start time (usually zero), and the initial events are placed in the event list.
- **Loop phase**: The cycle continues until the termination condition occurs. At each step, the clock is set equal to the start time of the first event in the queue, the event is simulated and removed from the queue, and finally, the statistics are updated.

- **End phase**: The simulation ends by generating a report with the statistics of the simulated system.

The simulation can be applied to all systems that can be modeled, respecting the characteristics mentioned in the previous paragraphs. This approach is particularly suitable for diagnosing problems related to complex processes. Understand where the system bottlenecks are. These areas are critical as they significantly decrease system performance and are usually associated with slower components. The only way to significantly improve the performance of a process is to improve it precisely in these critical areas, which, unfortunately, in many processes, are clouded by excess inventory, overproduction, diversity of processes, different sequencing, and stratification.

Therefore, by modeling these processes accurately and inserting them into a simulator, it is possible to obtain a more detailed view of the entire system, find the bottlenecks, and use performance indicators to analyze the system and improve its performance.

A simulation is a numerical model that mimics the operations of a real existing system, such as the daily operations of a bank, the process of a factory assembly line, and the assignment of personnel in a hospital or call center. For this reason, it must consider all the resources and constraints related to the system and their interactions over time. It must also coincide as much as possible with reality and therefore consider that events can have variable completion times, and this is done by introducing pseudorandom generators. In this way, the simulation is much closer to the real situation, and therefore through it, it is possible to predict the behavior of the system in the face of changes in the input data or its structure, just as if the simulation were real. Therefore, with this type of simulation, you can test your ideas much faster and at a much lower cost than real simulation.

Generally, any real system that is expressible through a flow of processes with events can be simulated. The processes that can be most beneficial are those with more changes over time and with a high level of randomness. A good example is a vehicle filling station: you cannot predict exactly when the next customer will arrive, nor the type and quantity of fuel they will require.

Consequently, the simulation, where applicable, should be preferred to the real simulation since a gain in terms of cost, time, and repeatability of the simulation is obtained. Just think of the cost associated with each real simulation, not only as regards the expense of hiring new employees or purchasing new components, but also the consequences that these decisions can have on the real system, which can lead to positive but also negative results. On the other hand, this is avoidable in numerical simulation in which the system has no connections with the real system, and therefore the test does not cause economic consequences.

Furthermore, it is more difficult to simulate the same real system twice with the same conditions, while in numerical simulation, you can test a system with the same conditions by modifying only the input. This gives greater certainty that one idea is better than another. Finally, as mentioned before, the time needed to carry out a numerical simulation is much shorter than that of a real simulation, especially if the simulated system takes a very long time. For example, if we want to analyze the throughput of customers in a month, the real simulation must pass at least 1 month, and the numerical one just a few seconds.

Summary

In this chapter, we learned what is meant by simulation modeling. We understood the difference between modeling and simulation, and we discovered the strengths of simulation models, such as defects. To understand these concepts, we clarified the meaning of the terms that appear most frequently when dealing with these topics.

We then analyzed the different types of models: static versus dynamic, deterministic versus stochastic, and continuous versus discrete. We then explored the workflow connected to a numerical simulation process and highlighted the crucial steps. Furthermore, we analyzed in detail the discrete event systems that simulate a dynamic system whose states can take on logical or symbolic, rather than numerical, values. Finally, we studied some practical modeling cases to understand how to elaborate on a model starting from the initial considerations.

In the next chapter, we will learn how to approach a stochastic process and understand the random number simulation concepts. Then, we will explore the differences between pseudo and non-uniform random numbers, as well as the methods we can use for random distribution evaluation

2
Understanding Randomness and Random Numbers

Frequently, in real life, it is useful to flip a coin in order to decide what to do. Many computers also use this procedure as part of their decision-making process. In fact, many problems can be solved in a very effective and relatively simple way by using probabilistic algorithms. In an algorithm of this type, decisions are made based on random contributions that remember the dice roll with the help of a randomly chosen value.

The generation of random numbers has ancient roots, but only recently has the process been speeded up, allowing it to be used on a large scale in scientific research as well. These generators are mainly used for computer simulations, statistical sampling techniques, or in the field of cryptography.

In this chapter, we're going to cover the following topics:

- Stochastic processes
- Random number simulation
- The **pseudorandom number generator**
- Testing uniform distribution
- Exploring generic methods for random distributions
- Random number generation using Python
- Randomness requirements for security
- Cryptographic random number generator

Technical requirements

In this chapter, we will introduce random number generation techniques. In order to understand these topics, a basic knowledge of algebra and mathematical modeling is needed.

To work with the Python code in this chapter, you need the following files (available on GitHub at https://github.com/PacktPublishing/Hands-On-Simulation-Modeling-with-Python-Second-Edition):

- `linear_congruential_generator.py`
- `learmouth_lewis_generator.py`
- `lagged_fibonacci_algorithm.py`
- `uniformity_test.py`
- `random_generation.py`
- `random_password_generator.py`
- `encript_decript_strings.py`

Stochastic processes

A **stochastic process** is a family of random variables that depends on a parameter, t. A stochastic process is specified using the following notation:

$$\{X_t, t \in T\}$$

Here, t is a parameter, and T is the set of possible values of t.

Usually, time is indicated by t, so a stochastic process is a family of time-dependent random variables. The variability range of t—that is, the set, T—can be a set of real numbers, possibly coinciding with the entire time axis, but it can also be a discrete set of values.

The random variables, X_t, are defined on the set, X, called the space of states. This can be a continuous set, in which case it is defined as a continuous stochastic process, or a discrete set, in which case it is defined as a discrete stochastic process.

Consider the following elements:

$$x_0, x_1, x_2, \ldots, x_n \in X$$

This means the values that the random variables, X_t, can take are called system states and represent the possible results of an experiment. The X_t variables are linked together by dependency relationships. We can know a random variable if we know both the values it can assume and the probability distribution. So, to understand a stochastic process, it is necessary not only to know the values that X_t can take but also the probability distributions of the variables and the joint distributions between the values. Simpler stochastic processes, in which the variability range of t is a discrete set of time values, can also be considered.

> **Important note**
>
> In practice, numerous phenomena are studied through the theory of stochastic processes. A classic application in physics is the study of the motion of a particle in each medium, the so-called **Brownian motion**. This study is carried out statistically using a stochastic process. There are processes where even by knowing the past and the present, the future cannot be determined, whereas, in other processes, the future is determined by the present without considering the past.

Types of stochastic processes

Stochastic processes can be classified according to the following characteristics:

- Space of states
- Time index
- Type of stochastic dependence between random variables

The state space can be **discrete** or **continuous**. In the first case, a stochastic process with discrete space is also called a **chain**, and space is often referred to as the set of non-negative integers. In the second case, the set of values assumed by the random variables is not finite or countable, and the stochastic process is in continuous space.

The time index can also be discrete or continuous. A discrete-time stochastic process is also called a **stochastic sequence** and is denoted as follows:

$$\{X_n \mid n \in T\}$$

Here, the set, T, is finite or countable.

In this case, the changes of state are observed only in certain instances: finite or countable. If state changes occur at any instant at a finite or infinite set of real intervals, then there is a continuous-time process, which is denoted as follows:

$$\{X(t) \mid t \in T\}$$

The stochastic dependence between random variables, $X(t)$, for different values of t characterizes a stochastic process and sometimes simplifies its description. A stochastic process is stationary in the strict sense that the distribution function is invariant with respect to a shift on the time axis, T. A stochastic process is stationary in the broad sense that the first two moments of the distribution are independent of the position on the T axis.

Examples of stochastic processes

The mathematical treatment of stochastic processes seems complex, yet we find cases of stochastic processes every day. For example, the number of patients admitted to a hospital as a function of time, observed at noon each day, is a stochastic process in which the space of states is discrete, being a finite subset of natural numbers, and time is discrete. Another example of a stochastic process is the temperature measured in a room as a function of time, observed at every instant, with continuous state space and continuous time. Let's now look at a number of structured examples that are based on stochastic processes.

The Bernoulli process

The concept of a random variable allows us to formulate models that are useful for the study of many random phenomena. An important early example of a probabilistic model is the **Bernoulli distribution**, named in honor of the Swiss mathematician, James Bernoulli (1654-1705), who made important contributions to the field of probability.

Some of these experiments consist of repeatedly performing a given test. For example, we want to know the probability of a head when throwing a coin 1,000 times.

In each of these examples, we look for the probability of obtaining x successes in n trials. If x indicates the successes, then $n - x$ will be the failures.

Bernoulli trials fit into the following hypotheses:

- There are only two possible mutually exclusive results for each trial, arbitrarily called **success** and **failure**
- The probability of success, p, is the same for each trial
- All tests are independent

Independence means that the result of a test is not influenced by the result of any other test. For example, the "the third test was successful" event is independent of the "the first test was successful" event.

Coin flipping is a Bernoulli trial: the "heads" event can be considered successful, and the "tails" event can be considered unsuccessful. In this case, the probability of success is $p = 1/2$.

> **Important note**
> Two events are said to be complementary when the occurrence of the first excludes the occurrence of the second but one of the two will certainly occur.

Let p denote the probability of success in a Bernoulli trial. The random variable, X, which counts the number of successes in n trials is called the binomial random variable of the n and p parameters. X can take integer values between 0 and n.

Random walk

The **random walk** is a discrete parameter stochastic process in which X_t, where X represents a random variable, describes the position taken at time t by a moving point. The term *random walk* refers to the mathematical formalization of statistics that describe the displacement of an object that moves randomly. This kind of simulation is extremely important for a physicist and has applications in statistical mechanics, fluid dynamics, and quantum mechanics.

Random walks represent a mathematical model that is used universally to simulate a path formalized by a succession of random steps. This model can assume a variable number of degrees of freedom, depending on the system we want to describe. From a physical point of view, the path traced over time will not necessarily simulate a real motion, but it will represent the trend of the characteristics of the system over time. Random walks find applications in chemistry, biology, and physics, but also other fields such as economics, sociology, and information technology.

Random one-dimensional walking is a model that is used to simulate the movement of a particle moving along a straight line. There are only two potential movements on the allowed path: either to the right (with a probability that is equal to **p**) or to the left (with a probability that is equal to **q**) of the current position. Each step has a constant length and is independent of the others, as shown in the following diagram:

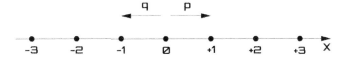

Figure 2.1: One-dimensional walking

The position of the point in each instant is identified by its abscissa, $X(n)$. This position, after n steps, will be characterized by a random term. Our aim is to calculate the probability of passing from the starting point after n movements. Obviously, nothing assures us that the point will return to the starting position. The variable, $X(n)$, returns the abscissa of the particle after n steps. It is a discrete random variable with a binomial distribution.

At each instance, the particle steps right or left based on the value returned by a random variable, $Z(n)$. This variable can take only two values: +1 and -1. It assumes a +1 value with a probability of p > 0 and a value of -1 with a probability that is equal to q. The sum of the two probabilities is p + q = 1. The position of the particle at instant n is given by the following equation:

$$X_n = X_{n-1} + Z_n \;;\; n = 1, 2, \ldots$$

This shows the average number of returns to the origin of the particle, named p. The probability of a single return is given by the following geometric series:

$$\mu = \sum_{n=0}^{\infty} n\, p^n (1-p) = \sum_{n=0}^{\infty} n\, p^n \frac{1}{\sqrt{n * \pi}} \to \infty$$

We assume that the probability of the particle returning to the origin tends to 1. This means that despite the frequency of the returns decreasing with the increase in the number of steps taken, they will always be an infinite value of steps taken. So, we can conclude that a particle with equal probability of left and right movement, which is left free to walk casually to infinity with great probability, returns infinite times to the point from which it started.

The Poisson process

There are phenomena in which certain events, with reference to a certain interval of time or space, rarely happen. The number of events that occur in that interval varies from 0 to n, and n cannot be determined a priori. For example, a number of cars passing through an uncrowded street in a randomly chosen 5-minute time frame can be considered a rare event. Similarly, the number of accidents at work that happen at a company in a week, or the number of printing errors on a page of a book, should be few and far between.

In the study of rare events, a reference to a specific interval of time or space is fundamental. For the study of rare events, the Poisson probability distribution is used, named in honor of the French mathematician, Siméon Denis Poisson (1781-1840), who first obtained the distribution. The Poisson distribution is used as a model in cases where the events or realizations of a process, distributed randomly in space or time, are counts—that is, discrete variables.

The binomial distribution is based on a set of hypotheses that define the Bernoulli trials; the same happens for the Poisson distribution. The binomial distribution originates from the repeated observation (n times) of a Bernoulli test, characterized by two outcomes that we will call "success" and "failure" with probability p and (1-p) respectively. The probability of success p does not change with each subsequent observation, which is therefore defined as independent; the count of successes in n sequences of observations determines the binomial random variable.

The following conditions describe the so-called Poisson process:

- The realizations of the events are independent, meaning that the occurrence of an event in a time or space interval has no effect on the probability of the event occurring a second time in the same, or another, interval
- The probability of a single realization of the event in each interval is proportional to the length of the interval
- In any arbitrarily small part of the interval, the probability of the event occurring more than once is negligible

An important difference between the Poisson distribution and the binomial distribution is the number of trials and successes. In a binomial distribution, the number, n, of trials is finite and the number, x, of successes cannot exceed n; in a Poisson distribution, the number of tests is essentially infinite and the number of successes can be infinitely large, even if the probability of having x successes becomes very small as x increases.

After analyzing the stochastic processes, we can see how to generate random numbers.

Random number simulation

The availability of random numbers is a necessary requirement in many applications. In some cases, the quality of the final application strictly depends on the possibility of generating good-quality random numbers. Think, for example, of applications such as video games, cryptography, generating visuals or sound effects, telecommunications, signal processing, optimizations, and simulations. In an algorithm of this type, decisions are made based on the pull of a virtual currency, which is based on a randomly chosen value.

There is no single or general definition of a random number since it often depends on the context. The concept of a random number itself is not absolute, as any number or sequence of numbers can appear to be random to an observer, but not to another who knows the law with which they are generated. Put simply, a random number is defined as a number selected in a random process from a finite set of numbers. With this definition, we focus on the concept of randomness in the process of selecting a sequence of numbers.

In many cases, the problem of generating random numbers concerns the random generation of a sequence of 0 and 1, from which numbers in any format can be obtained: integers, fixed points, floating points, or strings of arbitrary length. With the right functions, it is possible to obtain good-quality sequences that can also be used in scientific applications, such as a **Monte Carlo** simulation. These techniques should be easy to implement and be usable by any computer. In addition, as with all software solutions, they should be very versatile and quickly improved.

> **Important note**
> These techniques have a big problem that is inherent to the algorithmic nature of the process: the final string can be predicted from the starting seed. This is why we call this process **pseudorandom**.

Despite this, many problems of an algorithmic nature are solved very effectively and relatively simply using probabilistic algorithms. The simplest example of a probabilistic algorithm is perhaps the randomized quicksort. This is a probabilistic variant of the homonymous sorting algorithm, where, by choosing the pivot element, the algorithm manages to randomly guarantee optimal complexity in the average case, no matter the distribution of the input. Cryptography is a field in which randomness plays a fundamental role and deserves specific mention. In this context, randomness does not lead to computational advantages, but it is essential to guarantee the security of authentication protocols and encryption algorithms.

Probability distribution

It is possible to characterize a random process from different points of view. One of the most important characteristics is the **probability distribution**. The probability distribution is a model that associates a probability with each observable modality of a random variable.

The probability distribution can be either discrete or continuous, depending on whether the variable is random, discrete, or continuous. It is discrete if the phenomenon is observable with an integer number of modes. The throw of the dice is a discrete statistical phenomenon because the number of observable modalities is equal to 6. The random variable can take only six values (1, 2, 3, 4, 5, and 6). Therefore, the probability distribution of the phenomenon is discrete. The probability distribution is continuous when the random variable assumes a continuous set of values; in this case, the statistical phenomenon can be observed with an infinite or very high number of modalities. The probability distribution of body temperature is continuous because it is a continuous statistical phenomenon—that is, the values of the random variable vary continuously.

Let's now look at different kinds of probability distributions.

Uniform distribution

In many cases, processes characterized by a **uniform distribution** are considered and used. This means that each element is as likely as any of the others to be selected if an infinite number of extractions is performed. If you represent the elements and their respective probabilities of being extracted on a graph, you get a rectangular graph, as shown in the following diagram:

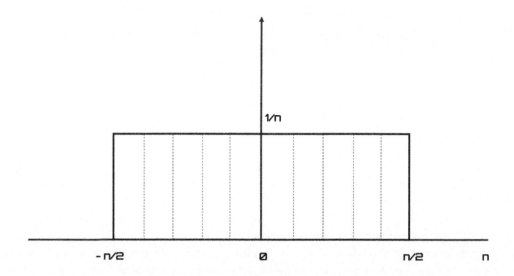

Figure 2.2: The probabilities of the elements

Since the probability is expressed as a real number between 0 and 1, where 0 represents the impossible event and 1 is the certain event, in a uniform distribution, each element will have a 1/n probability of being selected, where n is the number of items. In this case, the sum of all the probabilities must give a uniform result since, in an extraction, at least one of the elements is chosen for sure. A uniform distribution is typical of artificial random processes such as dice rolling, lotteries, and roulette and is also the most used in several applications.

Gaussian distribution

Another very common probability distribution is the **Gaussian** or "normal" distribution, which has a typical bell shape. In this case, the smaller values, or those that are closer to the center of the curve, are more likely to be extracted than the larger ones, which are far away from the center. The following diagram shows a typical Gaussian distribution:

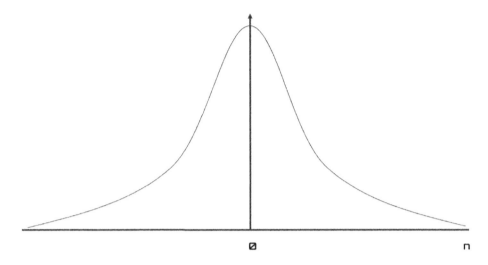

Figure 2.3: Gaussian distribution

Gaussian distribution is important because it is typical of natural processes. For example, it can represent the distribution of noise in many electronic components, or it can represent the distribution of errors in measurements. It is, therefore, used to simulate statistical distributions in the fields of telecommunications or signal processing.

Properties of random numbers

By random number, we refer to a random variable distributed in a uniform way between 0 and 1. The statistical properties that a sequence of random numbers must possess are as follows:

- Uniformity
- Independence

Suppose you divide an interval, [0.1], into n subintervals of equal amplitude. The consequence of the uniformity property is that if N observations of a random number are made, then the number of observations in each subinterval is equal to N/n. The consequence of the independence property is that the probability of obtaining a value in a particular range is independent of the values that were previously obtained.

Random numbers are an important resource for developers, but we can often settle for pseudorandom numbers. Let's see what they are and how to generate them.

The pseudorandom number generator

The generation of real random sequences using deterministic algorithms is impossible: at most, pseudorandom sequences can be generated. These are, apparently, random sequences that are actually perfectly predictable and can be repeated after a certain number of extractions. A pseudorandom number generator (PRNG) is an algorithm designed to output a sequence of values that appear to be generated randomly.

The pros and cons of a random number generator

A random number generation routine must have the following attributes:

- Be replicable
- Be fast
- Not have large gaps between two generated numbers
- Have a sufficiently long-running period
- Be able to generate numbers with statistical properties that are as close as possible to ideal ones

The most common cons of random number generators are as follows:

- Numbers not uniformly distributed
- Discretization of the generated numbers
- Incorrect mean or variance
- Presence of cyclical variations

Random number generation algorithms

The first to deal with the random number generation problem was John von Neumann, in 1949. He proposed a method called **middle-square**. This method allows us to understand some important characteristics of a random number generation process. To start, we need to provide an input as the seed or a value that starts the sequence. This is necessary to be able to generate different sequences each time. However, it is important to ensure that the good behavior of the generator does not depend on which seed is used. From here, the first flaw of the middle-square method appears—that is, when using the zero value as a seed, only a sequence of zeros will be obtained.

Another shortcoming of this method is the repetitiveness of the sequences. As in all of the other PRNGs that will be discussed, each value depends on the previous one, and, at most, on the internal state variables of the generator. Since this is a limited number of digits, the sequence can only be repeated from a certain point onward. The length of the sequence, before it begins to repeat itself, is called the period. A long period is important because many practical applications require a large amount of random data, and a repetitive sequence might be less effective. In such cases, it is important that the choice of the seed has no influence on the potential outcomes.

Another important aspect is the efficiency of the algorithm. The size of the output data values and internal state—and, therefore, the generator input (**seed**)—are often intrinsic features of the algorithm and remain constant. For this reason, the efficiency of a PRNG should be assessed not so much in terms of computational complexity, but in terms of the possibility of a fast and efficient implementation of the calculation architectures available. In fact, depending on the architecture you are working on, the choice of different PRNGs, or different design parameters of a certain PRNG, can result in a faster implementation by many orders of magnitude.

Linear congruential generator

One of the most common methods for generating random numbers is the **linear congruential generator** (**LCG**). The theory on which it rests is simple to understand and implement. It also has the advantage of being computationally light. The recursive relationship underlying this technique is provided by the following equation:

$$x_{k+1} = (a * x_k + c) \bmod m$$

Here, we can observe the following:

- a is the multiplier (non-negative integers)
- c is the increment (non-negative integers)
- m is the mode (non-negative integers)
- x_0 is the initial value (seed or non-negative integers)

The **modulo function**, denoted by *mod*, results in the remainder of the Euclidean division of the first number by the second. For example, 18 mod 4 gives 2 as it is the remainder of the Euclidean division between the two numbers.

The linear congruence technique has the following characteristics:

- It is cyclical with a period that is approximately equal to m
- The generated numbers are discretized

To use this technique effectively, it is necessary to choose very large m values. As an example, set the parameters of the method and generate the first 16 pseudorandom values. Here is the Python code that allows us to generate that sequence of numbers (linear_congruential_generator.py):

```
import numpy as np
```

Understanding Randomness and Random Numbers

```
a = 2
c = 4
m = 5
x = 3

for i in range (1,17):
    x= np.mod((a*x+c), m)
    print(x)
```

The following results are returned:

```
0
4
2
3
0
4
2
3
0
4
2
3
0
4
2
3
```

In this case, the period is equal to 4. It is easy to verify that, at most, m distinct integers, X_n, can be generated in the interval, $[0, m - 1]$. If $c = 0$, the generator is called **multiplicative**. Let's analyze the Python code line by line. The first line was used to import the library:

```
import numpy as np
```

numpy is a library of additional scientific functions of the Python language, designed to perform operations on vectors and dimensional matrices. numpy allows you to work with vectors and matrices more efficiently and faster than you can do with lists and lists of lists (matrices). In addition, it contains an extensive library of high-level mathematical functions that can operate on these arrays.

After importing the `numpy` library, we set the parameters that will allow us to generate random numbers using the LCG:

```
a = 2
c = 4
m = 5
x = 3
```

At this point, we can use the LCG formula to generate random numbers. We only generate the first 16 numbers, but we will see from the results that these can be enough to understand the algorithm. To do this, we use a `for` loop:

```
for i in range (1,17):
    x= np.mod((a*x+c), m)
    print(x)
```

To generate random numbers according to the LCG formula, we have used the `np.mod()` function. This function returns the remainder of a division when given a dividend and divisor.

Random numbers with uniform distribution

A sequence of numbers uniformly distributed between [0,1] can be obtained using the following formula:

$$U_n = \frac{X_n}{m}$$

The obtained sequence is periodic, with a period less than or equal to m. If the period is m, then it has a full period. This occurs when the following conditions are true:

- If m and c are prime numbers among them
- If m is divisible by a prime number, b, for which it must also be divisible
- If m is divisible by 4, then $a - 1$ must also be divisible by 4

> **Important note**
> By choosing a large value of m, you can reduce both the phenomenon of periodicity and the problem of generating rational numbers.

Furthermore, it is not necessary for simulation purposes that all numbers between [0,1] are generated, because these are infinite. However, it is necessary that as many numbers as possible within the range have the same probability of being generated.

Generally, a value of m is $m \geq 10^9$ so that the generated numbers constitute a dense subset of the interval, [0, 1].

An example of a multiplicative generator that is widely used in 32-bit computers is the **Lewis-Learmonth generator**. This is a generator in which the parameters assume the following values:

- a = 75
- c = 0
- m = $2^{31} - 1$

Let's analyze the code that generates the first 100 random numbers according to this method (learmouth_lewis_generator.py):

```
import numpy as np
a = 75
c = 0
m = 2**(31)-1
x = 0.1

for i in range(1,100):
    x= np.mod((a*x+c),m)
    u = x/m
    print(u)
```

The code we have just seen has been analyzed line by line in the *Linear congruential generator* section of this chapter. The difference, in addition to the values of the parameters, lies in the generation of a uniform distribution in the range of [0.1] through the following command:

```
u = x/m
```

The following results are returned:

1.0477378969303043e-07	0.4297038430486358	0.2719089376143687	0.6478998370691668
7.858034226977282e-06	0.22778821374652358	0.3931703117644276	0.5924877578357644
0.0005893525670232962	0.08411602260736563	0.48777336836223184	0.43658181719322775
0.044201442526747216	0.30870169275845477	0.5830026104035799	0.743636274590919
0.3151081876433958	0.15262694570823895	0.7251957597793991	0.7727205682418871
0.6331140620788159	0.44702092252998654	0.38968195830922664	0.9540425911331748
0.4835546335594517	0.526569173916508	0.22614685922216013	0.5531943014604944
0.26659750019507367	0.49268802511165294	0.9610144342114285	0.489572589979308
0.9948125053172989	0.9516018666101629	0.07608253232952325	0.7179442316842937
0.6109378643384845	0.3701399622345995	0.7061899219202762	0.8458173511763184
0.8203398039659205	0.7604971545564463	0.9642441198063288	0.4363013084215584
0.525485268573037	0.03728656472511895	0.3183089519470506	0.7225981167157172
0.4113951243513241	0.7964923534525988	0.873171384852925	0.19485872853307926
0.8546343123794693	0.7369264810052358	0.4878538332357322	0.6144046334616862
0.09757339865787579	0.269486049315652844	0.5890374759161088	0.08034748820604173
0.3180048956153937	0.21145368936073672	0.177810673219063633	0.02606161265916266
0.8503671599786575	0.8590266946046737	0.33580048491051445	0.954620948505877
0.7775369685969953	0.4270020655482086	0.1850363561813889	0.5965711044131644
0.3152726177662949	0.025154901214481752	0.8777267070849085	0.7428328104982306
0.6454463212962478	0.8866175901548088	0.829503000634491	0.7124607612902581
0.40847407486684345	0.49631923087701163	0.212272501871582356	0.4345570716236518
0.6355556010434197	0.223394229808074528	0.9543763962361852	0.5917803568727246
0.6666700559047377	0.7956723486053163	0.578229684186275	0.3835267449652435
0.0002541695722631037	0.6754261174590448	0.3672262934815261	0.7645058593547465
0.01906271791973278	0.6569587861452991	0.5419719980759415	

Figure 2.4: LCG output

Since we are dealing with random numbers, the output will be different from the previous one.

A comparison between the different generators must be made based on the analysis of the periodicity, the goodness of the uniformity of the numbers generated, and the computational simplicity. This is because the generation of very large numbers can lead to the use of expensive computer resources. Also, if the X_n numbers get too big, they are truncated, which can cause a loss of the desired uniformity statistics.

Lagged Fibonacci generator

The **Lagged Fibonacci algorithm** for generating pseudorandom numbers arises from the attempt to generalize the method of linear congruences. One of the reasons that led to the search for new generators is the need (which is useful for many applications, especially in parallel computing) to lengthen the generator period. The period of a linear generator when m is approximately 10^9 is enough for many applications, but not all of them.

One of the techniques developed is to make X_{n+1} dependent on the two previous values, X_n and X_{n-1}, instead of only on X_n, as is the case in the LCG method. In this case, the period may arrive close to the value, m^2, because the sequence will not repeat itself until the following equality is obtained:

$$(X_{n+\lambda}, X_{n+\lambda+1}) = (X_n, X_{n+1})$$

The simplest generator of this type is the Fibonacci sequence represented by the following equation:

$$X_{n+1} = (X_n + X_{n-1}) \bmod m$$

This generator was first analyzed in the 1950s and provides a period, m, but the sequence does not pass the simplest statistical tests. We then tried to improve the sequence using the following equation:

$$X_{n+1} = (X_n + X_{n-k}) \bmod m$$

This sequence, although better than the Fibonacci sequence, does not return satisfactory results. We had to wait until 1958 when Mitchell and Moore proposed the following sequence:

$$X_n = (X_{n-24} + X_{n-55}) \bmod m, \quad n \geq 55$$

Here, m is even and $X_0, \ldots X_{54}$ are arbitrary integers that are not all even. Constants 24 and 55 are not chosen at random but are numbers that define a sequence whose least significant bits ($X_n \bmod 2$) have a period of length $2^{55}-1$. Therefore, the sequence (X_n) must have a period of length of, at least, $2^{55}-1$. The succession has a period of $2^{M-1}(2^{55}-1)$ where $m = 2^M$.

Numbers 24 and 55 are commonly called lags, and the sequence (X_n) is called a **Lagged Fibonacci generator** (**LFG**). The LFG sequence can be generalized with the following equation:

$$X_n = (X_{n-l} \otimes X_{n-k}) \bmod 2^M, l > k > 0$$

Here, $\otimes$ refers to any of the following operations: $+$, $-$, $\times$, or $\otimes$ (exclusive or).

Only some pairs (k, l) give sufficiently long periods. In these cases, the period is $2^{M-1}(2^l-1)$. The pairs (k, l) must be chosen appropriately. The only condition on the first l values is that at least one of them must be odd; otherwise, the sequence will be composed of even numbers.

Let's look at how to implement a simple example of an additive LFG in Python using the following parameters: $x_0 = x_1 = 1$ and $m = 2^{32}$. Here is the code to generate the first 100 random numbers (lagged_fibonacci_algorithm.py):

```
import numpy as np
x0=1
x1=1
m=2**32

for i in range (1,101):
    x= np.mod((x0+x1), m)
    x0=x1
    x1=x
    print(x)
```

Let's analyze the Python code line by line. The first line was used to import the library:

```
import numpy as np
```

After importing the `numpy` library, we set the parameters that will allow us to generate random numbers using the LFG:

```
x0=1
x1=1
m=2**32
```

At this point, we can use the LFG formula to generate random numbers. We only generate the first 100 numbers. To do this, we use a `for` loop:

```
for i in range (1,101):
    x= np.mod((x0+x1), m)
    x0=x1
    x1=x
    print(x)
```

To generate random numbers according to the LFG formula, we use the `np.mod()` function. This function returns the remainder of a division when given a dividend and divisor. After generating the random number, the two previous values are updated as follows:

```
x0=x1
x1=x
```

The following random numbers are printed:

2	317811	1776683621	375819880
3	514229	368225352	3634140029
5	832040	2144908973	4009959909
8	1346269	2513134325	3349132642
13	2178309	363076002	3064125255
21	3524578	2876210327	2118290601
34	5702887	3239286329	887448560
55	9227465	1820529360	3005739161
89	14930352	764848393	3893187721
144	24157817	2585377753	2603959586
233	39088169	3350226146	2202180011
377	63245986	1640636603	511172301

610	102334155	695895453	2713352312
987	165580141	2336532056	3224524613
1597	267914296	3032427509	1642909629
2584	433494437	1073992269	572466946
4181	701408733	4106419778	2215376575
6765	1134903170	885444751	2787843521
10946	1836311903	696897233	708252800
17711	2971215073	1582341984	3496096321
28657	512559680	2279239217	4204349121
46368	3483774753	3861581201	3405478146
75025	3996334433	1845853122	3314859971
121393	3185141890	1412467027	2425370821
196418	2886509027	3258320149	1445263496

Table 2.1: Table of random numbers using the LFG

The initialization of an LFG is particularly complex, and the results of this method are very sensitive to the initial conditions. If extreme care is not taken when choosing the initial values, statistical defects may occur in the output sequence. These defects could harden the initial values along with subsequent values that have a careful periodicity. Another potential problem with LFG is that the mathematical theory behind the method is incomplete, making it necessary to rely on statistical tests rather than theoretical performance.

Having seen how to generate pseudorandom numbers, we will now focus on the testing procedures.

Testing uniform distribution

Test adaptation (that is, the goodness of fit), in general, has the purpose of verifying whether a variable under examination does or does not have a certain hypothesized distribution on the basis, as usual, of experimental data. It is used to compare a set of frequencies observed in a sample, with similar theoretical quantities assumed for the population. By means of the test, it is possible to quantitatively measure the degree of deviation between the two sets of values.

The results obtained in the samples do not always exactly agree with the theoretical results that are expected according to the rules of probability. Indeed, it is very rare for this to occur. For example, although theoretical considerations lead us to expect 100 heads and 100 tails from 200 flips of a coin, it is rare that these results are obtained exactly. However, despite this, we must not unnecessarily deduce that the coin is rigged.

Chi-squared test

The **chi-squared test** is a test of hypotheses that gives us back the significance of the relationship between two variables. It is a statistical inference technique that is based on the chi-squared statistic and its probability distribution. It can be used with nominal and/or ordinal variables, generally arranged in the form of contingency tables.

The main purpose of this statistic is to verify the differences between observed and theoretical values, called **expected** values, and to make an inference on the degree of deviation between the two. The technique is used with three different objectives that are all based on the same fundamental principle:

- The randomness of the distribution of a categorical variable
- The independence of two qualitative variables (nominal or ordinal)
- The differences with a theoretical model

For now, we will just consider the first aspect. The method consists of a comparison procedure between the observed empirical frequencies and the theoretical frequencies. Let's consider the following definitions:

- **H0**: Null hypothesis or the absence of a statistical relationship between two variables
- **H1**: Research hypothesis that supports the existence of the relationship—for instance, H_1 is true if H0 is false
- **Fo**: Observed frequencies—that is, the number of data of a cell detected
- **Fe**: Expected frequencies—that is, the frequency that should be obtained based on the marginal totals if there was no association between the two variables considered

The chi-squared test is based on the difference between observed and expected frequencies. If the observed frequency is very different from the expected frequency, then there is an association between the two variables.

As the difference between the observed frequency and the expected frequency increases, so does the value of chi-squared. The chi-squared value is calculated using the following equation:

$$\chi^2 = \sum \frac{(F_o - F_e)^2}{F_e}$$

Let's look at an example to understand how to calculate this value. We build a contingency table that shows student choices for specific courses divided by genres. These are the observed values:

	Male	Female	Tot
Biotechnology	411	574	985
Health Management	452	253	705
MBA	303	246	549
Tot	1166	1073	2239

Table 2.2: Table of student choices divided by genres

In addition, we calculate the representation of each value as a percentage of column totals (observed frequencies):

	Male	Female	Tot
Biotechnology	35,25%	53,49%	43,99%
Health Management	38,77%	23,58%	31,49%
MBA	25,99%	22,93%	24,52%
Tot	100,00%	100,00%	100,00%

Table 2.3: Table of choices in percentages of column totals

Now, we calculate the expected value, as follows:

$$\text{Expected value} = \frac{(Tot\ row) * (Tot\ column)}{Tot}$$

Let's calculate it for the first cell (**Biotechnology – Male**):

$$\text{Expected value} = \frac{(985) * (1166)}{2239} = 512.9567$$

The contingency matrix of expected values is as follows:

	Male	Female
Biotechnology	512.9567	472.0433
Health Management	367.1416	337.8584
MBA	285.9017	263.0983

Table 2.4: Table of contingency matrix

Let's calculate the contingency differences ($F_o - F_e$), as follows:

	Male	Female
Biotechnology	-101.957	101.9567
Health Management	84.85842	-84.8584
MBA	17.09826	-17.0983

Table 2.5: Table of contingency differences

Finally, we can calculate the chi-squared value, as follows:

$$\chi^2 = \sum \frac{(F_o - F_e)^2}{F_e} = \frac{(-101.957)^2}{512.9567} + \frac{(-101.957)^2}{472.0433} + \cdots = 84.35$$

If the two characters were independent, we would expect a chi-squared value of 0. On the other hand, random fluctuations are always possible. So, even in the case of perfect independence, we will never have 0. Therefore, even chi-squared values that are far from 0 could make the result compatible with the null hypothesis, H0, of independence between the variables.

> **Important note**
> One question that arises is this: is the value obtained only the result of a fluctuation, or does it arise from the dependence between the data?

Statistical theory tells us that if the variables are independent, the distribution of the chi-squared frequencies follows an asymmetric curve. In our case, we have a frequency distribution table of two features: course and gender—that is, the course feature with three modes, and the gender feature with two modes. In the case of independence, how much is the square value that leaves a 5% probability on the right?

To answer this question, we must first calculate the so-called degrees of freedom, n, which is defined as follows:

$$n = (number\ of\ rows - 1) * (number\ of\ columns - 1)$$

In our case, from the contingency table, we obtain the following:

$$n = (3 - 1) * (3 - 1) = 2 * 1 = 2$$

Now, we need the following chi-squared distribution table:

	Probability of exceeding the critical value		
Degrees of freedom	0.05	0.01	0.001
1	3.84	6.64	10.83
2	5.99	9.21	13.82
3	7.82	11.35	16.27
4	9.49	13.28	18.47
5	11.07	15.09	20.52
6	12.59	16.81	22.46
7	14.07	18.48	24.32
8	15.51	20.09	26.13
9	16.92	21.67	27.88
10	18.31	23.21	29.59

Table 2.6: Chi-squared distribution table

Table 2.6 provides the critical values for the chi-squared hypothesis tests (χ^2). The intersections of columns and rows are the critical right-tail values for a given probability and degrees of freedom. The column headings indicate the probability of χ^2 exceeding the critical value. The row headers define the degrees of freedom for the chi-squared test. The cells within the table represent the critical chi-squared value for a right-tailed test.

In the previous table, we look for the value, n = 2, in the first column, and then we scroll through the rows until we reach the column that is equal to 0.05. Here, we find the following:

$$\chi^2_{2,0.05} = 5.99$$

This means that if the data were independent, we would only have a 5% chance of getting $\chi^2 > 5.99$ from the calculations. Having obtained $\chi^2 = 84.35$, we can discard the null hypothesis, H_0, of independence from the data with a confidence of 5%. This means the possibility that H_0 is true is only 5%. Therefore, the research hypothesis, H_1, will be true, with 95% confidence.

Uniformity test

After having generated the pseudorandom numerical sequence, it is necessary to verify the goodness of the obtained sequence. It is a question of checking whether the sequence obtained, which constitutes a random sample of the experiment, follows a uniform distribution. To carry out this check, we can use the χ^2 test (chi-squared test). Let's demonstrate how to do this.

The first operation is to divide the interval, [0,1], into s subintervals of the same length. Then, we count how many numbers of the generated sequence are included in the i-th interval, as follows:

$$R_i = \{x_i \mid x_j \in s_i, \quad j = 1, \ldots N\}$$

The R_i values should be as close as possible to the N/s value. If the sequence were perfectly uniform, then each subinterval would have the same number of samples in the sequence.

We indicate, with V, the variable to perform the test. This variable is calculated using the following formula:

$$V = \sum_{i=1}^{s} \frac{\left(R_i - \frac{N}{s}\right)^2}{\frac{N}{s}}$$

After introducing the tools that allow us to perform a uniformity test, let's analyze a practical example that will help us to understand how to carry out this procedure. We generate a pseudorandom numerical sequence of 100 values, by means of the congruent linear generator, by fixing the parameters as follows:

- a = 75
- c = 0
- m = $2^{31} - 1$

This is the random number generator already seen in the *Lagged Fibonacci generator* section. We have already introduced the code that allows us to generate the sequence, so let's modify it for our new requirements by storing the sequence in an array (uniformity_test.py):

```
import numpy as np
a = 75
c = 0
m = 2**(31) -1
x = 0.1
u=np.array([])

for i in range(0,100):
    x= np.mod((a*x+c),m)
    u= np.append(u,x/m)
    print(u[i])
```

The following results are returned:

0.000349246	0.873963	0.405964	0.437041
0.0261934	0.547262	0.447335	0.778111
1.96451	0.044633	0.55013	0.358325
0.00147338	0.347471	0.259782	0.874402
0.110504	0.0603575	0.483628	0.580178
0.28777	0.526814	0.272083	0.513331
0.582786	0.511059	0.406251	0.499849
0.708915	0.329403	0.468846	0.488678
0.16862	0.705262	0.163429	0.650882
0.646474	0.89468	0.257203	0.816133
0.485546	0.10097	0.290253	0.209991
0.415936	0.572771	0.769002	0.74933
0.195233	0.95786	0.675151	0.199786
0.642455	0.839501	0.636331	0.983964
0.184134	0.962572	0.724838	0.797331
0.810025	0.19288	0.362847	0.799806
0.751858	0.465979	0.213493	0.985428
0.389342	0.948448	0.0119538	0.907136
0.200681	0.133636	0.896533	0.0351827

0.0510482	0.0226703	0.239982	0.638706
0.828618	0.700276	0.998665	0.902933
0.146342	0.520684	0.899845	0.719949
0.975639	0.0512744	0.488352	0.996165
0.172955	0.845583	0.626389	0.712364
0.971653	0.418746	0.979161	0.427272

Table 2.7: Output table of LFG random numbers

To better understand how the numbers are distributed in the range considered, we will divide the interval, [0,1], into 20 parts (s = 20), and then count how many values of the sequence fall into each interval of amplitude 0.05.

Finally, we calculate the V variable:

```
N=100
s=20
Ns =N/s
S = np.arange(0, 1, 0.05)
counts = np.empty(S.shape, dtype=int)
V=0
for i in range(0,20):
    counts[i] = len(np.where((u >= S[i]) & (u < S[i]+0.05))[0])
    V=V+(counts[i]-Ns)**2 / Ns

print("R = ",counts)
print("V = ", V)
```

Let's analyze the code line by line:

```
N=100
s=20
Ns =N/s
```

The first three lines set the variable, N (the number of random numbers), and s (the number of parts), and then we calculated the ratio.

After that, we divide the interval, [0,1], into 20 subintervals:

```
S = np.arange(0, 1, 0.05)
```

Now, we initialize the counts array, which contains how many values of the sequence fall into each interval, and the V variable, as follows:

```
counts = np.empty(S.shape, dtype=int)
V=0
```

To count how many values of the sequence fall into each interval, we will use a for loop:

```
for i in range(0,20):
    counts[i] = len(np.where((u >= S[i]) & (u < S[i]+0.05))[0])
    V=V+(counts[i]-Ns)**2 / Ns
```

First, we use the np.where() function to count how many values satisfy the following conditions: (u >= S[i]) & (u < S[i]+0.05); these are the extremes of each subinterval. Then, we calculate the V variable using the following equation:

$$V = \sum_{i=1}^{S} \frac{\left(R_i - \frac{N}{S}\right)^2}{\frac{N}{S}}$$

Finally, we will print the results:

```
print("R = ",counts)
print("V = ", V)
```

The following results are returned:

```
R =  [8 3 4 7 4 5 2 3 7 7 5 4 5 2 7 5 5 5 3 9]
V =  14.8
```

Before analyzing the meaning of the calculated V value, let's consider the sequence of counts obtained. To appreciate the distribution of the frequencies obtained, we draw a bar graph:

```
import matplotlib.pyplot as plt
Ypos = np.arange(len(counts))
plt.bar(Ypos,counts)
```

The following diagram is printed:

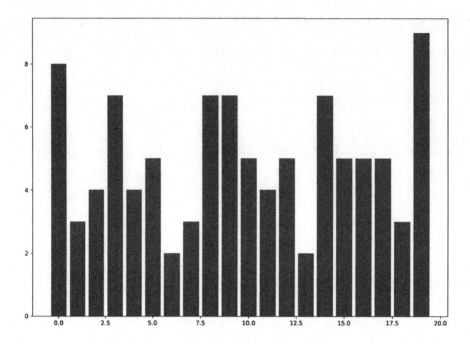

Figure 2.5: Distribution of frequencies

As you can see, all of the ranges are covered, with values ranging from a minimum of 2 to a maximum of 9.

However, now, let's analyze the value of V obtained. As we anticipated, V = 14.8, so what do we do with this value? First, let's calculate the so-called degrees of freedom, n:

$$n = (2 - 1) * (20 - 1) = 1 * 19 = 19$$

Now, we must compare the V value obtained with the probability of exceeding the critical value. To get this value, we must visualize the chi-squared distribution table:

d.f.	.995	.99	.975	.95	.9	.1	.05	.025	.01
1	0.00	0.00	0.00	0.00	0.02	2.71	3.84	5.02	6.63
2	0.01	0.02	0.05	0.10	0.21	4.61	5.99	7.38	9.21
3	0.07	0.11	0.22	0.35	0.58	6.25	7.81	9.35	11.34
4	0.21	0.30	0.48	0.71	1.06	7.78	9.49	11.14	13.28
5	0.41	0.55	0.83	1.15	1.61	9.24	11.07	12.83	15.09
6	0.68	0.87	1.24	1.64	2.20	10.64	12.59	14.45	16.81
7	0.99	1.24	1.69	2.17	2.83	12.02	14.07	16.01	18.48
8	1.34	1.65	2.18	2.73	3.49	13.36	15.51	17.53	20.09
9	1.73	2.09	2.70	3.33	4.17	14.68	16.92	19.02	21.67
10	2.16	2.56	3.25	3.94	4.87	15.99	18.31	20.48	23.21
11	2.60	3.05	3.82	4.57	5.58	17.28	19.68	21.92	24.72
12	3.07	3.57	4.40	5.23	6.30	18.55	21.03	23.34	26.22
13	3.57	4.11	5.01	5.89	7.04	19.81	22.36	24.74	27.69
14	4.07	4.66	5.63	6.57	7.79	21.06	23.68	26.12	29.14
15	4.60	5.23	6.26	7.26	8.55	22.31	25.00	27.49	30.58
16	5.14	5.81	6.91	7.96	9.31	23.54	26.30	28.85	32.00
17	5.70	6.41	7.56	8.67	10.09	24.77	27.59	30.19	33.41
18	6.26	7.01	8.23	9.39	10.86	25.99	28.87	31.53	34.81
19	6.84	7.63	8.91	10.12	11.65	27.20	30.14	32.85	36.19
20	7.43	8.26	9.59	10.85	12.44	28.41	31.41	34.17	37.57

Figure 2.6: Chi-squared distribution table

In the previous table, we look for the value, n = 19, in the first column, and then we scroll through the rows until we reach the column that is equal to 0.05. Here, we find the following:

$$\chi^2_{19,0.05} = 30.14$$

If the statistic of the V test is less (14.8 < 30.14), then we can accept the hypothesis of uniformity of the generated sequence.

After analyzing several examples of distribution testing, in the next section, we will see generic methods for random distributions.

Exploring generic methods for random distributions

Most programming languages provide users with functions for the generation of pseudorandom numbers with uniform distributions in the range of 0/1. These generators are, very often, considered to be continuous. However, in reality, they are discrete even if they have a very small discretization step. Any sequence of pseudorandom numbers can always be generated from a uniform distribution of random numbers. In the following sections, we will examine some methods that allow us to derive a generic distribution starting from a uniform distribution of random numbers.

The inverse transform sampling method

By having a PRNG with continuous and uniform distributions in the range of 0 ÷ 1, it is possible to generate continuous sequences with any probability distribution using the inverse transform sampling technique. Consider a continuous random variable, x, having a probability density function of f (x). The corresponding distribution function, F(x), is determined for this function, as follows:

$$F(x) = \int_0^x f(x) * dx$$

The distribution function, F(x), of a random variable indicates the probability that the variable assumes a value that is less than or equal to x. The analytical expression (if any) of the inverse function is then determined, such as x = F^{-1}. The determination of the sample of the variable, x, is obtained by generating a value between 0 and 1 and replacing it in the expression of the inverse distribution function.

This method can be used to obtain samples from many types of distribution functions, such as exponential, uniform, or triangular. It turns out to be the most intuitive, but not the most computationally effective, method.

Let's proceed by starting with a decreasing exponential distribution:

$$f(x) = \lambda * e^{-\lambda x}, \quad \lambda > 0$$

The corresponding distribution function, F(x), is determined for this function, as follows:

$$F(x) = \int_0^\infty \lambda * e^{-\lambda x} * dx = 1$$

The trend of the decreasing exponential distribution function is shown in the following diagram:

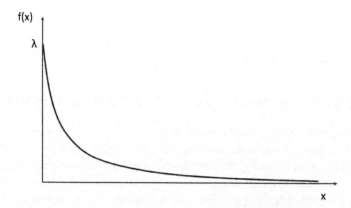

Figure 2.7: Representation of the decreasing exponential function

Get the distribution function by solving the integral:

$$F(x) = \int_0^\infty \lambda * e^{-\lambda x} * dx = 1 - e^{-\lambda x} = r$$

By operating the inverse transformation of the distribution function, we get the following:

$$x = -\frac{1}{\lambda} * ln(1 - r)$$

Here, r is within the range of [0 ÷ 1]. r is extracted from a uniform distribution by means of a uniform generator. λ represents the average inter-arrival frequency if x represents the time, and 1/λ is the average inter-arrival time.

The method of inverse transformation in the discrete case has a very intuitive justification. The interval, [0, 1], is divided into contiguous subintervals of amplitude p (x1), p (x2),... and X is assigned according to whether these intervals contain the U value that is being generated.

The acceptance-rejection method

The inverse transformation method is based on the calculation of the inverse transformation, F – 1, which cannot always be calculated (or, at least, not efficiently). In the case of law distributions defined on finite intervals [a, b], the rejection-acceptance method is used.

Suppose we know the probability density of the random variable, X, that we intend to generate: $f_X(x)$. This is defined on a finite interval, [a, b], and the image is defined in the range of [0, c]. In practice, the $f_X(x)$ function is all contained within the rectangle, [a, b] x [0, c], as shown in the following diagram:

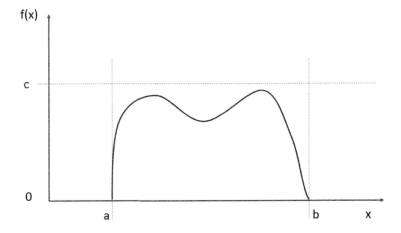

Figure 2.8: Representation of the f(x) function

We generate two uniform pseudorandom sequences between [0.1]: U_1 and U_2. Next, we derive two other uniform numerical sequences according to the following rule:

$$\begin{cases} X = a + (b-a) * U_1 \\ Y = c * U_2 \end{cases}$$

Each pair of values (u_1, u_2) will correspond to a pair (x, y) belonging to the rectangle, [a, b] x [0, c]. If the pair (x, y) falls within the area of the function, $f_X(x)$, it is accepted and will subsequently be used to create the desired pseudorandom sequence; otherwise, it will be discarded. In the latter case, the procedure is repeated until a new pair located in the area of $f_X(x)$ is found. The sequence of X values that is obtained is a pseudorandom sequence that follows the distribution law, $f_X(x)$, because we have chosen only values that fall in that area.

Random number generation using Python

So far, we have seen which methods can be used for generating random numbers. We have also proposed some solutions in Python code for the generation of random numbers through some universally used methods. These applications have been useful for understanding the basis on which random number generators have been made. In Python, there is a specific module for the generation of random numbers: this is the `random` module. Let's examine what it is.

Introducing the random module

The `random` module implements PRNGs for various distributions and is based on the **Mersenne Twister** algorithm. The Mersenne Twister algorithm is a PRNG that produces almost uniform numbers suitable for a wide range of applications. The `random` module was originally developed to produce inputs for Monte Carlo simulations.

It is important to note that random numbers are generated using repeatable and predictable deterministic algorithms. They begin with a certain seed value and, every time we ask for a new number, we get one based on the current seed. The seed is an attribute of the generator. If we invoke the generator twice with the same seed, the sequence of numbers that will be generated starting from that seed will always be the same. However, these numbers will be evenly distributed.

Let's analyze, in detail, the functions contained in the module through a series of practical examples.

The random.random() function

The `random()` function returns the next nearest floating-point value from the generated sequence. All return values are enclosed between 0 and 1.0. Let's explore a practical example that uses this function:

```
import random
for i in range(20):
```

```
        print('%05.4f' % random.random(), end=' ')
print()
```

We first imported the `random` module and then we used a `for` loop to generate 20 pseudorandom numbers. Each number is printed in a format that includes 5 digits, including 4 decimal places.

The following results are returned:

```
0.7916 0.2058 0.0654 0.6160 0.1003 0.3985 0.3573 0.9567 0.0193
0.4709 0.8573 0.2533 0.8461 0.1394 0.4332 0.7084 0.7994 0.3361
0.1639 0.4528
```

As you can see, the numbers are uniformly distributed in the range of [0.1].

By running the code repeatedly, you get sequences of different numbers. Let's try the following:

```
0.6918 0.8197 0.4329 0.2674 0.4118 0.1937 0.2267 0.8259 0.9081
0.4583 0.7300 0.7148 0.9814 0.2237 0.7419 0.7766 0.2626 0.1886
0.1328 0.0037
```

We have confirmed that every time we invoke the `random()` function, the generated sequence is different from the previous one.

The random.seed() function

As we have verified, the `random()` function produces different values each time it is invoked and has a very large period before any number is repeated. This is useful for producing unique values or variations, but there are times when it is useful to have the same set of data available to be processed in different ways. To do this, we can use the `random.seed()` function. This function initializes the basic random number generator. Let's look at an example:

```
import random
random.seed(1)
for i in range(20):
    print('%05.4f' % random.random(), end=' ')
print()
```

We used the same code from the previous example. However, this time, we set the seed (`random.seed(1)`). The number in parentheses is an optional argument and can be any object whose hash can be calculated. If this argument is omitted, the current system time is used. The current system time is also used to initialize the generator when the module is imported for the first time.

The following results are returned:

```
0.1344  0.8474  0.7638  0.2551  0.4954  0.4495  0.6516  0.7887  0.0939
0.0283  0.8358  0.4328  0.7623  0.0021  0.4454  0.7215  0.2288  0.9453
0.9014  0.0306
```

Let's see what happens if we launch this piece of code again:

```
0.1344  0.8474  0.7638  0.2551  0.4954  0.4495  0.6516  0.7887  0.0939
0.0283  0.8358  0.4328  0.7623  0.0021  0.4454  0.7215  0.2288  0.9453
0.9014  0.0306
```

The result is similar. The seed setting is particularly useful when you want to make the simulation repeatable.

The random.uniform() function

The `random.uniform()` function generates numbers within a defined numeric range. Let's look at an example:

```
import random
for i in range(20):
    print('%6.4f' % random.uniform(1, 100), end=' ')
print()
```

We asked it to generate 20 random numbers in the range of [1,100]. The following results are returned:

```
26.2741  84.3327  67.6382   9.2402   2.6524   2.4414  75.8031  25.7064
11.8394  62.8554  35.0979   7.8820  16.8029  53.2107  17.6463  28.0185
71.4474  46.0155  32.8782  47.9033
```

This function can be used when requesting random numbers at well-defined intervals.

The random.randint() function

This function generates random integers. The arguments for `randint()` are the values of the range, including the extremes. The numbers may be negative or positive, but the first value should be less than the second. Let's look at an example:

```
import random
for i in range(20):
    print(random.randint(-100, 100), end=' ')
print()
```

The following results are returned:

```
9 -85 88 -24 -68 -46 -88 -22 -82 -81 -21 -24 90 -60 6 44 -36
-67 -98 43
```

The entire range is represented in the sequence of randomly generated numbers.

A more generic form of selecting values from a range is obtained by using the `random.range()` function. In this case, the step argument is provided, in addition to the start and end values. Let's look at an example:

```
import random
for i in range(20):
    print(random.randrange(0, 100,5), end=' ')
print()
```

The following results are returned:

```
5 90 30 90 70 25 95 80 5 60 30 55 15 30 90 65 90 30 75 15
```

The returned sequence is a random distribution of the values expected from the passed arguments.

The random.choice() function

A common use for random number generators is to select a random element from a sequence of enumerated values, even if these values are not numbers. The `choice()` function returns a random element of the non-empty sequence passed as an argument. Let's look at an example:

```
import random
CitiesList = ['Rome','New York','London','Berlin','Moskov',
'Los Angeles','Paris','Madrid','Tokio','Toronto']
for i in range(10):
    CitiesItem = random.choice(CitiesList)
    print ("Randomly selected item from Cities list is - ",
CitiesItem)
```

The following results are returned:

```
Randomly selected item from Cities list is -  Paris
Randomly selected item from Cities list is -  Moskov
Randomly selected item from Cities list is -  Tokio
Randomly selected item from Cities list is -  Madrid
Randomly selected item from Cities list is -  Rome
```

```
Randomly selected item from Cities list is - Los Angeles
Randomly selected item from Cities list is - Toronto
Randomly selected item from Cities list is - Paris
Randomly selected item from Cities list is - Moskov
Randomly selected item from Cities list is - Rome
```

At each iteration of the cycle, a new element is extracted from the list containing the names of the cities. This function is suitable to use in extracting values from a predetermined list.

The random.sample() function

Many simulations require random samples from a population of input values. The `random.sample()` function generates samples without repeating the values and without changing the input sequence. Let's look at an example:

```
import random
DataList = range(10,100,10)
print("Initial Data List = ",DataList)
DataSample = random.sample(DataList,k=5)
print("Sample Data List = ",DataSample)
```

The following results are returned:

```
Initial Data List =  range(10, 100, 10)
Sample Data List =  [30, 60, 40, 20, 90]
```

Only five elements of the initial list were selected, and this selection was completely random.

Generating real-value distributions

The following functions generate specific distributions of real numbers:

- `betavariate (alpha, beta)`: This is the beta distribution. The conditions for the parameters are alpha> -1 and beta> -1. Return values are in the range of 0 to 1.
- `expovariate (lambd)`: This is the exponential distribution. `lambd` is 1.0 divided by the desired average (the parameter was supposed to be called "lambda," but this is a reserved word in Python). The return value is between 0 and positive infinity.
- `gammavariate (alpha, beta)`: This is the gamma distribution. The conditions on the parameters are alpha> 0 and beta> 0.
- `gauss (mu, sigma)`: This is the Gaussian distribution. `mu` is the mean and `sigma` is the standard deviation. This is slightly faster than the `normalvariate()` function, defined next.

- `lognormvariate (mu, sigma)`: This is the normal logarithmic distribution. If you take the natural logarithm of this distribution, you will get the normal distribution with mean mu and sigma standard deviation. mu can have any value, while sigma must be greater than zero.
- `normalvariate (mu, sigma)`: This is the normal distribution. mu is the mean and sigma is the standard deviation.
- `vonmisesvariate (mu, kappa)`: mu is the average angle, expressed in radians with a value between 0 and 2 * pi, while `kappa` is the concentration parameter, which must be greater than or equal to zero. If kappa is equal to zero, this distribution narrows to a constant random angle in a range between 0 and 2 * pi.
- `paretovariate (alpha)`: This is the Pareto distribution. `alpha` is the form parameter.
- `weibullvariate (alpha, beta)`: This is the Weibull distribution. Here, alpha is the scale parameter and `beta` is the form parameter.

Randomness requirements for security

With the onset of the Digital Age, the topic of information security has extended to almost all areas of today's society, affecting every aspect of international governance systems. With the spread of digitization, illegality and cybercrime have evolved considerably, and consequently, attention to cybersecurity has grown. All companies are put at serious risk of possible attacks by malicious people and malfunctions that could cause a slowdown or stop of the production and decision-making processes managed by an information system. To avoid such risks, it is necessary to carefully study their causes, analyze the possible behaviors of those who try to attack the hardware and software integrity of computer systems, and prepare appropriate countermeasures that can neutralize or minimize the danger of attacks or at least their consequences. The term *IT security* means the study of problems related to information security and the solutions that are devised to deal with them.

Cybersecurity is based on systems of identification of digital identities, which are called authentication. Authentication represents the process by which the veracity of a partial attribute or information that claims to be true by an entity is confirmed. Authentication, therefore, represents the first element that a user is faced with when interacting with a computer system. Creating a secure authentication system is a complex problem to face, due to the vastness of devices and services that can be combined with a specific application. However, giving little importance to this factor would inevitably lead to the crumbling of the entire security system.

Password-based authentication systems

A password is a string of alphanumeric characters used to identify a user to gain access to resources. Passwords are inextricably linked to information technology; in fact, they have been used since the first computers. More specifically, passwords are strictly linked to the presence of accounts or profiles, which are related to a person or a group of people and contain personal data and customizations relating to a specific site, device, or service. This data must not be accessible to anyone with access to that service or device for both privacy and security reasons and is therefore protected by passwords.

Even today, a password is the most used online authentication method; in fact, anyone who has a computer account needs to deal with passwords. The success of the password/username pair as authentication is due to their simplicity, as well as the simple fact that they can be kept, after an encryption operation by hashing, in a simple table, which in the case of websites can be found all inside a remote server, while for offline authentication it is inside the local device. When an authentication attempt is made, the hashed password entered by the user is compared with the one in the table where passwords and usernames are stored. If both match, the authentication is successful and you will be able to access the reserved area.

Since passwords have existed, however, there has also been a problem regarding their robustness to any attacks of any kind to avoid the loss of sensitive information, identity theft, or tampering with personal data. People tend to create a more important password for accounts that they think are more valuable, such as the account for their bank's website. Passwords, although still the most used method of authentication today, are not free from various problems that make them highly vulnerable, causing security risks. Their main problem is that users cannot easily remember strings of nonsensical alphanumeric characters, so they are led to choose words and numbers that are either very predictable or related to their daily life, often even written down to avoid being forgotten. To this must be added the fact that they are identifiable by various types of attacks that will be described in the following pages.

Random password generator

Random numbers are a powerful tool for automatically generating secure passwords. The random numbers we learned to generate in the previous sections can easily be transformed into strings of characters and numbers that represent the essential elements of a password. A random password generator allows us to create a strong and random password using a combination of numbers, letters, and special characters. These combinations can be modified to suit the requirements of the authentication system.

To understand how a random password generator works, let's analyze a practical example. Here is the Python code that allowed us to generate a password consisting of a number of random characters set by the user (random_password_generator.py):

```
import string
import random


char_set = list(string.ascii_letters + string.digits +
                                       "()!$%^&*@#")
random.shuffle(char_set)
password_length = int(input("How long should your
                                      password be?: "))
password = []
for i in range(password_length):
```

```
        password.append(random.choice(char_set))
random.shuffle(password)
print("".join(password))
```

The following results are returned:

```
How long should your password be?: 15
g3@X&ps*&56MeH2
```

In this case, when asked to enter the number of characters in the password, we typed 15, and an absolutely random password made up of 15 characters was returned. In it, we can see a combination of lowercase and uppercase characters, numbers, and, finally, some of the special characters that we have foreseen in the character set.

So, let's analyze the code line by line to understand how we approached the problem. To get started, we imported the libraries:

```
import string
import random
```

The `string` module holds some constants, utility functions, and classes for string manipulation. The `random` module provides PRNGs for various distributions.

Let's now move on to set the characters for the generation of the password:

```
char_set = list(string.ascii_letters + string.digits +
                                "()!$%^&*@#")
```

To do this, we have set a variable of the `list` (`char_set`) type that will contain these characters: To start, we have inserted the letters provided by the **American Standard Code for Information Interchange (ASCII)** code in uppercase ('ABCDEFGHIJKLMNOPQRSTUVWXYZ') and lowercase ('abcdefghijklmnopqrstuvwxyz'). We have added the numerical digits to them—namely, the lowercase letters '0123456789'. Finally, we added a series of special characters ('()! $% ^ & * @ #'). Now, the character set is ready for use.

Now, let's mix the characters:

```
random.shuffle(char_set)
```

The `random.shuffle()` method shuffles a list, rearranging the contents of the variable by simply changing the order of succession of its elements.

Before proceeding to the generation of the password, it is necessary to set the number of characters:

```
password_length = int(input("How long should your
                                password be?: "))
```

To make this procedure customizable, we have planned to require the user to type the number of characters into the command line.

Now, we can generate the password:

```
password = []
for i in range(password_length):
    password.append(random.choice(char_set))
```

First, we generate an empty `list` variable (`password`), then fill it with a number of elements equal to those typed by the user and contained in the `password_length` variable. To do this, we use a `for` loop. Each iteration will add a character chosen from the `char_set` list and generated by the `random.choice()` method. The `random.choice()` method returns a random element of the non-empty sequence passed as an argument.

At this point, we have our password, but to add even more randomness we mix the characters of our password by applying the `random.shuffle()` method again:

```
random.shuffle(password)
```

At this point we just have to print the password on the screen:

```
print("".join(password))
```

Our password is of the `list` type. To be able to see it as a `string` type, we have to transform it into a string. We did this using the `join()` method. This method joins all items in a list into a string, with no spaces between characters.

Cryptographic random number generator

The oldest and still most widely used mechanism in the IT field for data protection is encryption. Cryptography is a technique used by man since ancient times in response to the need to communicate in a secret and secure way, managing to send messages that can be read quickly by recipients and not decrypted by the enemy or anyone who is not authorized.

Introducing cryptography

The need to communicate in a secret and secure way is extremely topical in the field of information security, as the continuous development of electronic systems, on the one hand, makes communications easier: on the other hand, it makes them very vulnerable to external attacks, if not protected adequately.

Modern cryptography must therefore ensure the following:

- **Authentication**: Verify the identity of whoever sends or receives the data, and prevent an intruder from impersonating the sender or recipient

- **Confidentiality**: Prevent the data sent by one subject to another subject from being intercepted by a third party
- **Integrity**: Make sure that the data received is the same as that sent, preventing an intruder from tampering with the data during transmission
- **Non-repudiation**: Prevent those who send data from denying that they have sent it in the future and ensure that they have sent that data

The algorithms used for modern cryptography are listed here:

- **Secret key (symmetric) algorithm**: The key to encrypt is also used to decrypt the message.
- **Public key (asymmetric) algorithm**: The sender and the recipient of the communication both have two keys: a private one, which only the owner knows, and a public one, which is made known to everyone.
- **Hash function algorithm**: It does not require a key; a fixed-length hash value is computed, which makes it impossible to retrieve the plaintext content.

Cryptoanalysis is a science that deals with the analysis and validity of cryptographic algorithms. It is generally used in the design phase of an encryption algorithm to find any weaknesses within it. Cryptography is based on the open source philosophy, meaning an algorithm is considered more secure if its code is public. This is because its code is subjected to analysis while an algorithm that bases its degree of security on the secrecy of the code is considered unsafe as the presence of a possible bug, if discovered, would make the algorithm ineffective. In other words, to increase the security of the algorithm, it is necessary to assume that the cryptoanalyst knows in detail the functioning of the encryption and decryption algorithm. Security, therefore, resides in the key, which identifies a particular encryption among many possibilities. It is secret compared to the method that is publicly known, and moreover, it can be easily changed. This philosophy is the basis of Kerckhoff's principle: all algorithms must be public; only the keys are secret. This is why the concept of keys is so fundamental in the study of cryptography, and this is why it is important to carefully choose the space of the keys.

Randomness and cryptography

Randomness is essential in cryptography: we cannot consider an algorithm secure enough if it is based on a predictable generation of random numbers. The concept of predictability in the generation of random numbers, however, is of fundamental importance in numerical simulation: however, it loses its usefulness in computer security. But what do we mean by predictability associated with generating random numbers? This concept is inversely related to the concept of entropy. But what does entropy have to do with IT security? Technicians know that entropy measures the disorder of a thermodynamic system.

If I have two empty piggy banks and I fill them with coins by flipping the coin each time to choose the piggy bank to fill, it is extremely unlikely that after thousands of tosses one piggy bank is empty and the other full. The uncertainty in the result of an experiment with a finite number of possible results each with probability p₁, p₂, ..., p_i is given by the following equation:

$$H = -\sum p_i \, log_2 \, p_i$$

H is measured in bits and is called the entropy of the experiment, thus measuring its unpredictability. Each experiment that leads to an increase in information costs a greater increase in entropy; it follows that any knowledge that can bring order is paid for with no less physical disorder. We can therefore conclude that entropy measures the randomness contained in a sequence of bits.

Referring to a string of characters, we can rewrite the entropy formula in a new form:

$$H = L * log_2(N)$$

Here, the following applies:

- H is the entropy and is measured in bits
- L is the number of the characters
- N is the number of values that each character can assume

The greater the value of H, the greater its entropy and the greater the unpredictability. If H measures the uncertainty of a string used as a password to access a computer system, it follows that the greater the entropy, the greater the unpredictability of this password. In order to increase this unpredictability, it will therefore be necessary to work on L, then increase the number of characters that make up the password. We will also be able to work on N, increasing the number of characters to choose from, and adding special characters to the letters and digits.

This is what we did in the example proposed in the *Random password generator* section. We then calculated the entropy of a password of 15 (L) characters as for the proposed example. Recall that the character set consisted of lowercase and uppercase characters, numbers, and, finally, some special characters. In all, we counted 26 uppercase letters, 26 lowercase letters, 10 digits, and 10 special characters, for a total of 72 (N) possible values. We apply the formula and the entropy calculation:

$$H = L * log_2(N) = 15 * log_2(72) = 92,54887502 \; bits$$

The value of entropy is given by the knowledge of a certain value; the more this knowledge increases, the more its entropy decreases.

Encrypted/decrypted message generator

In the *Introducing cryptography* section, we said that the need to generate encrypted messages dates back to antiquity. The message encryption techniques used by the Roman emperor Julius Caesar (Caesar cipher) to protect the messages he sent to his troops in the field remain famous and documented. Currently, the development of electronic systems facilitates communications but makes them vulnerable if they are not adequately protected. With encryption, we can make communication between two people or entities secret and secure by hiding the meaning of the message from everyone else. The strategy to be adopted includes the following phases:

- User A wants to send user B a message, OM (called a clear message)
- User A encrypts the message m obtaining a message, CM (called an encrypted message), which they send to B
- User B receives CM and decrypts it, getting OM back

To perform these operations, it is necessary that the encryption process be reversed, in order to allow the original message to be found. Whoever receives the encrypted message must be able to interpret it (decrypt). Users A and B must first agree on encryption and decryption techniques, choosing an effective method.

Encryption provides effective methods for encrypting and decrypting messages. The transformation process from the unencrypted message to the encrypted message and vice versa is often not known, but it is based on specific information (called a key), without which it is not possible to operate. Encryption methods have evolved extremely over the course of history. Mathematics is the fundamental tool for cryptography, for the following reasons:

- It provides methods for encrypting, decrypting, signing, and checking messages
- It ensures the security of encryption
- It studies (together with information technology) how to quickly carry out cryptographic operations

Modern cryptography has been strongly influenced by computer science and mathematics and has contributed in an essential way to the development of computer science. The evolution of IT tools poses new challenges by making it possible to easily decipher ciphers in a way previously considered impossible. New applications and new techniques are always required: for example, key exchange, public key cryptography, and digital signatures. The evolution of cryptography techniques follows innovations in the hardware and, especially, software fields hand in hand.

To get an idea of the tools currently available, we will analyze a simple example. We will see how to encrypt/decrypt a character string. Here is the Python code (`encript_decript_strings.py`):

```
from cryptography.fernet import Fernet

title = "Simulation Modeling with Python"
title=title.encode()
secret_key = Fernet.generate_key()
fernet_obj = Fernet(secret_key)
enc_title = fernet_obj.encrypt(title)
print("My last book title = ", title)
```

```
print("Title encrypted = ", enc_title)
dec_title = fernet_obj.decrypt(enc_title)
dec_title = dec_title.decode()
print("Title decrypted = ", dec_title)
```

So, let's analyze the code line by line to understand how we approached the problem. To get started, we imported the libraries:

```
from cryptography.fernet import Fernet
```

We imported the `Fernet` module from the `cryptography` library. This is a package of cryptographic modules and functions. The library contains both high-level and low-level interfaces for common cryptographic algorithms such as symmetric ciphers, message digests, and key derivation functions. The `Fernet` module allows us to perform symmetric (secret key) cryptography by generating strings of text that cannot be read without having the key.

To start, let's set up the message we want to encrypt:

```
title = "Simulation Modeling with Python"
```

We just used the title of this book. Now, we need to transform the text string into a bit string:

```
title=title.encode()
```

This method converts the text string to a byte string; only then can it be encrypted.

Once this is done, let's move on to generating a secret key that represents the essential element of symmetric cryptography. A secret key is used for both encryption and decryption, so it is essential to exchange the key between users without it being intercepted. In secret key cryptography methods, substitutions and transpositions are used but combined in complex algorithms. To generate a secret key, we will use the `generate_key()` method:

```
secret_key = Fernet.generate_key()
```

The `generate_key()` method uses the `os.urandom()` method: The `os.urandom()` method is used as a cryptographically secure PRNG. This method returns random bytes from an operating system-specific source of randomness, which is quite unpredictable for cryptographic applications. However, its quality depends on the operating system.

In the *Introducing the random module* section, we have already learned how to generate pseudorandom numbers. The main distinction between the methods provided by the `random` module and the `urandom()` method lies in the reproducibility of the generation of pseudorandom numbers. The `random` module implements deterministic PRNGs. There are scenarios in which the reproducibility of the experiment is a required requirement—for example, in the case of numerical simulation in which the random element must be reproducible in order to test the model that is meant to be simulated.

In this case, you need a deterministic PRNG that can be seeded. On the contrary, the urandom() method cannot be reproduced (no-seed) and generates its entropy from many unpredictable sources, making the result more random.

> **Important note**
> This randomness also has limitations: absolute random is something else that needs a physical source of randomness such as atomic decay measures. This is truly random in a physical sense but would be overkill for most applications.

Now, we need to instantiate the `Fernet` class:

```
fernet_obj = Fernet(secret_key)
```

We created an object as an instantiation of the `Fernet` class using the secret key generated in the previous step. Now, the newly created object has the methods of the `Fernet` class. In order to use them, we will adopt dot notation. Let's see this right away using the `encrypt()` method:

```
enc_title = fernet_obj.encrypt(title)
```

With this line of code, we have encrypted the title of the book. The result is called a Fernet token and has strong guarantees of privacy and authenticity. To verify the operations carried out, we print the clear and encrypted title on the screen:

```
print("My last book title = ", title)
print("Title encrypted = ", enc_title)
```

The following lines will be printed on the screen:

```
My last book title =  Simulation Modeling with Python
Title encrypted =   b'gAAAAABi1md0brrT0xLorTvM-FQMnEA702ZIqK7LK
pepHwdHEZ6ryX5D100EOYNlhzg5_EeO_YHKhmadZOj_efISEtM5zIPUfRmFd_
lv2LcamDkQ5M0H1WM='
```

Now, we just have to decrypt our string:

```
dec_title = fernet_obj.decrypt(enc_title)
```

To do this, we used the instantiated object of the `Fernet` class by applying the `decrypt()` method. This method decrypts a `Fernet` token: if the message is successfully decrypted, you will receive the original plaintext as a result; otherwise, an exception will be raised. It is safe to use this data immediately as `Fernet` verifies that the data has not been tampered with before returning it. For the decryption, we had to transform the byte string into a text string by applying the `decode()` method:

```
dec_title = dec_title.decode()
```

Finally, let's check the result:

```
print("Title decrypted = ", dec_title)
```

The following text string will be printed on the screen:

```
Title decrypted =  Simulation Modeling with Python
```

The encrypt/decrypt operation has been performed successfully. On screen, we can read the title of the book.

Summary

In this chapter, we learned how to define stochastic processes and understood the importance of using them to address numerous real-world problems. For instance, the operation of slot machines is based on the generation of random numbers, as are many complex data encryption procedures. Next, we introduced the concepts behind random number generation techniques. We explored the main methods of generating random numbers using practical examples in Python code. The generation of uniform and generic distributions was discussed. We also learned how to perform a uniformity test using the chi-squared method. Then, we looked at the main functions available in Python for generating random numbers: random, seed, uniform, randint, choice, and sample. Finally, we explored the randomness requirements for security systems, and we analyzed an encrypted/decrypted message generator.

In the next chapter, we will learn about the basic concepts of probability theory. Additionally, we will learn how to calculate the probability of an event happening after it has already occurred, and then we will learn how to work with discrete and continuous distributions.

3
Probability and Data Generation Processes

The field of **probability calculation** was born in the context of gambling. It was then developed further, assuming a relevant role in the analysis of collective phenomena and becoming an essential feature of statistics and statistical decision theory. Probability calculation is an abstract and highly formalized mathematical discipline while maintaining relevance to its original and pertinent empirical context. The concept of probability is strongly linked to that of uncertainty. The probability of an event can, in fact, be defined as the quantification of the level of randomness of that event. What is not known or cannot be predicted with an absolute level of certainty is known as being **random**. In this chapter, we will learn how to distinguish between the different definitions of probabilities and how these can be integrated to obtain useful information in the simulation of real phenomena.

In this chapter, we're going to cover the following main topics:

- Explaining probability concepts
- Understanding Bayes' theorem
- Probability distributions
- Generating synthetic data
- Data generation with Keras
- Simulation of power analysis

Technical requirements

This chapter will discuss an introduction to the theory of probability. To deal with these topics, it is necessary that you have a basic knowledge of algebra and mathematical modeling.

To install a library not contained in your Python environment, use the `pip install` command. To work with the Python code in this chapter, you need the following files (available on GitHub at the following URL: `https://github.com/PacktPublishing/Hands-On-Simulation-Modeling-with-Python-Second-Edition`):

- `Uniform_distribution.py`
- `Binomial_distribution.py`
- `Normal_distribution.py`
- `image_augmentation.py`
- `power_analysis.py`

Explaining probability concepts

If we take a moment to reflect, we'll notice that our everyday lives are full of **probabilistic** considerations, although not necessarily formalized as such. Examples of probabilistic assessments include choosing to participate in a competition given the limited chance of winning, the team's predictions of winning the championship, and statistics that inform us about the probability of death from smoking or failure to use seat belts in the event of a road accident, and the chances of winning in games and lotteries.

In all situations of uncertainty, there is basically a tendency to give a measure of uncertainty that, although indicated in various terms, expresses the intuitive meaning of probability. The fact that probability has an intuitive meaning also means that establishing its rules can, within certain limits, be guided by intuition. However, relying completely on intuition can lead to incorrect conclusions. To avoid reaching incorrect conclusions, it is necessary to formalize the calculation of probabilities by establishing their rules and concepts logically and rigorously.

Types of events

We define an **event** as any result to which, following an experiment or an observation, a well-defined degree of truth can be uniquely assigned. In everyday life, some events happen with certainty, while others never happen. For example, if a box contains only yellow marbles, by extracting one at random, we are sure that it will be yellow, while it is impossible to extract a red ball. We call the events of the first type – that is, extracting a yellow marble – **certain events**, while we call those of the second type – that is, extracting a red marble – **impossible events**.

These two types of events – certain and impossible – are events that can happen, but without certainty. If the box contains both yellow and red balls, then extracting a yellow ball is a possible but not certain event, as is extracting a red ball. In other words, we cannot predict the color of the extracted ball because the extraction is random.

Something that may or may not happen at random is called a **random event**. In *Chapter 2, Understanding Randomness and Random Numbers*, we introduced random events. An example of such a random event is being selected in chemistry to check homework over a week's worth of lessons.

The same event can be certain, random, or impossible, depending on the context in which it is considered. Let's analyze an example: winning the Mega Millions jackpot game. This event can be considered certain if we buy all the tickets for the game; it is impossible if we do not buy even one, and it is random if we buy one or more than one, but not all.

Calculating probability

The succession of random events has led people to formulate bets on their occurrence. The concept of probability was born precisely because of gambling. Over 3,000 years ago, the Egyptians played an ancestor of the dice game. The game of dice was widespread in ancient Rome too, so much so that some studies have found that this game dates back to the age of Cicero. But the birth of the systematic study of the calculation of probabilities dates back to 1654, by the mathematician and philosopher Blaise Pascal.

Probability definition with an example

Before we analyze some simple examples of calculating the probability of the occurrence of an event, it is good to define the concept of probability. To start, we must distinguish between a classical approach to the definition of probability and the **frequentist** point of view.

A priori probability

The **a priori probability** $p(E)$ of a random event E is defined as the ratio between the number s of the favorable cases and the number n of the possible cases, which are all considered equally probable:

$$P(E) = \frac{\text{number of the favorable cases}}{\text{number of the possible cases}} = \frac{s}{n}$$

In a box, there are 14 yellow marbles and 6 red marbles. The marbles are similar in every way except for their color; they're made of the same material, are the same size, are perfectly spherical, and so on. We'll put a hand into the box without looking inside, pulling out a random marble. What is the probability that the pulled-out marble is red?

In total, there are 14 + 6 = 20 marbles. By pulling out a marble, we have 20 possible cases. We have no reason to think that some marbles are more privileged than others; that is, they are more likely to be pulled out. Therefore, the 20 possible cases are equally probable.

Of these 20 possible cases, there are only 6 cases in which the marble being pulled out is red. These are the cases that are favorable to the expected event.

Therefore, the red marble being pulled out has 6 out of 20 possible occurrences. Defining its probability as the ratio between the favorable and possible cases, we will get the following:

$$P(E) = P(\text{red marble pulled out}) = \frac{6}{20} = 0.3 = 30\%$$

Based on the definition of probability, we can say the following:

- The probability of an impossible event is 0
- The probability of a certain event is 1
- The probability of a random event is between 0 and 1

Previously, we introduced the concept of equally probable events. Given a group of events, if there is no reason to think that some event occurs more frequently than others, then all group events should be considered equally likely.

Complementary events

Complementary events are two events – usually referred to as E and $\bar{E}$ – that are mutually exclusive.

For example, when rolling some dice, we consider the event as E = number 5 comes out.

The complementary event will be $\bar{E}$ = number 5 does not come out.

E and $\bar{E}$ are mutually exclusive because the two events cannot happen simultaneously; they are exhaustive because the sum of their probabilities is 1.

For event E, there are 1 (5) favorable cases, while for event $\bar{E}$, there are 5 favorable cases, that is, all the remaining cases (1, 2, 3, 4, 6). So, the a priori probability is as follows:

$$P(E) = \frac{1}{6} \, ; P(\bar{E}) = \frac{5}{6}$$

Due to this, we can observe the following:

$$P(E) + P(\bar{E}) = \frac{1}{6} + \frac{5}{6} = 1$$

Relative frequency and probability

However, the classical definition of probability is not applicable to all situations. To affirm that all cases are equally probable is to make an a priori assumption about their probability of occurring, thus using the same concept in the definition that you want to define.

The relative frequency *f(E)* of an event subjected to *n* experiments, all carried out under the same conditions, is the ratio between the number *v* of the times the event occurred, and the number *n* of tests carried out:

$$f(E) = \frac{v}{n}$$

If we consider the toss of a coin and the event *E* = heads up, classical probability gives us the following value:

$$P(E) = \frac{1}{2}$$

If we perform many throws, we will see that the number of times the coin landed heads up is almost equal to the number of times a cross occurs. That is, the relative frequency of the event *E* approaches the theoretical value:

$$f(E) \cong P(E) = \frac{1}{2}$$

Given a random event *E*, subjected to *n* tests performed all under the same conditions, the value of the relative frequency tends to the value of the probability as the number of tests carried out increases.

> **Important note**
> The probability of a repeatable event coincides with the relative frequency of its occurrence when the number of tests being carried out is sufficiently high.

Note that in the classical definition, the probability is evaluated a priori, while the frequency is a value that's evaluated posteriori.

The frequency-based approach is applied, for example, in the field of insurance, to assess the average life span of an individual, the probability of theft, and the probability of accidents. It can also be applied in the field of medicine in order to evaluate the probability of contracting a certain disease or the probability that a drug is effective. In all these events, the calculation is based on what has happened in the past, that is, by evaluating the probability by calculating the relative frequencies.

Let's now look at another approach we can use to calculate probabilities that estimate the levels of confidence in the occurrence of a given event.

Understanding Bayes' theorem

From the Bayesian point of view, probability measures the degree of likelihood that an event will occur. It is an inverse probability in the sense that from the observed frequencies, we obtain the probability.

Bayesian statistics foresee the calculation of the probability of a certain event before carrying out the experiment; this calculation is made based on previous considerations. Using Bayes' theorem, by using the observed frequencies, we can calculate the a priori probability, and from this, we can determine the posterior probability. By adopting this method, the prediction of the degree of credibility of a given hypothesis is used before observing the data, which is then used to calculate the probability after observing the data.

> **Important note**
> In the frequentist approach, we determine how often the observation falls within a certain interval, while in the Bayesian approach, the probability of truth is directly attributable to the interval.

In cases where a frequentist result exists within the limit of a very large sample, the Bayesian and frequentist results coincide. There are also cases where the frequentist approach is not applicable.

Compound probability

Now, consider two events, *E1* and *E2*, where we want to calculate the probability *P(E1∩E2)* that both occur. Two cases can occur:

- *E1* and *E2* are stochastically independent
- *E1* and *E2* are stochastically dependent

The two events, *E1* and *E2*, are stochastically independent if they do not influence each other, that is, if the occurrence of one of the two does not change the probability of the second occurring. Conversely, the two events, *E1* and *E2*, are stochastically dependent if the occurrence of one of the two changes the probability of the second occurring.

Let's look at an example: you draw a card from a deck of 40 that contains the numbers 1 to 7, plus the 3 face cards for each suit. What is the probability that it is a face card and from the hearts suit?

To start, we must ask ourselves whether the two events are dependent or independent.

There are 12 faces, 3 for each symbol, so the probability of the first event is equal to 12/40, that is, 3/10. The probability that the card is from the hearts suit is not influenced by the occurrence of the event that the card is a face card; therefore, it is worth 10/40, that is, 1/4. Therefore, the compound probability will be 3/40.

Therefore, this is a case of independent events. The **compound probability** is given by the product of the probabilities of the individual events, as follows:

$$P(E_1 \cap E_2) = P(E_1) * P(E_2) = \frac{3}{10} * \frac{1}{4} = \frac{3}{40}$$

Let's look at a second example: we draw a card from a deck of 40 and, without putting it back in the deck, we draw a second one. What is the probability that they are two queens?

The probability of the first event is 4/40, that is, 1/10. But when drawing the second card, there are only 39 remaining, and there are only 3 queens. So, the probability that the second card is still a queen will have become 3/39, that is, 1/13. Therefore, the compound probability will be given by the product of the probability that the first card is a queen for the probability that the second is still a queen, that is, 1/130.

Thus, this is a case of dependent events; that is, the probability of the second event is conditioned by the occurrence of the first event. Similarly, the two events are considered dependent if the two cards are drawn simultaneously when there is no reintegration.

When the probability of an *E2* event depends on the occurrence of the *E1* event, we speak of the **conditional probability**, which is denoted by *P(E2|E1)*, and we see that the probability of *E2* is conditional on *E1*.

When the two events are stochastically dependent, the compound probability is given by the following equation:

$$P(E_1 \cap E_2) = P(E_1) * P(E_2|E_1) = \frac{1}{10} * \frac{1}{13} = \frac{1}{130}$$

From the previous equation, we can derive the equation that gives us the conditional probability:

$$P(E_2|E_1) = \frac{P(E_1 \cap E_2)}{P(E_1)}$$

After defining the concept of conditional probability, we can move on and analyze the heart of Bayesian statistics.

Bayes' theorem

Let's say that *E1* and *E2* are two dependent events. In the *Compound probability* section, we learned that the compound probability of the two events is calculated using the following equation:

$$P(E_1 \cap E_2) = P(E_1) * P(E_2|E_1)$$

By exchanging the order of succession of the two events, we can write the following equation:

$$P(E_1 \cap E_2) = P(E_2) * P(E_1|E_2)$$

The left-hand part of the two previous equations contains the same quantity, which must also be true for the right-hand part. Based on this consideration, we can write the following equation:

$$P(E_2|E_1) = \frac{P(E_2) * P(E_1|E_2)}{P(E_1)}$$

The same is true by exchanging the order of events:

$$P(E_1|E_2) = \frac{P(E_1) * P(E_2|E_1)}{P(E_2)}$$

The preceding equations represent the mathematical formulation of Bayes' theorem. The use of one or the other depends on the purpose of our work. Bayes' theorem is derived from two fundamental probability theorems: the compound probability theorem and the total probability theorem. It is used to calculate the probability of a cause that triggered the verified event.

In Bayes' theorem, we know the result of the experiment, and we want to calculate the probability that it is due to a certain cause. Let's analyze the elements that appear in the equation that formalizes Bayes' theorem in detail:

$$P(E_2|E_1) = \frac{P(E_2) * P(E_1|E_2)}{P(E_1)}$$

Here, we have the following:

- $P(E_2|E_1)$ is called **posterior probability** (what we want to calculate)
- $P(E_2)$ is called **prior probability**
- $P(E_1|E_2)$ is called **likelihood** (represents the probability of observing the E_1 event when the correct hypothesis is E_2)
- $P(E_1)$ is called **marginal likelihood**

Bayes' theorem applies to many real-life situations, such as in the medical field for finding false positives in one analysis or verifying the effectiveness of a drug.

Now, let's learn how to represent the probabilities of possible results in an experiment.

Exploring probability distributions

A **probability distribution** is a mathematical model that links the values of a variable to the probabilities that these values can be observed. Probability distributions are used to model the behavior of a phenomenon of interest in relation to the reference population, or to all the cases in which the researcher observes a given sample.

Based on the measurement scale of the variable of interest X, we can distinguish two types of probability distributions:

- **Continuous distributions**: The variable is expressed on a continuous scale
- **Discrete distributions**: The variable is measured with integer numerical values

In this context, the variable of interest is seen as a random variable whose probability law expresses the degree of uncertainty with which its values can be observed. Probability distributions are expressed by a mathematical law called the **probability density function** (*f(x)*) or **probability function** (*p(x)*) for continuous or discrete distributions, respectively. The following diagram shows a continuous distribution (to the left) and a discrete distribution (to the right):

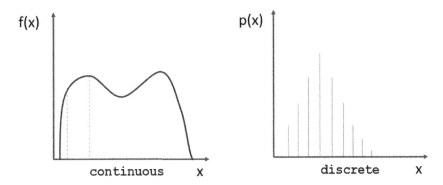

Figure 3.1 – A continuous distribution and a discrete distribution

To analyze how a series of data is distributed, which we assume can take any real value, it is necessary to start with the definition of the probability density function. Let's see how that works.

The probability density function

The **probability density function** (**PDF**) *P(x)* represents the probability *p(x)* that a given *x* value of the continuous variable is contained in the interval (*x*, *x* + Δ*x*), divided by the width of the interval Δ*x* when this tends to be zero:

$$P(x) = \lim_{\Delta x \to 0} \frac{p(x + \Delta x) - p(x)}{\Delta x} = \frac{dp(x)}{dx}$$

The probability of finding a given *x* value in the interval [*a*, *b*] is given by the following equation:

$$p(x) = \int_a^b P(x)dx$$

Since *x* takes a real value, the following property holds:

$$\int_{-\infty}^{+\infty} P(x)dx = 1$$

In practice, we do not have an infinite set of real values but rather a discrete set *N* of real numbers x_i. Then, we proceed by dividing the interval [*xmin*, *xmax*] into a certain number *Nc* of subintervals (bins) of amplitude Δx, considering the definition of probability as the ratio between the number of favorable cases and the number of possible cases.

The calculation for the PDF refers to dividing the interval [xmin, xmax] into Nc subintervals and counting how many xi values fall into each of these subintervals before dividing each value by Δx*N, as shown in the following equation:

$$P(x) = \frac{n_i}{\Delta x * N}$$

Here, we can see the following:

- P(x) is the PDF
- ni is the number of x values that fall into the i-th sub-interval
- Δx is the amplitude of each sub-interval
- N is the number of observations x

Now, let's learn how to determine the probability distribution of a variable in the Python environment.

Mean and variance

The expected value, which is also called the average of the distribution of a random variable, is a position index. The expected value of a random variable represents the expected value that can be obtained with a sufficiently large number of tests so that it is possible to predict, by probability, the relative frequencies of various events.

The expected value of a discrete random variable, if the distribution is finite, is a real number given by the sum of the products of each value of the random variable for the respective probability:

$$E(x) = x_1 p_1 + x_2 p_2 + \cdots + x_n p_n = \sum_{i=1}^{n} x_i p_i$$

The expected value is, therefore, a weighted sum of the values that the random variable assumes when weighted with the associated probabilities. Due to this, it can be either negative or positive.

After the expected value, the most used parameter to characterize the probability distributions of the random variables is the variance, which indicates how scattered the values of the random variable are relative to its average value.

Given a random variable X, whatever E(X) is is its expected value. Consider the random variable X– E (X), whose values are the distances between the values of X and the expected value E(X). Substituting a variable X for the variable X-E (X) is equivalent to translating the reference system that brings the expected value to the origin of the axes.

The variance of a discrete random variable X, if the distribution is finite, is calculated with the following equation:

$$\sigma^2 = \sum_{i=1}^{n}(x_i - E(x))^2 * p_i$$

The variance is equal to zero when all the values of the variable are equal, and therefore, there is no variability in the distribution; in any case, it is positive and measures the degree of variability of a distribution. The greater the variance, the more scattered the values are. The smaller the variance, the more the values of X are concentrated around the average value.

Uniform distribution

The simplest of the continuous variable probability distribution functions is the one in which the same degree of confidence is assigned to all the possible values of a variable defined in a certain range. Since the probability density function is constant, the distribution function is linear. The uniform distribution is used to treat measurement errors whenever they occur with the certainty that a certain variable is contained in a certain range, but there is no reason to believe some values are more plausible than others. Using suitable techniques, starting from a uniformly distributed variable, it is possible to build other variables that have been distributed at will.

Now, let's start practicing using it. We will start by generating a uniform distribution of random numbers contained within a specific range. To do this, we will use the numpy random.uniform() function. This function generates random values uniformly distributed over the half-open interval *[a, b)*; that is, it includes the first but excludes the second. Any value within the given interval is equally likely to be drawn by uniform distribution:

1. To start, we import the necessary libraries:

   ```
   import numpy as np
   import matplotlib.pyplot as plt
   ```

   The numpy library is a Python library that contains numerous functions that help us manage multidimensional matrices. Furthermore, it contains a large collection of high-level mathematical functions that we can perform on these matrices.

   The matplotlib library is a Python library for printing high-quality graphics. With matplotlib, it is possible to generate graphs, histograms, bar graphs, power spectra, error graphs, scatter graphs, and so on with a few commands. This is a collection of command-line functions like those provided by the MATLAB software.

2. After this, we define the extremes of the range and the number of values we want to generate:

   ```
   a=1
   b=100
   N=100
   ```

Now, we can generate the uniform distribution using the `random.uniform()` function, as follows:

```
X1=np.random.uniform(a,b,N)
```

With that, we can view the numbers that we generated. To begin, draw a diagram in which we report the values of the 100 random numbers that we have generated:

```
plt.plot(X1)
plt.show()
```

The following graph will be output:

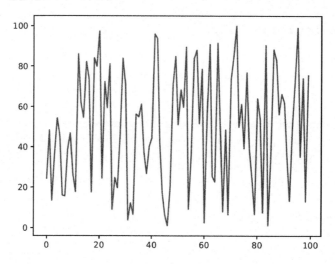

Figure 3.2 – Diagram plotting the 100 numbers

3. At this point, to analyze how the generated values are distributed within the interval considered, we draw a graph of the probability density function:

```
plt.figure()
plt.hist(X1, density=True, histtype='stepfilled',
alpha=0.2)
plt.show()
```

The `matplotlib.hist()` function draws a histogram, that is, a diagram of a continuous character shown in classes. It is used in many contexts, usually to show statistical data, when there is an interval of the definition of the independent variable divided into subintervals. These subintervals can be intrinsic or artificial, can be of equal or unequal amplitude, and are or can be considered constant. Each of these can either be an independent or dependent variable. Each rectangle has a non-random length equal to the width of the class it represents. The height of each rectangle is equal to the ratio between the absolute frequency associated with the class and the amplitude of the class, and it can be defined as **frequency density**. The following four parameters are passed:

- X1: Input values.
- density=True: This is a bool where, if True, the function returns the counts normalized to form a probability density.
- histtype='stepfilled': This parameter defines the type of histogram to draw. The stepfilled value generates a line plot that is filled by default.
- alpha=0.2: This is a float value that defines the characteristics of the content (0.0 transparent through 1.0 opaque).

The following graph will be output:

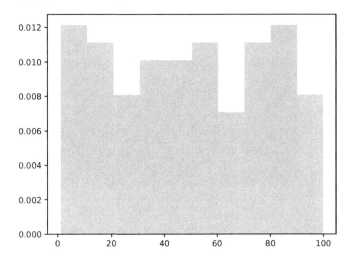

Figure 3.3 – Graph plotting the generated values

Here, we can see that the generated values are distributed almost evenly throughout the range. What happens if we increase the number of generated values?

4. Then, we repeat the commands we just analyzed to modify only the number of samples to be managed. We change this from 100 to 10000:

```
a=1
b=100
N=10000
X2=np.random.uniform(a,b,N)

plt.figure()
```

```
plt.plot(X2)
plt.show()

plt.figure()
plt.hist(X2, density=True, histtype='stepfilled',
alpha=0.2)
plt.show()
```

It is not necessary to reanalyze the piece of code line by line since we are using the same commands. Let's see the results, starting from the generated values:

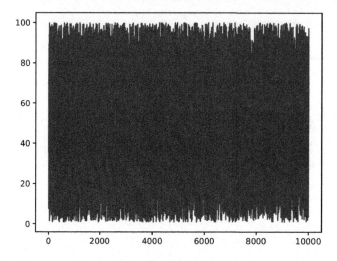

Figure 3.4 – Graph plotting the number of samples

Now, the number of samples that have been generated has increased significantly. Let's see how they are distributed within the range considered:

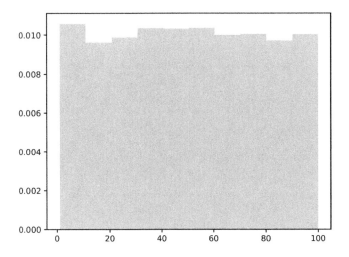

Figure 3.5 – Graph showing the sample distribution

Analyzing the previous histogram and comparing it with what we obtained in the case of N=100, we can see that this time, the distribution appears to be flatter. The distribution becomes flatter as N increases, increasing the statistics in each individual bin.

Binomial distribution

In many situations, we are interested in checking whether a certain characteristic occurs or not. This corresponds to an experiment with only two possible outcomes – also called dichotomous – that can be modeled with a random variable X that assumes value 1 (success) with probability p and value 0 (failure) with probability $1-p$, with $0 < p < 1$, as follows:

$$X = \begin{cases} 1, & p \\ 0, & 1-p \end{cases}$$

The expected value and variance of X are calculated as follows:

$$E(X) = 0 * (1-p) + 1 * p = p$$

$$\sigma^2 = E(X^2) - \big(E(X)\big)^2 = (0^2 * (1-p) + 1^2 * p) - p^2 = p * (1-p)$$

The binomial distribution is the probability of obtaining x successes in n independent trials. The probability density for the binomial distribution is obtained using the following equation:

$$P_x = \binom{n}{x} p^x q^{n-x}, 0 \leq x \leq n$$

Here, we have the following:

- Px is the probability density
- n is the number of independent experiments
- x is the number of successes
- p is the probability of success
- q is the probability of fail

Now, let's look at a practical example. We throw a dice n = 10 times. In this case, we want to study the binomial variable x = number of times a number <= 3 came out. We define the parameters of the problem as follows:

- $n = 10$
- $0 \leq x \leq n$
- $p = \dfrac{3}{6} = 0.5$
- $q = 1 - p = 0.5$

We then evaluate the probability density function with Python code as follows:

1. Let's start as always by importing the necessary libraries:

    ```
    import numpy as np
    import matplotlib.pyplot as plt
    ```

    Now, we set the parameters of the problem:

    ```
    N = 1000
    n = 10
    p = 0.5
    ```

    Here, N is the number of trials, n is the number of independent experiments in each trial, and p is the probability of success for each experiment.

2. Now, we can generate the probability distribution:

   ```
   P1 = np.random.binomial(n,p,N)
   ```

   The numpy random.binomial() function generates values from a binomial distribution. These values are extracted from a binomial distribution with the specified parameters. The result is a parameterized binomial distribution, in which each value is equal to the number of successes obtained in the n independent experiments. Let's take a look at the return values:

   ```
   plt.plot(P1)
   plt.show()
   ```

   The following graph is output:

   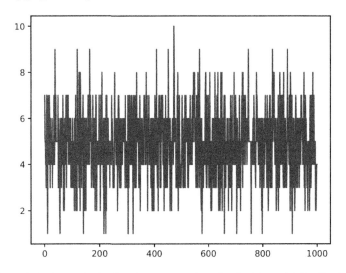

   Figure 3.6 – A graph plotting the return values for the binomial distribution

   Let's see how these samples are distributed within the range considered:

   ```
   plt.figure()
   plt.hist(P1, density=True, alpha=0.8, histtype='bar',
   color = 'green', ec='black')
   plt.show()
   ```

   This time, we used a higher alpha value to make the colors brighter, we used the traditional bar-type histogram, and we set the color of the bars.

3. Finally, we used the ec parameter to set the edge color of each bar. The following results are obtained:

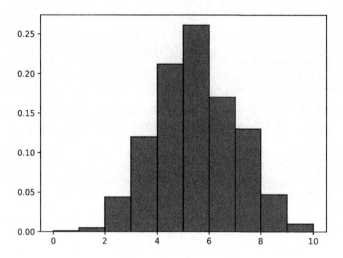

Figure 3.7 – Histogram plotting the return values

All the areas of the binomial distributions, that is, the sum of the rectangles, being the sum of probability, are worth 1.

Normal distribution

As the number of independent experiments that are carried out increases, the binomial distributions approach a curve called the **bell curve** or **Gauss curve**. The **normal distribution**, also called the **Gaussian distribution**, is the most used continuous distribution in statistics. Normal distribution is important in statistics for the following fundamental reasons:

- Several continuous phenomena seem to follow, at least approximately, a normal distribution
- The normal distribution can be used to approximate numerous discrete probability distributions
- The normal distribution is the basis of classical statistical inference by virtue of the central limit theorem (the mean of a large number of independent random variables with the same distribution is approximately normal, regardless of the underlying distribution)

Normal distribution has some important characteristics:

- The normal distribution is symmetrical and bell-shaped
- Its central position measures – the expected value and the median – coincide
- Its interquartile range is 1.33 times the mean square deviation
- The random variable in the normal distribution takes values between $-\infty$ and $+\infty$

In the case of a normal distribution, the normal probability density function is given by the following equation:

$$f(X) = \frac{1}{\sqrt{2\pi}\sigma} e^{-(1/2)[(X-\mu)/\sigma]^2}$$

Here, we have the following:

- μ is the expected value
- σ is the standard deviation

Note that, since e and π are mathematical constants, the probabilities of a normal distribution depend only on the values assumed by the parameters μ and σ.

Now, let's learn how to generate a normal distribution in Python. Let's start as always by importing the necessary libraries:

```
import numpy as np
import matplotlib.pyplot as plt
import seaborn as sns
```

Here, we have imported a new `seaborn` library. It is a Python library that enhances the data visualization tools of the `matplotlib` module. In the `seaborn` module, there are several features we can use to graphically represent our data. There are methods that facilitate the construction of statistical graphs with `matplotlib`.

Now, we set the parameters of the problem. As we've already mentioned, only two parameters are needed to generate a normal distribution: the expected value and the standard deviation. The μ value is also indicated as the center of the distribution and characterizes the position of the curve with respect to the ordinate axis. The σ parameter characterizes the shape of the curve since it represents the dispersion of the values around the maximum of the curve.

To appreciate the functionality of these two parameters, we will generate a normal distribution by changing the values of these parameters, as follows:

```
mu = 10
sigma =2
P1 = np.random.normal(mu, sigma, 1000)


mu = 5
sigma =2
P2 = np.random.normal(mu, sigma, 1000)
```

```
mu = 15
sigma =2
P3 = np.random.normal(mu, sigma, 1000)

mu = 10
sigma =2
P4 = np.random.normal(mu, sigma, 1000)

mu = 10
sigma =1
P5 = np.random.normal(mu, sigma, 1000)

mu = 10
sigma =0.5
P6 = np.random.normal(mu, sigma, 1000)
```

For each distribution, we have set the two parameters (μ and σ) and then used the `numpy.random.normal()` function to generate a normal distribution. Three parameters are passed: μ, σ, and the number of samples to generate. At this point, it is necessary to view the generated distributions. To do this, we will use the `histplot()` function of the `seaborn` library, as follows:

```
Plot1 = sns.histplot(P1,stat="density", kde=True, color="g")
Plot2 = sns.histplot(P2,stat="density", kde=True, color="b")
Plot3 = sns.histplot(P3,stat="density", kde=True, color="y")

plt.figure()
Plot4 = sns.histplot(P4,stat="density", kde=True, color="g")
Plot5 = sns.histplot(P5,stat="density", kde=True, color="b")
Plot6 = sns.histplot(P6,stat="density", kde=True, color="y")
plt.show()
```

The `histplot()` function allows us to flexibly plot a univariate or bivariate distribution of observations. Let's first analyze the results that were obtained in the first graph:

Exploring probability distributions   89

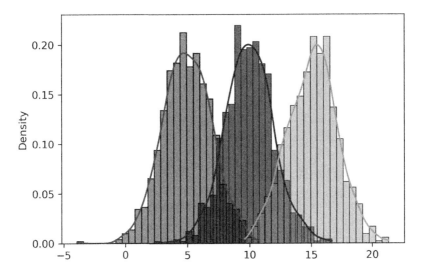

Figure 3.8   Seaborn plot of the samples

Three curves have been generated that represent the three distributions we have named: *P1*, *P2*, and *P3*. The only difference that we can notice lies in the value of μ, which assumes the values 5, 10, and 15. Due to the variation of μ, the curve moves along the *x*-axis, but its shape remains unchanged. Let's now see the graph that represents the remaining distributions:

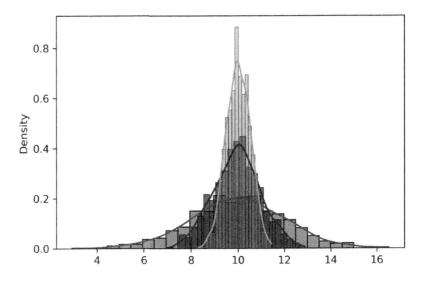

Figure 3.9 – Merged plots

In this case, by keeping the value of μ constant, we have varied the value of σ, which assumes the following values: 2, 1, and 0.5. As σ increases, the curve flattens and widens, while as σ decreases, the curve narrows and rises.

A specific normal distribution is one that's obtained with $\mu = 0$ and $\sigma = 1$. This distribution is called the **standardized normal distribution**.

Now that we've seen all the relevant kinds of probability distribution, let's learn how to generate data artificially.

Generating synthetic data

Machine learning-based systems have shown great progress and great results in real-world applications, but they have major limitations due to the quality of the data processed. In fact, the results and performances that these models return are strongly dependent on the data, on their quantity, and above all, on their quality. However, it is evident that the manual process of annotating and labeling data requires a very high level of work, which obviously increases with the amount of data generated.

Real data versus artificial data

The use of simulation systems as a method of data collection, then, becomes an effective solution that allows you to produce a large amount of data of better quality and with much less human effort. It is, in fact, possible to programmatically annotate the data that is produced and do so at a speed far superior to the real case. The precision of associating the correct annotation with data can make the difference in having an algorithm capable of interacting effectively with the surrounding environment. Usually, this is done by human annotators, resulting in an additional cost to be taken into consideration, in addition to the poor accuracy of the annotations. Further problems arise from the restrictions in terms of privacy that may occur if there is a need to have categories of sensitive data available. Generating annotations that are as accurate as possible when dealing with a massive amount of data is a challenge for the developer community. Although for some functions, more than satisfactory datasets are already available, in other contexts, you may have to deal with data that errs in the absence of precise annotations.

The availability of low-cost sensors and the ability to connect them together to create data collection networks is not always an advantage. Researchers find limitations when testing their methods because they need labeled data for the validation of their experiments. And even if information from sensors or sensor networks is available, it may not be accessible for legal reasons. Furthermore, obtaining data from a sensor network is associated with a double cost because the realization of a sensor network can be expensive in economic terms and in terms of time since the sensors must perform a sampling, which in some cases involves lengthening the time taken by weeks or even months, to have the data available and carry out the experimentation.

As a solution to such problems, there are various data repositories, both public and private. However, they are not always adapted to the needs of the problem they intend to deal with. The solution to all these limitations is the generation of synthetic data that emulates the behavior of the reality of the problem to be faced. This data is artificially created from detailed information on the process it is intended to represent. In this way, the synthetic data captures the structural and statistical characteristics of the original data without associating the personal information contained in the original dataset. By generating data on the basis of the statistical distributions derived from the original data, it is possible to obtain data that has similar behavior to the data returned by various sensors. The generation of pseudo-realistic character data cannot consist of obtaining random samples with a uniform probability distribution since the result would be a sequence of values devoid of any coherence. For this reason, several mechanisms are required to modify the original data. The generated data can be stored in the most popular database formats used for experimentation.

The use of synthetic data yields multiple advantages. For example, we obtain robustness since the sensors can return incorrect data in certain cases. Data obtained from a synthetic data generator lacks this problem. Another noteworthy feature is security, as synthetic data can be generated with a level of detail and realism that means you don't have to take any kind of risk.

Feature selection methodologies can be used to extract the characteristic properties of the original database. Feature selection is the process of identifying a set of data attributes that constitute the essential features of a dataset. Feature selection is commonly used in machine learning to limit the dimensionality of the input, remove irrelevant or redundant attributes, and filter out noisy data, with the aim of improving generalization, learning speed, or reducing the complexity of the template. The reduction of dimensionality through the selection of features has become even more important in recent times with the advent of big data.

Feature selection algorithms are commonly evaluated based on their stability or the mean probability of models trained on selected subsets of characteristics of the input data agreeing with their predictions. Feature selection can be associated with feature extraction, which creates new features as features of other features, where features found through selection directly map existing attributes. Feature extraction is most employed in data that has relatively few samples relative to the number of attributes.

In the context of synthetic data, feature selection techniques are used to identify attribute dependencies that should be modeled and reproduced in the output dataset. Additionally, high dimensionality in real-world data can be computationally challenging for generating synthetic data. Limiting the selection of modeled attributes to those that are important to the synthetic data use case can greatly improve performance and quality.

Features extracted into the real dataset can introduce other challenges to quality data generation if the features they derive from are reproduced as well, as this introduces additional feature relationships that should be maintained in the output.

Synthetic data generation methods

Synthetic data generation can be divided into two stages: kernel extraction and synthesis. In general, kernel extraction consists of analyzing real data and simulation requirements to identify the algorithms and parameters to be used. Synthesis is the invocation of the kernel algorithms to produce an output based on the derived parameters. So, let's see some of the methodologies used for the generation of synthetic data.

Structured data

Structured data is where its characteristics are easily understood and can be summarized by tables that allow them to be easily read and compared with other data. Thanks to this feature, it is much easier to analyze than unstructured data and typically takes the form of written text. It constitutes highly organized information.

The generation of structured synthetic data is not critical and can be generalized. There are several approaches to identifying and modeling value distributions and feature interdependencies. The most popular methods involve the adoption of independent statistical distributions of type t for each characteristic and the subsequent generation of data by imputation. For noise reduction, multiple imputations or aggregations of multiple synthetic datasets generated by the same kernel can be adopted.

The relationships between features can be reconstructed in several ways: For example, we can adopt synthetic reconstruction and combinatorial optimization, which apply conditional probability to impute different distributions based on feature relationships. The increase in computing power has led to the feasibility of more sophisticated machine learning techniques for maintaining feature relationships, such as the use of support vector machines and random forests.

Other forms of structured data include images and time series measurements, in which the relationships between features can be described spatially within a record. Technically, any image transformation method can be classified as generating new synthetic image data, but a particularly useful application of image synthesis is augmenting training images in a deep learning pipeline.

Semi-structured data

Unstructured or non-relational data is essentially the opposite of structured data. It has a structure within it, but it is difficult to summarize with schemes or models. Therefore, there is no schema, as in the case of multimedia objects or narrative text-only files. Semi-structured data is data with partial structure. It has characteristics of both structured and unstructured data. The email is a typical example of information containing structured and unstructured data; in fact, there are both personal data connotations and textual or multimedia content.

The main format for representing semi-structured data is XML. It can be used both to represent structured data, for example, for the purpose of exchanging it between different applications, and to represent semi-structured data, taking advantage of its flexibility and the possibility of indicating both the data and the schema. Synthesis of semi-structured documents such as XML can be challenging, particularly when the documents have complex nested structures that depend on values within the document. One approach is to treat document structures as templates and generate documents based on the frequency of template examples.

Rule-based procedural generation is another possibility, especially for recreating complex and plausible data structures. The extracted kernel could be condensed into a set of schema segments with rules for rebuilding, along with value distributions.

Having introduced the concepts behind the generation of synthetic data, let's now see some practical examples with the use of the Python Keras library.

Data generation with Keras

Keras is an open source library written in Python that contains algorithms based on machine learning and for training neural networks. The purpose of this library is to allow the configuration of neural networks. Keras does not act as a framework but as a simple interface (API) for accessing and programming various machine learning frameworks. Among these frameworks, Keras supports the TensorFlow, Microsoft Cognitive Toolkit, and Theano libraries. This library provides fundamental components on which complex machine learning models can be developed. Keras allows you to define even complex algorithms in a few lines of code, allowing you to define training and evaluation in a single line of code.

In the *Generating synthetic data* section, we saw that the generation of synthetic data represents a powerful tool to overcome the criticalities related to data scarcity or restrictions due to privacy that some types of data present.

Data augmentation

The lack of data represents a real problem in many applications. A widespread way to increase the number of available samples lies in the methodology called data augmentation. Data augmentation is a set of techniques that expands the dataset without adding new elements but applying random changes to the data that already exists. In applications that make use of machine learning, this is a widely used methodology because it reduces convergence times, increases the generalization capacity, prevents overfitting, and regularizes the model. This technique allows the enrichment of data through transformation procedures in the data and/or features.

In the case of an image dataset, there are several techniques that allow for the extraction of variations through artificial transformations of the images. The visual characteristics of an object in an image are different: brightness, focus, rotation, distance from the point of view, background, shape, and color. Let's see some of them in detail:

- **Flipping**: Both horizontal and vertical – the choice of one or both depends on the characteristics of the object. Car recognition does not benefit much from a vertical flip, as the overturned car will not be recognized by the algorithm. While for an object positioned on a horizontal plane, both flips can be considered useful. Similarly, the choice to rotate an object or not in the data augmentation phase is dictated by how it is supposed to be arranged in testing.

- **Cropping**: Random cropping into smaller pieces starting from an image allows the network to adapt to the case where the classification of an object of which only a part has been acquired is required. For example, for the recognition of animals, random crops allow a network to recognize them even if only the face is available. A typical way to proceed is to acquire images at a resolution higher than that required and take clippings with a resolution equal to that of the network input.
- **Rotation**: These are rotations made with respect to an axis of the image. The safety of the operation lies in the angular range of rotation performed.
- **Translation**: This represents a rigid translation of the image in one of the four main directions and is a very effective technique as it avoids positional bias in the data. For example, if all images in the dataset are centered, this means that the classifier should only be tested on perfectly centered images. To preserve the original spatial dimensions, the remaining space after the translation is filled either with a constant value (such as 0 or 255) or with Gaussian or random noise.

Color must be added to the spatial transformations. Color images are encoded as three stacked matrices, each height x width in size. These matrices represent the pixel values for each RGB channel. Brightness-related biases are among the most common challenges in recognition problems. Therefore, the meaning of the transformations in this context is understandable. A quick way to change the brightness is to add or subtract a constant value from the individual pixels or isolate the three channels. Another option is to manipulate the color histogram to apply filters that modify the characteristics of the color space. The disadvantages related to these techniques concern the required memory increase, transformation costs, and training time. They can also lead to the loss of information about the colors themselves and therefore represent an unsafe transformation. So, let's see some transformations in the color space:

- **Brightness and contrast**: The lighting conditions can significantly affect the recognition of an object. Therefore, in addition to acquiring images in the field with different light conditions, it is possible to act by randomly changing the brightness and contrast of the image.
- **Distortion**: This property is interesting to consider for objects subjected to various types of distortion (stretching) or for objects acquired from slightly different perspectives.

The choice of ideal transformations to consider is given by the characteristics of the dataset. The general rule to consider is to maximize the variation of the transformations of the objects within each class and minimize the variation between different classes. The use of data augmentation causes a slower model training convergence, a factor that is irrelevant in the face of greater accuracy in the testing phase.

The Python Keras library allows you to perform data augmentation. The `ImageDataGenerator` class has methods for generating images starting from a source and applying spatial and color space transformations. To understand how this class is used, we will analyze an example in detail. Here is the Python code (`image_augmentation.py`):

```
from keras.preprocessing.image import load_img
from keras.preprocessing.image import img_to_array
```

```
from keras.preprocessing.image import ImageDataGenerator

image_generation = ImageDataGenerator(
      rotation_range=10,
      width_shift_range=0.1,
      height_shift_range=0.1,
      shear_range=0.1,
      zoom_range=0.1,
      horizontal_flip=True,
      fill_mode='nearest')

source_img = load_img('colosseum.jpg')
x = img_to_array(source_img)
x = x.reshape((1,) + x.shape)

i = 0
for batch in image_generation.flow(x, batch_size=1,
            save_to_dir='AugImage', save_prefix='new_image',
            save_format='jpeg'):
   i += 1
   if i > 50:
       break
```

Now let's analyze the code just proposed line by line to understand all the commands. Let's start with the import of the libraries:

```
from keras.preprocessing.image import load_img
from keras.preprocessing.image import img_to_array
from keras.preprocessing.image import ImageDataGenerator
```

Three libraries were imported to process the data. All libraries belong to the Keras preprocessing module. This module contains utilities for data preprocessing and augmentation and provides utilities for working with image, text, and sequence data. To get started we imported the load_img () function, which allows you to load an image from a file as a PIL image object. **PIL** stands for **Python Imaging Library**, and it provides the Python interpreter with image editing capabilities. The second function we imported is img_to_array(), which converts a PIL image into a NumPy array. Finally, we have imprinted the ImageDataGenerator() function, which will allow us to generate new images from that source.

Let's start by defining the generator object by setting a series of parameters:

```
image_generation = ImageDataGenerator(
        rotation_range=10,
        width_shift_range=0.1,
        height_shift_range=0.1,
        shear_range=0.1,
        zoom_range=0.1,
        horizontal_flip=True,
        fill_mode='nearest')
```

In defining the object, we set the following parameters:

- `rotation_range`: An integer value to define the range of degrees in random rotations
- `width_shift_range`: Shifts the image left or right (horizontal shifts)
- `height_shift_range`: Shifts the image up or down (vertical shifts)
- `shear_range`: Sets a value for the random application of shear transformations
- `zoom_range`: Sets a value to randomly enlarge images
- `horizontal_flip`: Sets a value to randomly flip half of the image horizontally
- `fill_mode`: Defines the strategy used to fill the newly created pixels, which can appear after a rotation or a shift in width/height

At this point we can load the source image:

```
source_img = load_img('colosseum.jpg')
```

It is an image of the famous Flavian Amphitheater, known by the name the Colosseum, located in Rome (Italy), and famous because it represents the largest Roman amphitheater in the world. To load the image, we used the `load_img()` function, which loads an image from a file as a PIL image object. Let's now turn this image by doing the following:

```
x = img_to_array(source_img)
x = x.reshape((1,) + x.shape)
```

First, we transformed a GDP image (783.1156) into a NumPy matrix (783.1156.3). Next, we added an extra dimension to the NumPy array to make it compatible with the `ImageDataGenerator()` function.

Now, we can finally proceed with the creation of the images:

```
i = 0
for batch in image_generation.flow(x, batch_size=1,
        save_to_dir='AugImage', save_prefix='new_image',
        save_format='jpeg'):
    i += 1
    if i > 50:
        break
```

To do this we used a `for` loop that exploits the potential of the `flow()` method. This method takes advantage of data and labels, and generates batches of augmented data. We passed the following topics:

- Input data (x): A `Numpy` array of rank 4 or a tuple.
- `batch_size`: An integer value.
- `save_to_dir`: Optionally, specify a directory in which to save the generated augmented images. In our case, we have defined the `'AugImage'` folder, which must be previously created and added to the path.
- `save_prefix`: A prefix to be used for the filenames of saved images.
- `save_format`: The format of the generated image file. Available formats are PNG, JPEG, BMP, PDF, PPM, GIF, TIF, and JPG.

Finally, we set a break to end the `for` loop, which otherwise would have continued indefinitely.

If all goes well, we will find 50 new images generated by our script in the `AugImage` folder. In the following screenshot, a summary of augmented images is shown:

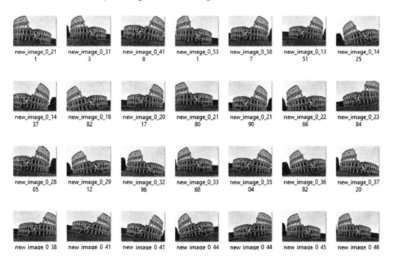

Figure 3.10 – Augmented images examples

After seeing how to practically generate data distributions, let's turn our attention to test statistics.

Simulation of power analysis

In the *Testing uniform distribution* section of *Chapter 2, Understanding Randomness and Random Numbers*, we introduced the concept of statistical testing. A statistical test is a procedure that allows us to verify with a high degree of confidence our initial hypothesis, which is also called a working hypothesis or null hypothesis. This is a calculation procedure based on the analysis of the numerical data we have available, data that is interpreted as observed values of a certain random variable.

The power of a statistical test

The power of a statistical test provides us with important information on the probability that the hypotheses formulated are confirmed by the data. In this way, we will be able to evaluate the reliability of the results of a study and, more importantly, to evaluate the sample size necessary to obtain statistically significant results.

Hypothesis tests allow us to verify whether, and to what extent, a given hypothesis is supported by experimental results. The goal is to decide whether a certain hypothesis formulated on a specific population is true or false. The phenomenon studied must be representable by means of a distribution. A hypothesis test begins with defining the problem in terms of a hypothesis about the parameter under study. In a hypothesis test, two types of errors can be made:

- **Type I error**: Reject a hypothesis when it is true. It occurs when the hypothesis testing procedure tells us which data supports the research hypothesis. However, this research hypothesis is false.
- **Type II error**: Accept a hypothesis when it is false. It occurs when the hypothesis testing procedure tells us that the results are inconclusive. However, the research hypothesis is true.

The probability of a type I error is called the level of significance and is denoted by the Greek letter α: it is like the degree of confidence.

We can conclude that a statistical test leads to an exact conclusion in two cases:

- If it does not reject the null hypothesis when it is true
- If it rejects the null hypothesis when it is false

There is a sort of competition between errors of the first type and errors of the second type. If the level of significance is lowered, that is, the probability of committing type I errors, the level of type II error increases and vice versa. It is a question of seeing which of the two is more harmful in the choice that must be made. The only way to reduce both is to increase the amount of data. However, it is not always possible to expand the size of the sample, either because it has already been collected or because the costs and time required become excessive due to the availability of the researcher.

The power of a test measures the probability that a statistical test has of falsifying the null hypothesis when the null hypothesis is actually false. In other words, the power of a test is its ability to detect differences when those differences exist. The statistical test is constructed in such a way as to keep the level of significance constant, regardless of the sample size. But this result is achieved at the expense of the power of the test, which increases as the sample size increases.

There are six factors that affect the power of a test:

- **Level of significance**: A test is significant only when the probability estimated by the test is lower than the pre-established conventional critical value, usually chosen in the range of 0.05 to 0.001.

  The size of the difference between the observed value and the expected value in the null hypothesis is the second factor affecting the power of a test. Frequently, tests are about the difference between means. The power of a statistical test is an increasing function of the difference, taken as an absolute value.

- **Variability of the data**: The power of a test is a decreasing function of the variance.

- **Direction of the hypothesis**: The alternative hypothesis H1, to be verified with a test, can be bilateral or unilateral.

- **Sample size**: This is the parameter that has the most important effect on the power of a test in the planning phase of the experiment and evaluation of the results, as it is closely linked to the behavior of the researcher.

- **Characteristics of the test**: Starting from the same data, not all tests have the same ability to reject the null hypothesis when it is false. It is therefore very important to choose the most suitable test.

Results obtained with insufficient potency will lead to incorrect conclusions. That is why only results with an acceptable power level should be taken into account. It is quite common to design experiments with an 80% power level, which results in a 20% chance of making a type II error. The main way to achieve adequate power is to plan an adequate sample size in the study protocol.

Power analysis

Power analysis is used at the beginning of a research project to determine the correct sample size. In fact, every time you start researching, you need to make several decisions: one of these is precisely the size of the sample.

Power analysis is generally used for two purposes:

- **A posteriori to determine the power of a test**: Since the search is carried out on a certain sample (of amplitude N) and using a certain level, and from the results obtained we can calculate the size, we can estimate the power of a test after the fact, such as the probability of having made the right choice.

- **A priori to determine the size of the sample**: If we want to do research that has a certain power, once a certain level of significance has been established, and a certain sample size assumed, what should the size of the sample be?

Power analysis is based on the following metrics:

- Effect size
- Power
- Significance level
- Sample size

These quantities are related to each other: A decrease in significance level can lead to a decrease in power, while a larger sample may make the effect easier to detect. We can then evaluate any of these values once the other three are known. In designing an experiment, by setting the level of significance, power, and size of the desired effect, we can estimate the size of the sample that we must collect in order for the experiment to return valid results. Or, in the validation procedure of an experiment, having set the sample size, the effect size, and the significance level, we can calculate the power, which is the probability of making a type II error.

Furthermore, we can carry out a sensitivity analysis by performing the power analysis several times and showing the results on a graph. In this way, we can see how the required sample size changes with an increase or decrease in the level of significance.

To see how to perform a power analysis in a Python environment, we can use the `statsmodels.stats.power` package. This package contains various statistical tests and tools. Some can be used independently of any model while others are intended as an extension of the models and model results. As always, to understand how it works, we will analyze in detail a practical example. Here is the Python code (`power_analysis.py`):

```python
import numpy as np
import matplotlib.pyplot as plt
import statsmodels.stats.power as ssp

stat_power = ssp.TTestPower()
sample_size = stat_power.solve_power(effect_size=0.5,
            nobs = None, alpha=0.05, power=0.8)
print('Sample Size = {:..2f}'.format(sample_size))

power = stat_power.solve_power(effect_size = 0.5,nobs=33,
                    alpha = 0.05, power = None)
print('Power = {:..2f}'.format(power))
```

```
effect_sizes = np.array([0.2, 0.5, 0.8,1])
sample_sizes = np.array(range(5, 500))
stat_power.plot_power(dep_var='nobs', nobs=sample_sizes,
            effect_size=effect_sizes)
plt.xlabel('Sample Size')
plt.ylabel('Power')
plt.show()
```

As always, we analyze the code line by line. Let's start by importing the libraries:

```
import numpy as np
import matplotlib.pyplot as plt
import statsmodels.stats.power as ssp
```

To start, we imported the `numpy` library: `numpy` is a library of additional scientific functions of the Python language designed to perform operations on vectors and dimensional matrices. `numpy` allows you to work with vectors and matrices more efficiently and faster than you can do with lists and lists of lists (matrices). In addition, it contains an extensive library of high-level mathematical functions that can operate on these arrays. We then imported the `matplotlib.pyplot` library: the `matplotlib` library is a Python library for printing high-quality graphics. Finally, we imported `statsmodels.stats.power`.

As a first step, we carry out the power analysis:

```
stat_power = ssp.TTestPower()
sample_size = stat_power.solve_power(effect_size=0.5,
            nobs = None, alpha=0.05, power=0.8)
print('Sample Size = {:.2f}'.format(ss))
```

The first command creates an instance for the calculation of the power of a t-test for two independent samples using the aggregate variance. The second command uses the `solve_power()` function, which solves any parameter of the power of a sample t-test. As previously stated, four parameters are available: `effect_size`, `nobs`, `alpha`, and `power`. One of these parameters must be set as None; all the others need numerical values. In our case, `nobs = None` has been set, which represents the sample size. In this way, the calculation will give us the number of samples necessary to obtain a power equal to 0.8, with `effect_size` equal to 0.5 and a significance level equal to 0.05. Finally, we printed the result on the screen:

```
Sample Size = 33.37
```

We then calculated the minimum number of samples to be used to obtain the required power (80%). Now let's see how to get the power available when we form the number of samples as input:

```
power = stat_power.solve_power(effect_size = 0.5,nobs=33,
                    alpha = 0.05, power = None)
print('Power = {:.2f}'.format(power))
```

Once again we used the `solve_power()` function, but this time we set the power as None. We have printed the following result on the screen:

```
Power = 0.80
```

As a third operation, we will draw a diagram to see how the power varies as the number of samples varies:

```
effect_sizes = np.array([0.2, 0.5, 0.8, 1])
sample_sizes = np.array(range(5, 500))
stat_power.plot_power(dep_var='nobs', nobs=sample_sizes,
            effect_size=effect_sizes)
plt.xlabel('Sample Size')
plt.ylabel('Power')
plt.show()
```

To start, we have set up two vectors; the first is to set `effect_sizes`. According to Cohen, 0.2 corresponds to a small effect size, 0.5 to medium, and 0.8 to large. Jacob Cohen introduced a measure of the distance between two proportions to quantify the difference between them. Therefore, Cohen believes that differences between means of less than 20% of the standard deviation should be considered irrelevant, even if potentially significant. The second vector will contain the variation in the number of samples that we will analyze.

Subsequently, we used the `plot_power()` function to plot a graph of the power trend as the number of samples changes. Four parameters were passed:

- `dep_var`: Specifies which variable is used for the horizontal axis
- `nobs`: Specifies the values of the number of observations in the chart
- `effect_size`: Specifies the effect size values in the graph

The following chart is returned:

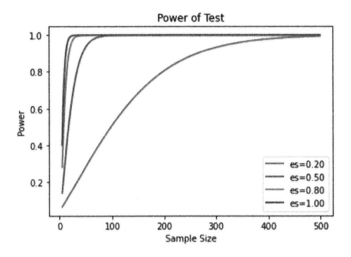

Figure 3.11 – Power of test

From the analysis of *Figure 3.11*, we can see that as the effect size decreases, the number of samples required to reach a power of at least 80% increases significantly.

Summary

Knowing the basics of probability theory in depth helps us to understand how random phenomena work. We discovered the differences between a priori, compound, and conditioned probabilities. We also saw how Bayes' theorem allows us to calculate the conditional probability of a cause of an event, starting from the knowledge of the a priori probabilities and the conditional probability. Next, we analyzed some probability distributions and how such distributions can be generated in Python.

In the final part of the chapter, we introduced the basics of synthetic data generation by analyzing a practical case of data augmentation with the Keras library. Finally, we explored power analysis for statistical tests.

In the next chapter, we will learn about the basic concepts of Monte Carlo simulation and explore some of its applications. Then, we will discover how to generate a sequence of numbers that have been randomly distributed according to Gaussian. Finally, we will take a look at the practical application of the Monte Carlo method in order to calculate a definite integral.

Part 2: Simulation Modeling Algorithms and Techniques

In this part of the book, we will analyze some of the most used algorithms in numerical simulation. We will see the basics of how these techniques work and how to apply them to solve real problems.

This part covers the following chapters:

- *Chapter 4, Exploring Monte Carlo Simulations*
- *Chapter 5, Simulation-Based Markov Decision Processes*
- *Chapter 6, Resampling Methods*
- *Chapter 7, Using Simulation to Improve and Optimize Systems*
- *Chapter 8, Introducing Evolutionary Systems*

4
Exploring Monte Carlo Simulations

Monte Carlo simulation is used to reproduce and numerically solve a problem in which random variables are also involved, and whose solution by analytical methods is too complex or impossible. In addition, the use of simulation allows you to test the effects of changes in the input variables or the output function more easily and with a high degree of detail. Starting from modeling the processes and generating random variables, simulations composed of multiple runs capable of obtaining an approximation of the probability of certain results are performed.

This method has assumed great importance in many scientific and engineering areas, above all for its ability to deal with complex problems that previously could only be solved through deterministic simplifications. It is mainly used in three distinct classes of problems: optimization, numerical integration, and generating probability functions. In this chapter, we will explore various techniques based on Monte Carlo methods for process simulation. First, we will learn about the basic concepts and then learn how to apply them to practical cases.

In this chapter, we're going to cover the following main topics:

- Introducing the Monte Carlo simulation
- Understanding the central limit theorem
- Applying the Monte Carlo simulation
- Performing numerical integration using Monte Carlo
- Exploring sensitivity analysis concepts
- Explaining the cross-entropy method

Technical requirements

In this chapter, we will provide an introduction to Monte Carlo simulation. To deal with the topics in this chapter, it is necessary to have a basic knowledge of algebra and mathematical modeling.

To work with the Python code in this chapter, you'll need the following files (available on GitHub at the following URL: `https://github.com/PacktPublishing/Hands-On-Simulation-Modeling-with-Python-Second-Edition`):

- `simulating_pi.py`
- `central_limit_theorem.py`
- `numerical_integration.py`
- `sensitivity_analysis.py`
- `cross_entropy.py`
- `cross_entropy_loss_function.py`

Introducing the Monte Carlo simulation

In simulation procedures, the evolution of a process is followed, but at the same time, forecasts of possible future scenarios are made. A simulation process consists of building a model that closely imitates a system. From the model, numerous samples of possible cases are generated and subsequently studied over time. After this, the results are analyzed over time, all while highlighting the alternative decisions that can be made.

The term Monte Carlo simulation was born at the beginning of the Second World War by J. von Neumann and S. Ulam as part of the Manhattan project at the Los Alamos nuclear research center. They replaced the parameters of the equations that describe the dynamics of nuclear explosions with a set of random numbers. The choice of the name Monte Carlo was due to the uncertainty of the winnings that characterize the famous casino of the Principality of Monaco.

Monte Carlo components

To obtain a simulation with satisfactory results, applications that use the Monte Carlo method are based on the following components:

- **Probability density functions** (**PDFs**) of the physical system
- Methods for estimating and reducing statistical error
- A uniform random number generator that allows us to obtain a uniform function distributed in a range between [0.1]
- An inversion function that allows one random uniform variable to be passed to a population variable

- Sampling rules so that we can divide the space into specific volumes of interest
- Parallelization and optimization algorithms for efficiently implementing the available computing architecture

The Monte Carlo simulation calculates a series of possible realizations of the phenomenon in question, along with the weight of the probability of a specific occurrence, while trying to explore the whole space of the parameters of the phenomenon.

Once this random sample has been calculated, the simulation gathers measurements of the quantities of interest in the sample. It is well executed if the average value of these measurements on the system realizations converges to the true value.

> **Important note**
> The functionality of Monte Carlo simulation can be summarized as follows: a phenomenon is observed n times, recording the methods adopted in each event, to identify the statistical distribution of the character.

First Monte Carlo application

The primary objective of the Monte Carlo method is to estimate a parameter that's representative of a population. To do this, the calculator generates a series of n random numbers that make up the sample of the population in question.

For example, suppose we want to evaluate a parameter, A, that's currently unknown, which can be interpreted as the average value of a random variable. The Monte Carlo method consists of, in this case, estimating this parameter by calculating the average of a sample consisting of N values of X. This is obtained using a procedure that involves the use of random numbers, as shown in the following diagram:

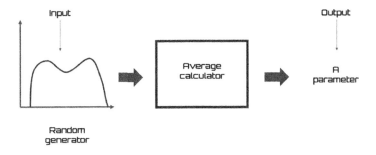

Figure 4.1 – Process of a random generator

In the Monte Carlo simulation, a series of possible realizations of a phenomenon are calculated to explore all the available parameters.

> **Important note**
> In this calculation, the weight of the probability of each event assumes importance. When the representative sample is calculated, the simulation measures the quantities of interest in this sample.

Monte Carlo simulation works if the average value of these measurements on the system converges to the real value.

Monte Carlo applications

The Monte Carlo simulation proves to be a valid tool for addressing the following problems:

- Intrinsically probabilistic problems involving phenomena related to the stochastic fluctuation of random variables
- Problems of an essentially deterministic nature, completely devoid of random components, but whose solution strategy can be treated as an expectation value of a function of stochastic variables

The necessary conditions to apply the method are the independence and analogy of the experiments. For independence, it is understood that the results of each repetition of the experiment must not be able to influence each other. By analogy, however, reference is made to the fact that, when observing the character, the same experiment is repeated n times.

Applying the Monte Carlo method for Pi estimation

The Monte Carlo method is a problem-solving strategy that uses statistics. If we use P to indicate the probability of a certain event, then we can randomly simulate this event and obtain P by making the ratio between the number of times our event occurred and the number of total simulations, as follows:

$$P = \frac{\text{number of event occurrences}}{\text{number of total simulations}}$$

We can apply this strategy to get an approximation of Pi. Pi (π) is a mathematical constant indicating the relationship between the circumference of a circle and its diameter. If we denote the length of a circumference with C and its diameter with d, we know that $C = d * \pi$. The circumference of a circle with a diameter equal to 1 is π.

> **Important note**
> Usually, we approximate the value of Pi with 3.14 to simplify the accounts. However, π is an irrational number; that is, it has an infinite number of digits after the decimal point that never repeats regularly.

Given that a circle has a radius of 1, it can be inscribed in a square with a side equal to 2. For convenience, we will only consider a fraction of the circle, as shown in the following figure:

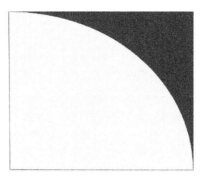

Figure 4.2 – A fraction of the circle

By analyzing the previous figure, we can see that the area of the square in blue is 1 and that the area of the circular sector in yellow (1/4 of the circle) is Pi/4. We randomly place a very large number of points inside the square. Thanks to the very large number and random distribution, we can approximate the size of the areas with the number of points contained in them.

If we generate *N* random numbers inside the square as the number of points that fall in the circular sector, which we will denote with *M*, divided by the total number of generated numbers, *N*, we will have to approximate the area of the circular sector, in which case it will be equal to *Pi/4*. From this, we can derive the following equation:

$$\pi = \frac{4 * M}{N}$$

The greater the number of points generated, the more precise the approximation of Pi will be.

Now, let's analyze the code line by line to understand how we have implemented the simulation procedure for estimating Pi:

1. To start, we import the necessary libraries:

    ```
    import math
    import random
    import numpy as np
    import matplotlib.pyplot as plt
    ```

    The `math` library provides access to the mathematical functions defined by the C standard library. The `random` library implements pseudo-random number generators for various distributions. The random module is based on the Mersenne Twister algorithm. The `numpy` library offers additional scientific functions of the Python language, designed to perform operations on vectors and dimensional matrices. Finally, the matplotlib library is a Python library for printing high-quality graphics.

2. Let's move on and initialize the parameters:

```
N = 10000
M = 0
```

As we mentioned previously, N represents the number of points that we generate – that is, those that we are going to position. Instead, M will be the points that fall within the circular sector. To start, these points will be zero and as we generate them, we will try to perform a check. In a positive scenario, we will gradually increase this number.

3. Let's proceed and initialize the vectors that will contain the coordinates of the points that we will generate:

```
XCircle=[]
YCircle=[]
XSquare=[]
YSquare=[]
```

Here, we have defined two types of points: Circle and Square. Circle is a point that falls within the circular sector, while Square is a point that falls within the space of the square outside the circular sector. Now, we can generate the points:

```
for p in range(N):
    x=random.random()
    y=random.random()
```

Here, we used a for loop that iterates the process several times equal to the number, *N*, of samples we want to generate. Then, we used the random() function of the random library to generate the points. The random() function returns the next nearest floating-point value from the generated sequence. All the return values fall between 0 and 1.0.

4. Now, we can check where the point we just generated falls:

```
if(x**2+y**2 <= 1):
    M+=1
    XCircle.append(x)
    YCircle.append(y)
else:
    XSquare.append(x)
    YSquare.append(y)
```

The `if` loop allows us to check the position of the points. Recall that the points of a circumference are defined by the following equation:

$$(x - x_0)^2 + (y - y_0)^2 = r^2$$

If $x_0=y_0=0$ and $r=1$, the previous equation turns into the following:

$$x^2 + y^2 = 1$$

This makes us understand that the necessary condition for a point to fall within the circular sector is that the following equation is verified:

$$x^2 + y^2 \leq 1$$

If this condition is satisfied, the value of M is increased by 1 unit and the values of the x and y values that are generated are stored in the `Circle` point vector (`XCircle`, `YCircle`). Otherwise, the value of M is not updated, and the values of x and y that are generated are stored in the vector of the `Square` point vector (`XSquare`, `YSquare`).

5. Now that we've iterated this procedure for the 10,000 points that we have decided to generate, we can estimate `Pi`:

```
Pi = 4*M/N
print("N=%d M=%d Pi=%.2f" % (N,M,Pi))
```

In this way, we can calculate `Pi` and print the results, as follows:

```
N=10000 M=7857 Pi=3.14
```

The estimate that we've obtained is acceptable. Usually, we stop at the second decimal place, so this is okay. Now, let's draw a graph, where we will draw the generated points. To start, we will generate the points of the circumference arc:

```
XLin=np.linspace(0,1)
YLin=[]
for x in XLin:
    YLin.append(math.sqrt(1-x**2))
```

The `linspace()` function of the `numpy` library allows us to define an array composed of a series of N numerical elements equally distributed between two extremes (0,1). This will be the x of the arc of the circumference (`XLin`). On the other hand, the y numerical elements (`YLin`) will be obtained from the equation of the circumference while solving them concerning y, as follows:

$$y = \sqrt{1 - x^2}$$

To calculate the square root, we used the `math.sqrt()` function.

6. Now that we have all the points, we can draw the graph:

```
plt.axis    ("equal")
plt.grid    (which="major")
plt.plot    (XLin , YLin, color="red" , linewidth="4")
plt.scatter(XCircle, YCircle, color="yellow",
marker     =".")
plt.scatter(XSquare, YSquare, color="blue"  ,
marker     =".")
plt.title   ("Monte Carlo method for Pi estimation")
plt.show()
```

The `scatter()` function allows us to represent a series of points not closely related to each other on two axes. The following diagram is printed:

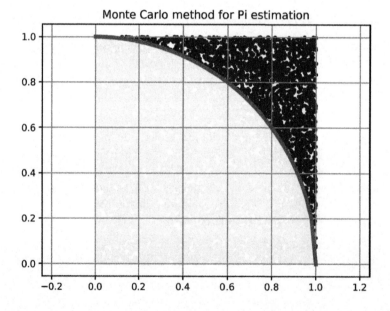

Figure 4.3 – Plot of the Pi estimation

Consistent with what we established at the beginning of this chapter, we plotted the points inside the circular sector in yellow, while those outside the circular sector are in blue. To highlight the separation line, we have drawn the circumference arc in red.

Now that we've applied the Monte Carlo method to estimate Pi, the time has come to deepen some fundamental concepts for simulation based on generating random numbers.

Understanding the central limit theorem

The Monte Carlo method is essentially a numerical method for calculating the expected value of random variables; that is, an expected value that cannot be easily obtained through direct calculation. To obtain this result, the Monte Carlo method is based on two fundamental theorems of statistics: the law of large numbers and the central limit theorem.

Law of large numbers

This theorem states the following: considering a very large number of variables, x ($N \to \infty$), the integral that defines the average value is approximate to the estimate of the expected value. Let's try to give an example so that you can understand this. We flip a coin 10 times, 100 times, and 1,000 times and check how many times we get heads. We can put the results we obtained into a table, as follows:

Number of coin flips	Number of heads	Head output frequency
10	4	40%
100	44	44%
1,000	469	46.9%

Table 4.1 – Table showing the results for a coin toss

Analyzing the last column of the previous table, we can see that the value of the frequency approaches that of the probability (50%). Therefore, we can say that as the number of tests increases, the frequency value tends to the theoretical probability value. The latter value can be achieved using the hypothesis of several throws that tend to infinity.

> **Important note**
>
> The use of the law of large numbers is different. The law of large numbers allowed us, in the *Monte Carlo method for Pi estimation* section, to equal the number of launches with the area of the circular sector. In this way, we were able to estimate the value of Pi simply by generating random numbers. Also, in this case, the greater the number of random variables generated, the closer the estimate of Pi is to the expected value.

The law of large numbers allows you to determine the centers and weights of a Monte Carlo analysis to estimate definite integrals but does not say how large the number, N, must be. You do not have an estimate to understand with what order of magnitude you can perform a simulation so that you can consider the numbers large enough. To answer this question, it is necessary to resort to the central limit theorem.

The central limit theorem

Monte Carlo not only allows us to obtain an estimate of the expected value, as established by the law of large numbers, but also allows us to estimate the uncertainty associated with it. This is possible thanks to the central limit theorem, which returns an estimate of the expected value and the reliability of that result.

> **Important note**
> The central limit theorem can be summarized with the following definition: given a dataset with an unknown distribution, the sample's mean will approximate the normal distribution.

If the law of large numbers tells us that the random variable allows us to evaluate the expected value, the central limit theorem provides information on its distribution.

An interesting feature of the central limit theorem is that there are no constraints on the distribution of the function that's used to generate the N samples from which the random variable is formed. It is not important what the distribution associated with the random variable is, but when the average is characterized by a finite variance and is obtained for a very large number of samples, it can be described through a Gaussian distribution.

Let's take a look at a practical example. We generate 10,000 random numbers with a uniform distribution. Then, we extract 100 samples from this population, also taken randomly. We repeat this operation a consistent number of times and for each time, we evaluate its average and store this value in a vector. In the end, we draw a histogram of the distribution that we have obtained. Here is the Python code:

```
import random
import numpy as np
import matplotlib.pyplot as plt
a=1
b=100
N=10000
DataPop=list(np.random.uniform(a,b,N))
plt.hist(DataPop, density=True, histtype='stepfilled',
alpha=0.2)
plt.show()

SamplesMeans = []
for i in range(0,1000):
    DataExtracted = random.sample(DataPop,k=100)
    DataExtractedMean = np.mean(DataExtracted)
```

```
        SamplesMeans.append(DataExtractedMean)
plt.figure()
plt.hist(SamplesMeans, density=True, histtype='stepfilled',
alpha=0.2)
plt.show()
```

Now, let's analyze the code line by line to understand how we have implemented the simulation procedure to understand the central limit theorem:

1. To start, we import the necessary libraries:

   ```
   import random
   import numpy as np
   import matplotlib.pyplot as plt
   ```

   The random library implements pseudo-random number generators for various distributions. The numpy library offers additional scientific functions of the Python language and is designed to perform operations on vectors and dimensional matrices.

   Finally, the matplotlib library is a Python library for printing high-quality graphics.

2. Let's move on and initialize the parameters:

   ```
   a=1
   b=100
   N=10000
   ```

   The a and b parameters are the extremes of the range, and N is the number of values we want to generate.

   Now, we can generate the uniform distribution using the NumPy random.uniform() function, as follows:

   ```
   DataPop=list(np.random.uniform(a,b,N))
   ```

3. At this point, we draw a histogram of the data to verify that it is a uniform distribution:

   ```
   plt.hist(DataPop, density=True, histtype='stepfilled',
   alpha=0.2)
   plt.show()
   ```

   The matplotlib.hist() function draws a histogram; that is, a diagram in classes of a continuous character.

   This is used in many contexts, usually to show statistical data when there is an interval of the definition of the independent variable divided into subintervals.

   The following diagram is printed:

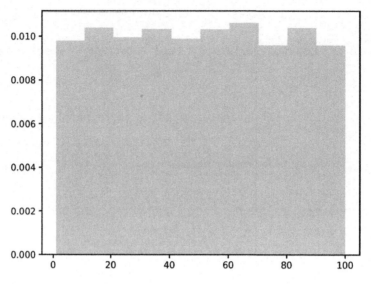

Figure 4.4 – Plot of the data distribution

The distribution appears to be uniform – we can see that each bin is populated with an almost constant frequency.

4. Now, let's pass the values to the extraction of the samples from the generated population:

```
SamplesMeans = []
for i in range(0,1000):
    DataExtracted = random.sample(DataPop,k=100)
    DataExtractedMean = np.mean(DataExtracted)
    SamplesMeans.append(DataExtractedMean)
```

First, we initialized the vector that will contain the samples. To do this, we used a `for` loop to repeat the operations 1,000 times. At each step, we extracted 100 samples from the population generated using the `random.sample()` function. The `random.sample()` function extracts samples without repeating the values and without changing the input sequence.

5. Next, we calculated the average of the extracted samples and added the result to the end of the vector containing the samples. Now, all we need to do is view the results:

```
plt.figure()
plt.hist(SamplesMeans, density=True,
histtype='stepfilled', alpha=0.2)
plt.show()
```

The following histogram is printed:

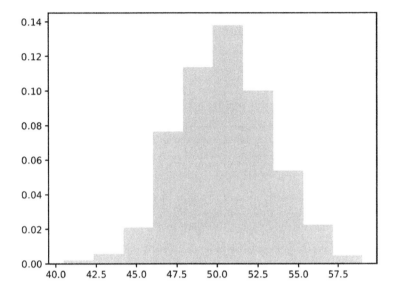

Figure 4.5 – Plot of the extracted samples

The distribution has now taken on the typical bell-shaped curve characteristic of the Gaussian distribution. This means that we have proved the central limit theorem.

Now, we are ready to apply the newly learned Monte Carlo simulation concepts to real cases.

Applying the Monte Carlo simulation

Monte Carlo simulation is used to study the response of a model that's used randomly generated inputs. The simulation process takes place in the following three phases:

1. N inputs are generated randomly.
2. A simulation is performed for each of the N inputs.
3. The outputs of the simulations are aggregated and examined. The most common measures include estimating the average value of a certain output and distributing the output values, as well as the minimum or maximum output value.

Monte Carlo simulation is widely used for analyzing financial, physical, and mathematical models.

Generating probability distributions

Generating probability distributions that cannot be found with analytical methods can easily be addressed with Monte Carlo methods. For example, let's say we want to estimate the probability distribution of the damage caused by earthquakes in a year in Japan.

> **Important note**
> In this type of analysis, there are two sources of uncertainty: how many earthquakes there will be in a year and how much damage each earthquake will do. Even if it is possible to assign a probability distribution to these two logical levels, it is not always possible to put this information together with analytical methods to derive the distribution of the annual losses.

It is easier to do a Monte Carlo simulation of this type like so:

1. A random number is extracted from the distribution of the number of annual events.
2. If events occur from the previous point, extractions are made from the distribution of losses.
3. Finally, we add the values of the extractions we performed to obtain a value that represents an annual loss caused by events.

By cyclically repeating these three points, a sample of annual losses is generated, from which it is possible to estimate the probability distribution that could not be obtained analytically.

Numerical optimization

Various algorithms can be used to find the local minima of a function. Typically, these algorithms proceed according to the following steps:

1. Start from an assigned point.
2. They control in which direction the function tends to have values smaller than the current one.
3. Moving in this direction, they find a new point where the function has a lower value than the previous one.

They keep repeating these steps until they reach a minimum. In the case of a function with only one minimum, this method allows us to achieve a result. But what if we have a function with many local minima and we want to find the point that minimizes the function globally? The following diagram shows the two cases just mentioned; that is, a distribution with only one minimum (left) and a distribution with several minimums (right):

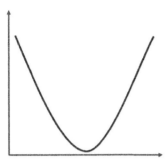 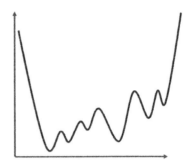

Figure 4.6 – Graphs of the two distributions

A local search algorithm could stop at any of the many local minima of the function. How would you know if you found one of the many local minimums or the global minimum? There is no way to strictly establish this. The only practical possibility is to explore different areas of the search domain to increase the probability of finding, among the various local minima, the global one.

> **Important note**
> Different methods have been developed to explore domains that can be very complicated, with many dimensions and with constraints to be respected.

Monte Carlo methods provide a solution to this problem; that is, an initial population of points belonging to the domain is created, which is then evolved by defining coupling algorithms between the points in which random genetic mutations also occur. When simulating different generations of points, a selection process intervenes that maintains only the best points – that is, those that give lower values of the function to be minimized.

Each generation keeps track of which point represents the best specimen ever. Continuing with this process, the points tend to move to local lows, but at the same time, they explore many areas of the optimization domain. This process can continue indefinitely, though at some point, it is stopped, and the best specimen is taken as an estimate of the global minimum.

Project management

Monte Carlo methods allow you to simulate the behavior of an event of interest and, in general, return a random variable as a result whose properties, such as mean, variance, probability density function, and so on, provide us with important information about the quality of the simulation.

This is a statistical analysis technique that can be applied to all those situations in which we are faced with very uncertain project estimates to reduce the level of uncertainty through a series of simulations.

In this sense, it can be applied when analyzing the times, costs, and risks associated with a project and, therefore, when evaluating the impact that this project may have on the community.

> **Important note**
> For each of these variables, the simulations do not provide a single estimate but a range of possible estimates, along with, associated with each estimate, the level of probability that that estimate is accurate.

For example, this technique can be used to determine the overall cost of a project through a discrete series of simulation cycles. In the planning phase of a project, the activities that make up the project are identified, and the cost associated with each activity is estimated. In this way, the total cost of the project can be determined. Since, however, we rely on cost estimates, we cannot be sure that this overall cost, and therefore also the completion costs, are certain. In such cases, Monte Carlo simulation can be carried out.

Now, let's learn how to apply the Monte Carlo simulation to compute integrals.

Performing numerical integration using Monte Carlo

Monte Carlo simulations represent numerical solutions for calculating integrals. In fact, with the use of the Monte Carlo algorithm, it is possible to adopt a numerical procedure to solve mathematical problems, with many variables that do not present an analytical solution. The efficiency of the numerical solution increases compared to other methods when the size of the problem increases.

> **Important note**
> Let's analyze the problem of a definite integral. In the simplest cases, there are methods for integration that foresee the use of techniques such as integration by parts, integration by replacement, and so on. In more complex situations, however, it is necessary to adopt numerical procedures that involve the use of a computer. In these cases, the Monte Carlo simulation provides a simple solution that's particularly useful in cases of multidimensional integrals.

However, it is important to highlight that the result that's returned by this simulation approximates the integral and not its precise value.

Defining the problem

In the following equation, we use I to denote the definite integral of the function, f, in the limited interval, $[a, b]$:

$$I = \int_{a}^{b} F(x)dx$$

In the interval, *[a, b]*, we identify the maximum of the function, *f*, and indicate it with *U*. To evaluate the approximation that we are introducing, we draw a base rectangle, *[a, b]*, and the height, *U*. The area under the function, *f (x)*, which represents the integral of *f(x)*, will surely be smaller than the area of the base rectangle, *[a, b]*, and the height, *U*. The following diagram shows the area subtended by the function, *f* – which represents the integral of *f(x)* – and the area, *A*, of the rectangle with the base, *[a, b]*, and the height, *U*, which represents our approximation:

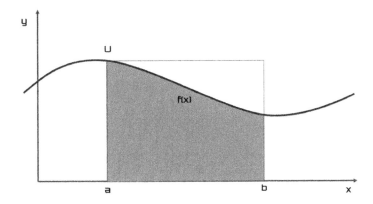

Figure 4.7 – Plot of the function

By analyzing the previous diagram, we can identify the following intervals:

- x ϵ [a, b]
- y ϵ [0, U]

In Monte Carlo simulation, *x* and *y* both represent random numbers. At this point, we can consider a point in the plane of the Cartesian coordinates (*x*, *y*). Our goal is to determine the probability that this point is within the area highlighted in the previous diagram; that is, that it is *y ≤ f(x)*. We can identify two areas:

- The area subtended by the function, *f*, which coincides with the definite integral, *I*
- The area, *A*, of the rectangle with a base of *[a, b]* and a height of *U*

Let's try to write a relationship between the probability and these two areas:

$$P(y \leq f(x)) = \frac{I}{A} = \frac{I}{(b-a) * U}$$

It is possible to estimate the probability, *P (y <= f (x))*, through Monte Carlo simulation. In fact, in the *Monte Carlo method for Pi estimation* section, we faced a similar case. To do this, *N* pairs of random numbers (x_i, y_i) are generated, as follows:

$$x_i \in [a, b]$$

$$y_i \in [0, U]$$

Generating random numbers in the intervals considered will certainly determine conditions in which $yi \leq f(xi)$ will result. If we number this quantity and denote it with M, we can analyze its variation. This is an approximation whose accuracy increases as the number of random number pairs (x_i, y_i) generated increases. The approximation of the calculation of the probability, $P(y \leq f(x))$, will therefore be equal to the following value:

$$\mu = \frac{M}{N}$$

After calculating this probability, it will be possible to trace the value of the integral using the previous equation, as follows:

$$I \cong \mu * (b - a) * U = \frac{M}{N} * (b - a) * U = \frac{M}{N} * A$$

This is the mathematical representation of the problem. Now, let's see the numerical solution.

Numerical solution

We will begin by setting up the components that we will need for the simulation, starting with the libraries that we will use to define the function and its domain of existence. The Python code for numerical integration through the Monte Carlo method is shown here:

```
import random
import numpy as np
import matplotlib.pyplot as plt

random.seed(2)
f = lambda x: x**2
a = 0.0
b = 3.0
NumSteps = 1000000
XIntegral=[]
YIntegral=[]
XRectangle=[]
YRectangle=[]
```

Now, let's analyze the code line by line to understand how we have implemented the simulation procedure to understand the central limit theorem:

1. To start, we import the necessary libraries:

   ```
   import random
   import numpy as np
   import matplotlib.pyplot as plt
   ```

   The `random` library implements pseudo-random number generators for various distributions. The `numpy` library offers additional scientific functions of the Python language, designed to perform operations on vectors and dimensional matrices. Finally, the `matplotlib` library is a Python library for printing high-quality graphics. Let's set the seed:

   ```
   random.seed(2)
   ```

   The `random.seed()` function is useful if we wish to have the same set of data available to be processed in different ways as it makes the simulation reproducible. This function initializes the basic random number generator. If you use the same seed in two successive simulations, you will always get the same sequence of pairs of numbers.

2. Now, we define the function that we want to integrate:

   ```
   f = lambda x: x**2
   ```

   We know that to define a function in Python, we must use the **def** clause, which automatically assigns a variable to it. Functions can be treated like other Python objects, such as strings and numbers. These objects can be created and used at the same time (on the fly) without us resorting to creating and defining variables that contain them.

   In Python, functions can also be used in this way, using a syntax called **lambda**. The functions that are created in this way are anonymous. This approach is often used when you want to pass a function as an argument for another function. The lambda syntax requires the lambda clause, followed by a list of arguments, a colon character, the expression to evaluate the arguments, and finally the input value.

3. Let's move on and initialize the parameters:

   ```
   a = 0.0
   b = 3.0
   NumSteps = 1000000
   ```

   As we mentioned in the *Defining the problem* section, a and b represent the ends of the range in which we want to calculate the integral. NumSteps represents the number of steps in which we want to divide the integration interval. The greater the number of steps, the better the simulation will be, even if the algorithm becomes slower.

4. Now, we must define four vectors so that we can store the pairs of generated numbers:

   ```
   XIntegral=[]
   YIntegral=[]
   XRectangle=[]
   YRectangle=[]
   ```

Whenever the generated y value is less than or equal to *f (x)*, this value and the relative x value will be added at the end of the `XIntegral`, `YIntegral` vectors. Otherwise, they will be added at the end of the `XRectangle`, `YRectangle` vectors.

Min-max detection

Before using this method, it is necessary to evaluate the minimum and maximum of the function:

> **Important note**
> Recall that if the function has only one minimum/maximum, the procedure is simple. If there are repeated minimums/maximums, then the procedure becomes more complex.

1. In the following Python code, we are extracting the min/max of the distribution:

   ```
   ymin = f(a)
   ymax = ymin
   for i in range(NumSteps):
       x = a + (b - a) * float(i) / NumSteps
       y = f(x)
       if y < ymin: ymin = y
       if y > ymax: ymax = y
   ```

2. To understand all these cases, even complex ones, we will look for the minimum/maximum for each step in which we have divided the interval, *[a, b]*. First, we must initialize the minimum and maximum with the value of the function in the far left of the range, (a):

   ```
   ymin = f(a)
   ymax = ymin
   ```

3. Then, we must use a `for` loop to check the value at each step:

   ```
   for i in range(NumSteps):
       x = a + (b - a) * float(i) / NumSteps
   ```

```
y = f(x)
```

4. For each step, the x value is obtained by increasing the left end of the interval (a) by a fraction of the total number of steps provided by the current value of i. Once this is done, the function at that point is evaluated. Now, you can check this, as follows:

```
if y < ymin: ymin = y
if y > ymax: ymax = y
```

5. The two if statements allow us to verify if the current value of f is less than/greater than the value chosen so far as the minimum/maximum and if so, to update these values. Now, we can apply the Monte Carlo method.

The Monte Carlo method

Now, we will apply the Monte Carlo method, as follows:

1. Now that we've set and calculated the necessary parameters, it is time to proceed with the simulation:

```
A = (b - a) * (ymax - ymin)
N = 1000000
M = 0
for k in range(N):
    x = a + (b - a) * random.random()
    y = ymin + (ymax - ymin) * random.random()
    if y <= f(x):
        M += 1
        XIntegral.append(x)
        YIntegral.append(y)
    else:
        XRectangle.append(x)
        YRectangle.append(y)
NumericalIntegral = M / N * A
print ("Numerical integration = " +
str(NumericalIntegral))
```

2. To start, we will calculate the area of the rectangle, as follows:

```
A = (b - a) * (ymax - ymin)
```

3. Then, we will set the numbers of random pairs we want to generate:

   ```
   N = 1000000
   ```

4. Here, we initialize the M parameter, which represents the number of points that fall under the curve that represents $f(x)$:

   ```
   M = 0
   ```

5. Now, we can calculate this value. To do this, we will use a `for` loop that iterates the process N times. First, we must generate the two random numbers, as follows:

   ```
   for k in range(N):
       x = a + (b - a) * random.random()
       y = ymin + (ymax - ymin) * random.random()
   ```

6. Both x and y fall within the rectangle of the area, A; that is, $x \in [a, b]$ and $y \in [0, maxy]$. Now, we need to determine whether the following is true:

$$y \leq f(x)$$

   We can do this with an `if` statement, as follows:

   ```
   if y <= f(x):
       M += 1
       XIntegral.append(x)
       YIntegral.append(y)
   ```

7. If the condition is true, then the value of M is incremented by one unit and the current values of x and y are added to the `XIntegral` and `YIntegral` vectors. Otherwise, the points will be stored in the `XRectangle` and `YRectangle` vectors:

   ```
   else:
       XRectangle.append(x)
       YRectangle.append(y)
   ```

8. After iterating N times, we can estimate the integral:

   ```
   NumericalIntegral = M / N * A
   print ("Numerical integration = " + str(NumericalIntegral))
   ```

   The following result is printed:

   ```
   Numerical integration = 8.996787006398996
   ```

The analytical solution for this simple integral is as follows:

$$I = \int_0^3 x^2 dx = \left[\frac{x^3}{3} + c\right]_0^3 = 9$$

The percentual error we made is equal to the following:

$$\frac{9 - 8.996787006398996}{9} * 100 = 0.03\,\%$$

This is a negligible error that defines our reliable estimate.

Visual representation

Now, let's plot the results using the following code:

1. Finally, we can visualize what we have achieved in the numerical integration by plotting scatter plots of the generated points. For this reason, we have memorized the pairs of points that were generated in the four vectors:

   ```
   XLin=np.linspace(a,b)
   YLin=[]
   for x in XLin:
       YLin.append(f(x))
   
   
   plt.axis    ([0, b, 0, f(b)])
   plt.plot    (XLin,YLin, color="red" , linewidth="4")
   plt.scatter(XIntegral, YIntegral, color="blue",
   marker    =".")
   plt.scatter(XRectangle, YRectangle, color="yellow",
   marker    =".")
   plt.title   ("Numerical Integration using Monte Carlo
   method")
   plt.show()
   ```

2. To start, we will generate the points we need to draw the representative curve of the function:

   ```
   XLin=np.linspace(a,b)
   YLin=[]
   for x in XLin:
       YLin.append(f(x))
   ```

The `linspace()` function of the `numpy` library allows us to define an array composed of a series of *N* numerical elements equally distributed between two extremes (0,1). This will be the x of the function, while the y of the function (`YLin`) will be obtained from the equation of the function solving them concerning y, as follows:

$$y = x^2$$

3. Now that we have all the points, we can draw the graph:

```
plt.axis    ([0, b, 0, f(b)])
plt.plot    (XLin,YLin, color="red" , linewidth="4")
plt.scatter(XIntegral, YIntegral, color="blue", 
marker   =".")
plt.scatter(XRectangle, YRectangle, color="yellow",
marker   =".")
plt.title   ("Numerical Integration using Monte Carlo 
method")
plt.show()
```

First, we set the length of the axes using the `plt.axis()` function. So, we plotted the curve of the x2 function, which, as we know, is a convex increasing the monotone function in the range of values considered, [0,3].

Then, we plotted two scatter plots:

- One for the points that are under the curve (points in blue)
- One for the points that are above the function (points in yellow)

The `scatter()` function allows us to represent a series of points not closely related to each other on two axes.

The following diagram is returned:

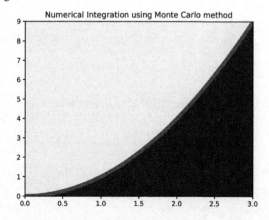

Figure 4.8 – Plot of numerical integration results

As we can see, all the points in blue are positioned below the curve of the function (curve in red), while all the points in yellow are positioned above the curve of the function.

Often, when performing a numerical calculation, it is necessary to evaluate what effect the input variables have on the output. In this case, it is possible to use sensitivity analysis. Let's see how.

Exploring sensitivity analysis concepts

The variability, or uncertainty, associated with a parameter propagates throughout the model, making it a strong contribution to the variability of the model's outputs. The model results can be highly correlated with an input parameter so that small changes in the input cause significant changes in the output. A widely used methodology in the field of data analytics is sensitivity analysis. It studies the correlation between the uncertainty of the output of a mathematical model and the various sources of randomness present in the input: we speak of uncertainty analysis when we focus on the quantitative aspect of the problem. There are many objectives of this type of study; here are some examples:

- Understand the complex relationships that exist between the input and output variables
- Identify the most influential risk factors (factor prioritization)
- Check the robustness of the model output to even minor variations of the input
- Identify areas of the model that need improvement

In the context of sensitivity analysis, we can distinguish between local and global methodologies. Local methodologies focus on a particular point in the domain of the input space when we are interested in understanding how the output behaves from this point. Global methods focus not on a single point but on a range of values in the input factor space. In general, for evaluations in the stochastic context, this type of methodology is used.

In sensitivity analysis, a change in the input of the model is required, which we can do by using a certain scenario, and the variation of the output due to this change is identified. The success of this technique derives from the possibility of studying the functioning of complex models with simplicity: this complexity prevents us from analyzing the behavior of the model through simple intuition. It follows that an operational methodology is needed to overcome these difficulties. You can think of the model as a black box: the system is described through inputs and outputs and its precise internal functioning is not visible.

Sensitivity analysis returns indices (sensitivity coefficients) that represent the importance of each parameter and thus allow you to rank the parameters. Therefore, the analysis aims to identify the parameters that require additional research to strengthen knowledge and therefore reduce the uncertainty of the output, ensuring calibration of the model. It allows you to identify insignificant parameters that can be eliminated from the model, thus allowing you to reduce the model. It also highlights how much the predictions of the model depend on the values of the parameters by carrying out a robustness analysis, and which parameters are most highly correlated with the output through adequate control of the system. Once a model is in use, it gives us the consequences of changing a given input parameter.

The methodology is based on the following tasks:

- Quantifying the uncertainty in each input by identifying the intervals and probability distributions
- The variation of parameters, one at a time or in a combination of parameters
- Identifying the outputs of the model to be analyzed
- Simulating the model to be used several times for each parameterization
- Calculating the sensitivity coefficients of interest, starting from the outputs of the obtained model

Ultimately, this analysis makes it possible to evaluate to what extent the uncertainty surrounding each of the independent variables may affect the value assumed by the valuation base. This impact essentially depends on two elements:

- The range of variability of each variable; for example, the relative degree of uncertainty
- The nature of the analytic relationships; for example, the type of decision problem under consideration

While the analyst cannot intervene in the latter, as it depends on the problem being faced, in the former, it is possible to intervene, in the sense that it can be reduced by taking on additional information that's useful for reducing the uncertainty surrounding the variable in question. However, any survey supplement aimed at improving the accuracy of the estimates involves additional calculations for the analyst. It should be noted, however, that this intervention makes sense, especially for the variables whose deviations from the base case may change the outcome of the assessment.

Therefore, sensitivity analysis provides useful information on the riskiness of a project and the sources from which it originates. Concerning the latter, it should be emphasized that it is not so much a sensitivity in an absolute sense that interests us, but rather that we wish to verify if there is the possibility that the objective function changes its sign. Consequently, determining the range of variability of each variable is critical, since it is incorrect and risky to assume a similar interval for each variable for simplicity. In doing so, completely unlikely scenarios can be assumed as possible.

Local and global approaches

In the local approach, the impact of small input perturbations on the model output is studied. These small perturbations occur around nominal values, such as the average of a random variable. This deterministic approach consists of calculating or estimating the partial derivatives of the model at a specific point in the space of the input variables. Using adjoint-based methods allows you to process models with a large number of input variables. These methods are affected by problems due to linearity and normality assumptions and local variations.

Global methods have been developed to overcome these limitations. With this approach, we do not distinguish any initial set of input values of the model, but we consider the numerical model in the whole domain of the possible variations of the input parameters. Therefore, global sensitivity analysis is a tool that's used to study a mathematical model as a whole rather than one of its solutions around specific parameter values.

Sensitivity analysis methods

Sensitivity analysis can be performed using different techniques. Let's see some of them:

- **Sensitivity analysis using slopes**: An intuitive approach to sensitivity analysis is to study the input-output functions expressed as slopes. In this way, we can calculate how many units the output will change by when increasing one unit of the input. This means studying the derivatives or partial derivatives in the case of a multidimensional input. Alternatively, we can look at the relative change, such as how many times the output changes with a 10% increase in input.

- **Direct methods**: If the model is not too complex, it is possible to use direct methods to calculate the sensitivity measurements directly from the relationships with purely mathematical methods. In many cases, however, there are problems with direct methods due to the complexity of the model.

- **Variance-based sensitivity analysis**: With this approach, we study the amount of variation in the output and how it can be explained by different inputs. If the total variation can be spread across the sources of variation, we can determine if the output can be stabilized by better controlling the inputs without having to know the exact input-output relationships. This sensitivity analysis can be very useful in the case of nonlinear response functions and categorical inputs.

Sensitivity analysis in action

Now that we've adequately introduced sensitivity analysis, let's look at a practical case in a Python environment. As we mentioned previously, with sensitivity analysis, we see how the outputs change over the entire range of possible inputs. It does not return any probability distribution of the results, instead providing a range of possible output values associated with each set of inputs. In the following code, we will learn how to use the tools available to perform sensitivity analysis on artificially generated data. Here is the Python code (sensitivity_analysis.py):

```
import numpy as np
import math
from sensitivity import SensitivityAnalyzer

def my_func(x_1, x_2,x_3):
    return math.log(x_1/ x_2 + x_3)
```

```
x_1=np.arange(10, 100, 10)
x_2=np.arange(1, 10, 1)
x_3=np.arange(1, 10, 1)

sa_dict = {'x_1':x_1.tolist(),'x_2':x_2.tolist(),'x_3':x_3.
tolist()}

sa_model = SensitivityAnalyzer(sa_dict, my_func)
plot = sa_model.plot()
styled_df = sa_model.styled_dfs()
```

As always, we will analyze the code line by line:

1. Let's start by importing the libraries:

   ```
   import numpy as np
   import math
   from sensitivity import SensitivityAnalyzer
   ```

   To start, we imported the numpy library, a library of additional scientific functions of the Python language, designed to perform operations on vectors and dimensional matrices. numpy allows you to work with vectors and matrices more efficiently and faster than you can do with lists and lists of lists (matrices). In addition, it contains an extensive library of high-level mathematical functions that can operate on these arrays. Then, we imported the math library, which provides access to the mathematical functions defined by the C standard. Finally, we imported the SensitivityAnalyzer function from the sensitivity library. This library contains the tools to perform sensitivity analysis in a Python environment.

2. Now, we must create a function that defines the output of our example:

   ```
   def my_func(x_1, x_2,x_3):
       return math.log(x_1/ x_2 + x_3)
   ```

   Here, we have created a simple three-variable function that, using the logarithm function and a fraction function, creates a wide variability of the output from the different inputs. This is because we aim to highlight the variability of the output from the different inputs.

   Now, let's define the variable domain of our function:

   ```
   x_1=np.arange(10, 100, 10)
   x_2=np.arange(1, 10, 1)
   x_3=np.arange(1, 10, 1)
   ```

Here, we have created three numpy arrays using the `np.arange()` function. This function creates an array with equidistant values within a given range.

3. As anticipated, to run sensitivity analysis, we must take advantage of the `sensitivity` library, so we must format the data according to the standards required by the tool. Such a library, for example, requires input such as a Python dictionary:

```
sa_dict = {'x_1':x_1.tolist(),'x_2':x_2.tolist(),'x_3':x_3.tolist()}
```

To convert numpy arrays into lists, we used the `tolist()` function. This function is used to convert a certain array into a normal list with the same elements, elements, or values.

4. Now, we can perform sensitivity analysis:

```
sa_model = SensitivityAnalyzer(sa_dict, my_func)
```

The `SensitivityAnalyzer` function performs sensitivity analysis based on the passed function and possible values for each argument. We only passed two arguments: `sa_dict` and `my_func`. The first is a Python dictionary that contains the values of the three input paths. The second argument is the function that defines the output. The `SensitivityAnalyzer` function executes the passed function with the Cartesian product of the possible values for each argument.

5. Now, we can see the results:

```
plot = sa_model.plot()
styled_df = sa_model.styled_dfs()
```

The following plots are returned:

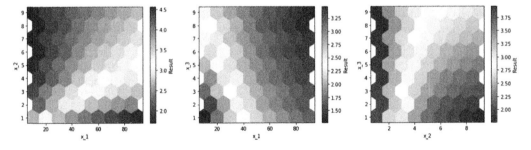

Figure 4.9 – Plot of sensitivity analysis

Three graphs have been drawn; this is because there are three input variables. To be able to appreciate the variability of the output as a function of the variability of the inputs, three graphs were drawn that couple the input variables into pairs.

By analyzing the third graph on the right, we can see, for example, that for low values of x_2, the x_3 variable seems irrelevant: when it changes, the output remains unchanged. We can appreciate this from the hexadecimal color map, which shows the same color for the entire range of variability of x_3. As the values of x_2 increase, there is a variability of the output, even as x_3 varies.

In the next section, we will learn how to evaluate the average information contained in data distributions. To do this, we will introduce the concepts of Cross-entropy.

Explaining the cross-entropy method

In *Chapter 2, Understanding Randomness and Random Numbers*, we introduced the entropy concepts in computing. Let's recall these concepts.

First, there's Shannon entropy. For a probability distribution, $P=\{p1, p2, ..., pN\}$, where p_i is the probability of the *N* extractions, x_i, of a random variable, *X*, Shannon defined the following measure, *H*, in probabilistic terms:

$$H = -\sum p_i \, log_2 \, p_i$$

This equation has the same form as the expression of thermodynamic entropy and for this reason, it was defined as entropy upon its discovery. The equation establishes that *H* is a measure of the uncertainty of an experimental result or a measure of the information obtained from an experiment that reduces the uncertainty. It also specifies the expected value of the amount of information transmitted from a source with a probability distribution. Shannon's entropy could be seen as the indecision of an observer trying to guess the result of an experiment or as the disorder of a system that can be under different configurations. This measure, defined as an information index, considers the only possibility that an event will happen or not, not its meaning or its value. This is the main limitation of the concept of entropy.

This equation would lead to maximum entropy if all probabilities were equal. The maximum entropy can be considered a measure of the total uncertainty. The statistically most probable state in which a system can find itself is that which corresponds to the maximum entropy. If there are two probability distributions with equally likely results, then it is possible to determine the differences in the information content of the two distributions.

The measure of entropy allows us to obtain a positive number. Since the probability is a number between 0 and 1, the logarithm of p_i will take on a negative value, reaching the maximum value of 0 when the probability is equal to 1. For this reason, the minus sign before the sum allows us to obtain a value of positive entropy. From this, we can understand the use of Shannon's entropy as a measure of uncertainty when distributing a random variable.

Introducing cross-entropy

Cross-entropy measures the accuracy of probabilistic forecasts, which is fundamental for modern forecasting systems since it allows us to produce very high-level estimates, even in the case of alternative indicators. Cross-entropy is very useful because it allows us to model even the rarest events, which are computationally expensive.

Cross-entropy measures the difference between two probability distributions for a given random variable or set of events. The information associated with an event quantifies the number of bits needed to encode and transmit it. Lower probability events provide more information; higher probability events provide less information. Cross-entropy tells us how likely an event is to happen based on its probability: if it is very probable, we have a small cross-entropy, while if it is not probable, we have a high cross-entropy.

At this point, we can define the cross-entropy: given two probability distributions, p and q, we can define the cross-entropy, $H(p, q)$, with the following equation:

$$H(p, q) = - \sum_{x} p(x) * log_2 \, q(x)$$

Here, x is the total number of values, $p(x)$ is the probability of the real distribution (actual values), and $q(x)$ is the probability of the distribution that was calculated, starting from the statistical model (predicted values).

If the expected values are equal to the current ones, then the cross-entropy and entropy are equal. In reality, this does not happen and the cross-entropy is obtained from the sum of the entropy and a term that takes the divergence into account.

Cross-entropy is commonly used in optimization procedures as a loss function. A loss function allows us to evaluate the performance of a simulation model: the better the model can predict the behavior of the real system, the smaller the values returned by the loss function. If we correct the algorithm to improve its predictions, then the loss function will give us a measure of the direction in which we are heading: if the loss function increasesm we are heading in the wrong direction, while if it decreases, we are heading in the right direction.

An algorithm based on cross-entropy provides for an iterative procedure in which each iteration can be divided into two phases:

1. A sample of random data is generated according to a specific mechanism.
2. We update the parameters of the random mechanism based on the data to produce a better sample in the next iteration.

Now, let's learn how to calculate cross-entropy in a Python environment.

Cross-entropy in Python

Let's start practicing with cross-entropy by applying the equation defined in the *Introducing cross-entropy* section to two artificially created distributions. Here is the Python code (cross_entropy.py):

```python
from matplotlib import pyplot
from math import log2

events = ['A', 'B', 'C','D']
p = [0.70, 0.05,0.10,0.15]
q = [0.45, 0.10, 0.20,0.25]
print(f'P = {sum(p):.3f}',f'Q = {sum(q):.3f}')

pyplot.subplot(2,1,1)
pyplot.bar(events, p)
pyplot.subplot(2,1,2)
pyplot.bar(events, q)
pyplot.show()

def cross_entropy(p, q):
    return -sum([p*log2(q) for p,q in zip(p,q)])

h_pq = cross_entropy(p, q)
print(f'H(P, Q) =  {h_pq:.3f} bits')
```

As always, we will analyze the code line by line:

1. Let's start by importing the libraries:

   ```python
   from matplotlib import pyplot
   from math import log2
   ```

   The matplotlib library is a Python library for printing high-quality graphics. Next, we imported the log2() function from the math library. This library provides access to the mathematical functions defined by the C standard.

2. Now, let's define the probability distributions:

   ```python
   events = ['A', 'B', 'C','D']
   p = [0.70, 0.05,0.10,0.15]
   ```

```
q = [0.45, 0.10, 0.20,0.25]
print(f'P = {sum(p):.3f}',f'Q = {sum(q):.3f}')
```

To start, we have defined four event labels so that we can identify them when plotted. Therefore, we have defined two lists with the probabilities associated with these events: recall that according to the definition of cross-entropy, with p, we indicate the probability of real distribution (actual values), while with q, we define the probability of the distribution that was calculated, starting from the statistical model (predicted values). Finally, after defining the two probability distributions, we made the sum to verify that it was equal to 1.

The following result is printed on the screen:

```
P = 1.000 Q = 1.000
```

3. Now, let's draw the diagrams of the two distributions:

    ```
    pyplot.subplot(2,1,1)
    pyplot.bar(events, p)
    pyplot.subplot(2,1,2)
    pyplot.bar(events, q)
    pyplot.show()
    ```

We have plotted two bar graphs for the two probability distributions. A bar chart represents a series of data from different categories: it displays data using multiple bars of the same width, each representing a category. The height of each bar is proportional to the probability value. The following diagram is returned:

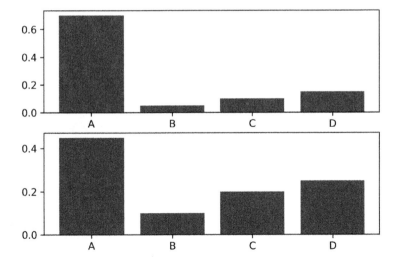

Figure 4.10 – Bar plot of probability distributions

4. Let's define a function that calculates the cross-entropy by applying the equation we defined in the *Introducing cross-entropy* section:

```
def cross_entropy(p, q):
    return -sum([p*log2(q) for p,q in zip(p,q)])
```

To do this, we used the `zip()` function, which returns a `zip` object. This is a tuple iterator. The two arguments we passed are coupled, element by element.

5. Now, let's look at a new application in which we will use cross-entropy as a loss function. We can apply the function we just defined to the two previously defined probability distributions:

```
h_pq = cross_entropy(p, q)
print(f'H(P, Q) = {h_pq:.3f} bits')
```

The following result is printed on the screen:

```
H(P, Q) = 1.505 bits
```

This represents the crossed entropy in bits. To verify its value, we can perform a new calculation simply by reversing the order of the distributions.

Binary cross-entropy as a loss function

Now, let's look at a new application in which we will use cross-entropy as a loss function. We will do this for a binary classification case where the outputs belong to only two classes, (0,1). In this case, the loss is equal to the mean of the categorical cross-entropy loss on many two-category tasks. Cross-entropy loss is used to measure the performance of a classification model. The loss is calculated between 0 and 1, where 0 is a perfect model. The goal is generally to get the model as close to 0 as possible. The formula for calculating the cross-entropy changes slightly:

$$Loss = -\frac{1}{N}\sum_{i} y_i * log(p(y_i)) + (1 - y_i) * log(1 - p(y_i))$$

Here, N is the number of observations, y is the label (that is, the actual value), and p is the estimated probability.

In binary classification, the probability is modeled as the Bernoulli distribution for the class 1 label: the probability for class 1 is predicted directly by the model, and the probability for class 0 is given as 1 minus the predicted probability.

Here is the Python code (`cross_entropy_loss_function.py`):

```
import numpy as np
```

```
y = np.array([1.0, 0.0, 1.0, 0.0, 1.0, 0.0, 1.0, 0.0, 1.0,
0.0])
p = np.array([0.8, 0.1, 0.9, 0.2, 0.8, 0.1, 0.7, 0.3, 0.6,
0.4])

ce_loss = -sum(y*np.log(p)+(1-y)*np.log(1-p))

ce_loss = ce_loss/len(p)
print(f'Cross-entropy Loss = {ce_loss:.3f} nats')
```

As always, we will analyze the code line by line:

1. Let's start by importing the libraries:

   ```
   import numpy as np
   ```

   Here, we imported the numpy library, which offers additional scientific functions of the Python language, designed to perform operations on vectors and dimensional matrices.

2. Now, let's define the probability distributions:

   ```
   y = np.array([1.0, 0.0, 1.0, 0.0, 1.0, 0.0, 1.0, 0.0,
   1.0, 0.0])
   p = np.array([0.8, 0.1, 0.9, 0.2, 0.8, 0.1, 0.7, 0.3,
   0.6, 0.4])
   ```

   As mentioned previously, y is the label (that is, the actual value) and p is the estimated probability.

3. Let's calculate the cross-entropy by applying the equation we defined previously:

   ```
   ce_loss = -sum(y*np.log(p)+(1-y)*np.log(1-p))
   ce_loss = ce_loss/len(p)
   ```

4. We just have to print the result:

   ```
   print(f'Cross-entropy Loss = {ce_loss:.3f} nats')
   ```

   The following result is returned:

   ```
   Cross-entropy Loss = 0.272 nats
   ```

Note that the result is in nats and not in bits since we used the natural logarithm. nat is a logarithmic unit of information or entropy based on natural logarithms, rather than the base 2 logarithms that define the bit.

Summary

In this chapter, we addressed the basic concepts of Monte Carlo simulation. We explored the Monte Carlo components used to obtain a simulation with satisfactory results. Hence, we used Monte Carlo methods to estimate the value of Pi.

Then, we tackled two fundamental concepts of Monte Carlo simulation: the law of large numbers and the central limit theorem. For example, the law of large numbers allows us to determine the centers and weights of a Monte Carlo analysis to estimate definite integrals. The central limit theorem is of great importance, and it is thanks to this that many statistical procedures work.

Next, we analyzed practical applications of using Monte Carlo methods in real life: numerical optimization and project management. Finally, we learned how to perform numerical integration using Monte Carlo techniques.

Finally, sensitivity analysis concepts and cross-entropy methods were explained using some practical examples.

In the next chapter, we will learn the basic concepts of the Markov process. We will understand the agent-environment interaction process and how to use Bellman equations as consistency conditions for the optimal value functions to determine the optimal policy. Finally, we will learn how to implement Markov chains to simulate random walks.

5
Simulation-Based Markov Decision Processes

Markov decision processes (**MDPs**) model decision-making in situations where outcomes are partly random and partly under the control of a decision maker. MDP is a stochastic process characterized by five elements: decision epochs, states, actions, transition probability, and reward. The characteristic elements of a Markovian process are the states in which the system finds itself and the available actions that the decision maker can carry out in that state. These elements identify two sets: the set of states in which the system can be found and the set of actions available for each specific state. The action chosen by the decision maker determines a random response from the system, which brings it into a new state. This transition returns a reward that the decision maker can use to evaluate the merit of their choice. In this chapter, we will learn how to deal with decision-making processes with Markov chains. We will analyze the concepts underlying Markovian processes and then analyze some practical applications to learn how to choose the right actions for the transition between different states of the system.

In this chapter, we're going to cover the following main topics:

- Introducing agent-based models
- Overview of Markov processes
- Introducing Markov chains
- Markov chain applications
- The Bellman equation explained
- Multi-agent simulation
- Schelling's model of segregation

Technical requirements

In this chapter, MDPs will be introduced. In order to deal with the topics in this chapter, it is necessary that you have a basic knowledge of algebra and mathematical modeling.

To work with the Python code in this chapter, you'll need the following files (available on GitHub at the following URL: https://github.com/PacktPublishing/Hands-On-Simulation-Modeling-with-Python-Second-Edition):

- `simulating_random_walk.py`
- `weather_forecasting.py`
- `schelling_model.py`

Introducing agent-based models

Agent-based simulation (**ABM**) is a class of computational models aimed at the computer simulation of actions and interactions of autonomous agents to evaluate their effects on the overall system. ABMs incorporate intelligent elements capable of learning, adapting, evolving, or following unpredictable logic. One of the key features, which turns out to be one of the strengths of this methodology, is the bottom-up approach, which, considering the interactions of the individual elements of the system, tries to determine the emergent characteristics produced by these interactions. These characteristics are not explicit of the system but emerge from the interaction of the group of agents who act independently, are oriented towards achieving their specific goal, and interact in a shared environment.

The applications of ABM are many and particularly useful. They are usually used to study phenomena that are too complex to be faithfully reproduced and/or expensive in terms of cost and effort. For example, the study of the behavior of a population bounded in a certain area, as well as the simulation of reactions between molecules in a chemical solution, or even the management of distributed systems that can be represented as nodes of a network that request and offer services. The reason for the increasing use of ABMs is to be found in the growing complexity of the networks with which a system can be modeled. The organizational, temporal, and economic costs for the analytical study of such systems would be too high; moreover, the experiment itself could prove to be risky. This tool allows us to study and analyze the dynamic behavior of a real system through the behavior of an artificial system.

An ABM is characterized by three fundamental components:

- **Agents**: Autonomous entities with their own characteristics and behaviors
- **Environment**: A field in which agents interact with each other
- **Rules**: They define the actions and reactions that occur during the simulation

An ABM assumes that only local information is available to agents. The ABM is a decentralized system; there is no central entity that globally disseminates the same information to all agents or controls their behavior with the aim of increasing the performance of the system. Agents interact with each other, but they don't all interact at the same time together. The interaction between them typically occurs with a sub-group of similar entities close to each other (neighbors). Similarly, the interaction between them and the environment is also localized; in fact, it is not certain that the agents interact with every part of the environment.

Therefore, the information obtained by the agents is obtained from the sub-group with which they interact and from the portion of the environment in which they are located. If the model is not static, the sub-group and the environment with which an agent interacts vary over time. The agent of an ABM interacts with others of its kind and with the environment. This plays a fundamental role in the simulation and is represented in such a way as to reproduce the context in which the real event occurs. It usually keeps the positional information of the agents over time to be able to keep track of all the movements made during the execution of the simulation.

There may also be objects in the field: they represent specific areas of the environment that have different characteristics depending on their location. Agents can interact with them to manipulate, produce, destroy, or be influenced by them and change behavior.

Some essential characteristics of ABMs are summarized in the following points:

- They define the decision-making processes of individual agents, considering both social and economic aspects
- They incorporate the influence of decisions made at the micro level by linking them to the consequences at the macro level
- They study the overall response generated by a change in the environment or the policies implemented

The ABM-based approach must be built and validated with the participation of the various stakeholders present within the model. In this way, the ABM promotes the collective learning process by integrating local and scientific knowledge, providing further information and assistance to policymakers in defining the most suitable policies for solving a specific problem.

Through ABM, it is possible to identify structures and patterns that emerge at the macroscopic level because of the interactions between agents at the microscopic level. No agent possesses the emergent properties of the system; these properties cannot be obtained as the sum of the individual behaviors of the single agents.

The complexity of a system is due to the interaction between the agents that develops at the microscopic level, which determines their characteristics at the macroscopic level. These interactions develop over time by linking the states of the system to each other in a chain of events. In Markov processes, the information obtainable from the observation of the history of the process up to the present can be useful for making inferences about the future state.

In the next section, we will study these processes to fully understand how agent-based models are structured.

Overview of Markov processes

Markov's decision-making process is defined as a discrete-time stochastic control process. In *Chapter 2, Understanding Randomness and Random Numbers*, we said that stochastic processes are numerical models used to simulate the evolution of a system according to random laws. Natural phenomena are characterized by random factors both in their very nature and observation errors. These factors introduce a random number into the observation of the system. This random factor determines an uncertainty in the observation since it is not possible to predict with certainty what the result will be. In this case, we can only say that it will assume one of the many possible values with a certain probability.

If starting from an instant *t* in which an observation of the system is made, the evolution of the process will depend only on *t*, while it will not be influenced by the previous instants. Here, we can say that the stochastic process is Markovian.

> **Important note**
> A process is called **Markovian** when the future evolution of the process depends only on the instant of observation of the system and does not depend in any way on the past.

Characteristic elements of a Markovian process include the states in which the system finds itself and the available actions that the decision maker can carry out in that state. These elements identify two sets: the set of states in which the system can be found and the set of actions available for each specific state. The action chosen by the decision maker determines a random response from the system, which brings it into a new state. This transition returns a reward that the decision maker can use to evaluate the goodness of their choice.

The agent-environment interface

A Markovian process takes on the characteristics of an interaction problem between two elements to achieve a goal. The two characteristic elements of this interaction are the **agent** and the **environment**. The agent is the element that must reach the goal, while the environment is the element that the agent must interact with. The environment corresponds to everything that is external to the agent.

The agent is a piece of software that performs the services necessary for another piece of software in a completely automatic and intelligent way. They are known as intelligent agents.

The essential characteristics of an agent are listed here:

- The agent continuously monitors the environment, and this action causes a change in the state of the environment
- The available actions belong to a continuous or discrete set

- The agent's choice of action depends on the state of the environment
- This choice requires a certain degree of intelligence as it is not trivial
- The agent has a memory of the choices made – intelligent memory

The agent's behavior is characterized by attempting to achieve a specific goal. To do this, it performs actions in an environment it does not know a priori, or at least not completely. This uncertainty is filled through the interaction between the agent and the environment. In this phase, the agent learns to know the state of the environment by measuring it, in this way, planning its future actions.

The strategy adopted by the agent is based on the principles of error theory: proof of the actions and memory of the possible mistakes made to make repeated attempts until the goal is achieved. These actions by the agent are repeated continuously, causing changes in the environment that change their state.

> **Important note**
> Crucial to the agent's future choices is the concept of reward, which represents the environment's response to the action taken. This response is proportional to the weight that the action determines in achieving the objective: it will be positive if it leads to correct behavior, while it will be negative in the case of an incorrect action.

The decision-making process that leads the agent to achieve their objective can be summarized in three essential points:

- The objective of the agent
- Their interaction with the environment
- The total or partial uncertainty of the environment

In this process, the agent receives stimuli from the environment through the measurements made by sensors. The agent decides what actions to take based on the stimuli received from the environment. As a result of the agent's actions, determining a change in the state of the environment will receive a reward.

The crucial elements in the decision-making process are shown in the following diagram:

Figure 5.1 – The agent's decision-making process

While choosing an action, it is crucial to have a formal description of the environment. This description must return essential information regarding the properties of the environment and not a precise representation of the environment.

Exploring MDPs

The agent-environment interaction we discussed in the previous section is approached as a Markov decision-making process. This choice is dictated by loading problems and computational difficulties. As anticipated in the *Overview of Markov processes* section, Markov's decision-making process is defined as a discrete-time stochastic control process.

Here, we need to perform a sequence of actions, with each action leading to a non-deterministic change regarding the state of the environment. By observing the environment, we know its state after performing an action. On the other hand, if the observation of the environment is not available, we do not know the state, even after performing the action. In this case, the state is a probability distribution of all the possible states of the environment. In such cases, the change process can be viewed as a snapshot sequence.

The state at time *t* is represented by a random variable *st*. Decision-making is interpreted as a discrete-time stochastic process. A discrete-time stochastic process is a sequence of random variables *xt*, with $t \in \mathbb{N}$. We can define some elements as follows:

- **State space**: Set of values that random variables can assume
- **History of a stochastic process (path)**: The realization of the sequence of random variables

The response of the environment to a certain action is represented by the reward. The agent-environment interaction in an MDP can be summarized by the following diagram:

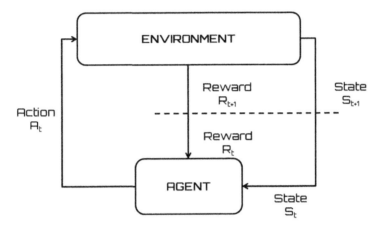

Figure 5.2 – The agent-environment interaction in MDP

The essential steps of the agent-environment interaction, schematically represented in the previous diagram, are listed here:

1. The interaction between the agent and the environment occurs at discrete instants over time.
2. In every instant, the agent monitors the environment by obtaining its state $st \in S$, where S is the set of possible states.
3. The agent performs an action $a \in A(s_t)$, where $A(s_t)$ is the set of possible actions available for the state s_t.
4. The agent chooses one of the possible actions according to the objective to be achieved.
5. This choice is dictated by the policy $\pi(s, a)$, which represents the probability that the action a is performed in the state *s*.
6. At time $t + 1$, the agent receives a numerical reward $r_{t+1} \in R$ corresponding to the action previously chosen.

7. Because of the choice, the environment passes into the new state.
8. The agent must monitor the state of the environment and perform a new action.
9. This iteration repeats until the goal is achieved.

In the iterative procedure we have described, the state s_{t+1} depends on the previous state and the action taken. This feature defines the process as an MDP, which can be represented by the following equation:

$$s_{t+1} = \delta(s_t, a)$$

In the previous equation, δ represents the state function. We can summarize an MDP as follows:

- The agent monitors the state of the environment and has a series of actions
- At a discrete time t, the agent detects the current state and decides to perform an action $a_t \in A$
- The environment reacts to this action by returning a reward rt = r (st, at) and moving to the state $s_{t+1} = \delta(s_t, a_t)$

> **Important note**
> The r and δ functions are characteristics of the environment that depend only on the current state and action. The goal of MDP is to learn a policy that, for each state of the system, provides the agent with an action that maximizes the total reward accumulated during the entire sequence of actions.

Now, let's analyze some of the terms that we introduced previously. They represent crucial concepts that help us understand Markovian processes.

The reward function

A reward function identifies the target in a Markovian process. It maps the states of the environment detected by the agent by enclosing it in a single number, which represents the reward. The purpose of this process is to maximize the total reward that the agent receives over the long term because of their choices. Then, the reward function collects the positive and negative results obtained from the actions chosen by the agent and uses them to modify the policy. If an action selected based on the indications provided by the policy returns a low reward, then the policy will be modified to select other actions. The reward function performs two functions: it stimulates the efficiency of the decisions and determines the degree of risk aversion of the agent.

Policy

A policy determines the agent's behavior in terms of decision-making. It maps both the states of the environment and the actions to be chosen in those states, representing a set of rules or associations that respond to a stimulus. The policy is the fundamental part of a Markovian agent as it determines its behavior. In a Markov decision-making model, a policy provides a solution that associates a recommended action with each state potentially achievable by the agent. If the policy provides the highest expected utility among the possible actions, it is called an optimal policy (π^*). In this way, the agent does not have to keep their previous choices in memory. To make a decision, the agent only needs to execute the policy associated with the current state.

The state-value function

The value function provides us with the information necessary to evaluate the quality of a state for an agent. It returns the value of the expected goal that was obtained following the policy of each state, which is represented by the total expected reward. The agent depends on the policy to choose the actions to be performed.

Understanding the discounted cumulative reward

The goal of MDP is to learn a policy that guides an agent in choosing the actions to be performed for each state of the environment. This policy aims to maximize the total reward received during the entire sequence of actions performed by the agent. Let's learn how to maximize this total reward. The total reward that's obtained from adopting a policy is calculated as follows:

$$R_T = \sum_{i=0}^{T} r_{t+1} = r_t + r_{t+1} + \cdots + r_T$$

In the preceding equation, r_T is the reward of the action that brings the environment into the terminal state s_T.

To get the maximum total reward, we can select the action that provides the highest reward for each individual state, which leads to the choice of the optimal policy that maximizes the total reward.

> **Important note**
>
> This solution is not applicable in all cases, for example, when the goal or terminal state is not achieved in a finite number of steps. In this case, both R_T and the sum of the rewards you want to maximize tend to infinity.

An alternative technique uses the discounted cumulative reward, which tries to maximize the following amount:

$$R_T = \sum_{i=0}^{\infty} \gamma^i * r_{t+1} = r_t + \gamma * r_{t+1} + \gamma^2 * r_{t+2} + \cdots$$

In the previous equation, γ is called the discount factor and represents the importance of future rewards. The discount factor is $0 \leq \gamma \leq 1$ and has the following conditions:

- $\gamma < 1$: The sequence r_t converges to a finite value

- **γ = 0**: The agent does not consider future rewards, thereby trying to maximize the reward only for the current state
- **γ = 1**: The agent will favor future rewards over immediate rewards

The value of the discount factor may vary during the learning process to take special actions or states into account. An optimal policy may include individual actions that return low rewards, provided that the total reward is higher.

Comparing exploration and exploitation concepts

Upon reaching the goal, the agent looks for the most rewarded behavior. To do this, they must link each action to the reward returned. In the case of complex environments with many states, this approach is not feasible due to many action-reward pairs.

> **Important note**
> This is the well-known exploration-exploitation dilemma: for each state, the agent explores all possible actions, exploiting the action most rewarded in achieving the objective.

Decision-making requires a choice between the two available approaches:

- **Exploitation**: The best decision is made based on current information
- **Exploration**: The best decision is made by gathering more information

The best long-term strategy can impose short-term sacrifices as this approach requires collecting adequate information to reach the best decisions.

In everyday life, we often find ourselves having to choose between two alternatives that, at least theoretically, lead to the same result: this approach is an exploration-exploitation dilemma. For example, let's say we need to decide whether to choose what we already know (exploitation) or choose something new (exploration). Exploitation keeps our knowledge unchanged, while exploration makes us learn more about the system. It is obvious that exploration exposes us to the risk of wrong choices.

Let's look at an example of using this approach in a real-life scenario: we must choose the best path to reach our trusted restaurant:

- **Exploitation**: Choose the path you already know
- **Exploration**: Try a new path

In complex problems, converging toward an optimal strategy can be too slow. In these cases, a solution to this problem is represented by a balance between exploration and exploitation.

An agent who acts exclusively based on exploration will always behave randomly in each state with convergence to an optimal strategy that is practically impossible. On the contrary, if an agent acts exclusively based on exploitation, they will always use the same actions, which may not be optimal.

Let's now introduce the Markov chains that describe a stochastic process whose evolution depends exclusively on the immediately preceding state.

Introducing Markov chains

Markov chains are discrete dynamic systems that exhibit characteristics attributable to Markovian processes. These are finite state systems – finite Markov chains – in which the transition from one state to another occurs on a probabilistic, rather than deterministic, basis. The information available about a chain at the generic instant t is provided by the probabilities that it is in any of the states, and the temporal evolution of the chain is specified by specifying how these probabilities update by going from the instant t at instant $t + 1$.

> **Important note**
> A Markov chain is a stochastic model in which the system evolves over time in such a way that the past affects the future only through the present: Markov chains have no memory of the past.

A random process characterized by a sequence of random variables $X = X_0, ..., X_n$ with values in a set $j_0, j_1, ..., j_n$ is given. This process is Markovian if the evolution of the process depends only on the current state, that is, the state after n steps. Using conditional probability, we can represent this process with the following equation:

$$P(X_{n+1} = j | X_0 = i_0, ..., X_n = i_n) = P(X_{n+1} = j | X_n = i_n)$$

If a discrete-time stochastic process X has a Markov property, it is called the Markov chain. A Markov chain is said to be homogeneous if the following transition probabilities do not depend on n and only on i and j:

$$P(X_{n+1} = j | X_n = i)$$

In such hypotheses, let's assume we have the following:

$$p_{ij} = P(X_{n+1} = j | X_n = i)$$

All joint probabilities can be calculated by knowing the numbers p_{ij} and the following initial distribution:

$$p_i^0 = P(X_0 = i)$$

This probability represents the distribution of the process at zero time. The probabilities p_{ij} are called transition probabilities, and p_{ij} is the probability of transitioning from I to j in each time phase.

Transition matrix

The application of homogeneous Markov chains is made easy by adopting the matrix representation. Through this, the formula expressed by the previous equation becomes much more readable. We can represent the structure of a Markov chain through the following transition matrix:

$$\begin{matrix} p_{11} & p_{12} & \cdots & p_{1n} \\ p_{21} & p_{22} & \cdots & p_{2n} \\ \cdots & \cdots & \cdots & \cdots \\ p_{n1} & p_{2n} & \cdots & p_{nn} \end{matrix}$$

This is a positive matrix in which the sum of the elements of each row is unitary. In fact, the elements of the ith row are the probabilities that the chain, being in the state S_i at the instant t, passes through S_1 or S_2,... or S_n at the next instant. Such transitions are mutually exclusive and exhaustive of all possibilities. Such a positive matrix with unit sum lines is stochastic. We will call each vector positive line x stochastic such that T = [x_1 x_2... x_n], in which the sum of the elements assumes a unit value:

$$\sum_{i=1}^{n} x_i = 1$$

The transition matrix has the position (i, j) to pass from result i to result j by performing a single experiment.

Transition diagram

The transition matrix is not the only solution for describing a Markov chain. An alternative is an oriented graph called a transition diagram, in which the vertices are labeled by the states S_1, S_2, ..., S_n, and there is a direct edge connecting the vertex S_i to the vertex S_j, if and only if, the probability of transition from S_i to S_j is positive.

> **Important note**
> The transition matrix and the transition diagram contain the same information necessary for representing the same Markov chain.

Le"s take a look at an example: consider a Markov chain with three possible states, 1, 2, and 3, represented by the following transition matrix:

$$\begin{bmatrix} \frac{1}{3} & \frac{1}{3} & \frac{1}{3} \\ \frac{2}{3} & 0 & \frac{1}{3} \\ \frac{1}{4} & \frac{1}{4} & \frac{2}{4} \end{bmatrix}$$

As mentioned previously, the transition matrix contains the same information as the transition diagram. Let's learn how to draw this diagram. There are three possible states, 1, 2, and 3, and the direct boundary from each state to other states shows the probabilities of transition p_{ij}. When there is no arrow from state i to state j, this means that $p_{ij} = 0$:

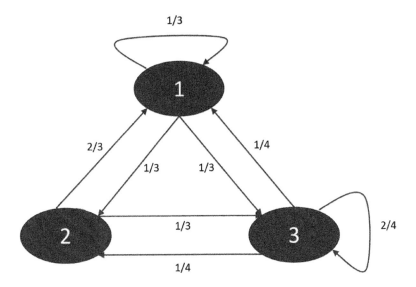

Figure 5.3 – Diagram of the transition matrix

In the preceding transition diagram, the arrows that come out of a state always add up to exactly 1, just like for each row in the transition matrix.

After having introduced the Markov chains, we now see some practical cases of the application of this methodology.

Markov chain applications

Now, let's look at a series of practical applications that can be made using Markov chains. We will introduce the problem and then analyze the Python code that will allow us to simulate how it works.

Introducing random walks

Random walks identify a class of mathematical models used to simulate a path consisting of a series of random steps. The complexity of the model depends on the system features we want to simulate, which are represented by the number of degrees of freedom and the direction. The authorship of the term is attributed to Karl Pearson, who, in 1905, first referred to the term casual walk. In this model, each step has a random direction that evolves through a random process involving known quantities that follow a precise statistical distribution. The path that's traced over time will not necessarily be descriptive of real motion: it will simply return the evolution of a variable over time. This is the reason for the widespread use of this model in all areas of science: chemistry, physics, biology, economics, computer science, and sociology.

One-dimensional random walk

The one-dimensional casual walk simulates the movement of a punctual particle that is bound to move along a straight line, thus having only two movements: right and left. Each movement is associated with a random shift of one step to the right with a fixed probability p or to the left with a probability q. Every single step is the same length and is independent of the others. The following diagram shows the path to which the punctual particle is bound, along with the direction and the two verses allowed:

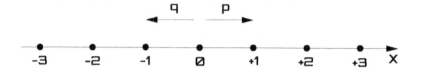

Figure 5.4 – One-dimensional walk

After n passes, the position of the point will be identified by its abscissa X (n), characterized by a random term. Our aim is to calculate the probability with which the particle will return to the starting point, after n steps.

> **Important note**
>
> The idea that the point will actually return to its starting position is not assured. To represent the position of the point on the straight line, we will adopt the $X(n)$ variable, which represents the abscissa of the line after the particle has moved n steps. This variable is a discrete random variable with a binomial distribution.

The path of the point particle can be summarized as follows: for each instant, the particle moves one step to the right or left according to the value returned by a random variable $Z(n)$. This random variable takes only two dichotomous values:

- +1 with probability p > 0
- -1 with probability q

The two probabilities are related to each other through the following equation:

$$p + q = 1$$

Let's consider random variables Z_n with $n = 1, 2, \ldots$. Suppose that these variables are independent and with equal distribution. The position of the particle at instant n will be represented by the following equation:

$$X_n = X_{n-1} + Z_n \; ; \quad n = 1, 2, \ldots$$

In the previous formula, X_n is the next value in the walk, X_{n-1} is the observation in the previous time phase, and Z_n is the random fluctuation in that step.

> **Important note**
> The X_n variable identifies a Markov chain; that is, the probability that the particle in the next moment is in a certain position depends only on the current position, even if we know all the moments preceding the current one.

Simulating a 1D random walk

The simulation of a casual walk does not represent a trivial succession of random numbers since the next step to the current one represents its evolution. The dependence between the next two steps guarantees a certain consistency from one passage to the next. This is not guaranteed in a banal generation of independent random numbers, which instead return big differences from one number to another. Let's learn how to represent the sequence of actions to be performed in a simple casual walking model through the following pseudocode:

1. Start from the 0 position.
2. Randomly select a dichotomous value (-1, 1).
3. Add this value to the previous time step.
4. Repeat *step 2* onward.

This simple iterative process can be implemented in Python by processing a list of 1,000 time steps for the random walk. Let's take a look:

1. Let's start by loading the necessary libraries:

    ```
    from random import seed
    from random import random
    from matplotlib import pyplot
    ```

    The random module implements pseudo-random number generators for various distributions. The random module is based on the Mersenne Twister algorithm. Mersenne Twister is a pseudorandom number generator. Originally developed to produce inputs for Monte Carlo simulations, almost uniform numbers are generated via Mersenne Twister, making it suitable for a wide range of applications.

From the `random` module, two libraries were imported: `seed` and `random`. In this code, we will generate random numbers. To do this, we will use the `random()` function, which produces different values each time it is invoked. It has a very large period before any number is repeated. This is useful for producing unique values or variations, but there are times when it is useful to have the same dataset available for processing in different ways. This is necessary to ensure the reproducibility of the experiment. To do this, we can use the `random.seed()` function contained in the `seed` library. This function initializes the basic random number generator.

The `matplotlib` library is a Python library for printing high-quality graphics. With `matplotlib`, it is possible to generate graphs, histograms, bar graphs, power spectra, error graphs, scatter graphs, and so on with a few commands. This is a collection of command-line functions like those provided by the MATLAB software.

2. Now, we will investigate the individual operations. Let's start with setting the seed:

```
seed(1)
```

The `random.seed()` function is useful if we wish to have the same set of data available to be processed in different ways, as this makes the simulation reproducible.

> **Important note**
> This function initializes the basic random number generator. If you use the same seed in two successive simulations, you will always get the same sequence of pairs of numbers.

3. Let's move on and initialize the crucial variable of the code:

```
RWPath= list()
```

The `RWPath` variable represents a list that will contain the sequence of values representative of the random walk. A list is an ordered collection of values, which can be of various types. It is an editable container, meaning that we can add, delete, and modify existing values. For our purposes, where we want to continuously update our values through the subsequent steps of the path, the list represents the most suitable solution. The `list()` function accepts a sequence of values and converts them into lists. With the preceding command, we simply initialized the list that is currently empty, and with the following code, we start to populate it:

```
RWPath.append(-1 if random() < 0.5 else 1)
```

The first value that we add to our list is a dichotomous value. It is simply a matter of deciding whether the value to be added is 1 or -1. The choice, however, is made on a random basis. Here, we generate a random number between 0 and 1 using the `random()` function and then check if it is <0.5. If it is, then we add -1, otherwise, we add 1. At this point, we will use an iterative cycle with a `for` loop, which will repeat the procedure for 1,000 steps:

```
for i in range(1, 1000):
```

At each step, we will generate a random term as follows:

```
ZNValue = -1 if random() < 0.5 else 1
```

> **Important note**
> As we did when we chose the first value to add to the list, we generate a random value with the `random()` function, so if the value that's returned is lower than 0.5, the `ZNValue` variable assumes the value -1; otherwise, 1.

4. Now, we can calculate the value of the random walk at the current step:

```
XNValue = RWPath[i-1] + ZNValue
```

The `XNValue` variable represents the value of the abscissa in the current step. It is made up of two terms: the first represents the value of the abscissa in the previous state, while the second is the result of generating the random value. This value must be added to the list:

```
RWPath.append(XNValue)
```

This procedure will be repeated for the 1,000 steps that we want to perform. At the end of the cycle, we will have the entire sequence stored in the list.

5. Finally, we can visualize it through the following piece of code:

```
pyplot.plot(RWPath)
pyplot.show()
```

The `pyplot.plot()` function plots the values contained in the `RWPath` list on the *y* axis using x as an index array with the following value: 0..N-1. The `plot()` function is extremely versatile and will take an arbitrary number of arguments.

Finally, the `pyplot.show()` function displays the graph that's created, as follows:

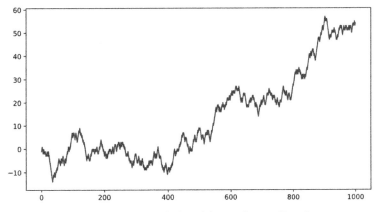

Figure 5.5 – The trend plot of the random walk path

In the previous graph, we can analyze the path followed by the point particle in a random process. This curve can describe the trends of a generic function, not necessarily associated with a road route. As anticipated, this process is configured as a Markovian process in that the next step is independent of the position from the previous step and depends only on the current step. The casual walk is a mathematical model widely used in finance. In fact, it is widely used to simulate the efficiency of information deriving from the markets: the price varies for the arrival of new information, which is independent of what we already know.

Simulating a weather forecast

Another potential application of Markov chains is in the development of a weather forecasting model. Let's learn how to implement this algorithm in Python. To start, we can work with a simplified model: we will consider only two climatic conditions/states, that is, sunny and rainy. Our model will assume that tomorrow's weather conditions will be affected by today's weather conditions, making the process take on Markovian characteristics. This link between the two states will be represented by the following transition matrix:

$$P = \begin{bmatrix} 0.80 & 0.20 \\ 0.25 & 0.75 \end{bmatrix}$$

The transition matrix returns the conditional probabilities P (A | B), which indicate the probability that event A occurs after event B has occurred. This matrix, therefore, contains the following conditional probabilities:

$$P = \begin{bmatrix} P(Sunny|Sunny) & P(Sunny|Rainy) \\ P(Rainy|Sunny) & P(Rainy|Rainy) \end{bmatrix}$$

In the previous transition matrix, each row contains a complete distribution. Therefore, all the numbers must be non-negative, and the sum must be equal to 1. The climatic conditions show a tendency to resist change. For this reason, after a sunny day, the probability of another sunny-P (Sunny | Sunny) day is greater than a rainy-P (Sunny | Rainy) day. The climatic conditions of tomorrow are not directly related to those of yesterday; it follows that the process is Markovian. The previous transition matrix is equivalent to the following:

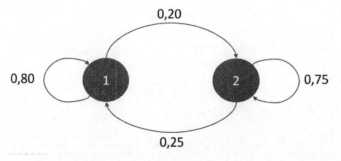

Figure 5.6 – Transition diagram

The simulation model we want to elaborate on will have to calculate the probability that it will rain in the next few days. It will also have to allow you to recover a statistic of the proportion between sunny and rainy days in a certain period of time. The process, as mentioned previously, is Markovian, and the tools we analyzed in the previous sections allow us to obtain the requested information. Let's get started:

1. Let's see the Python code that alternates sunny and rainy days, starting from a specific initial condition. As always, we will analyze it line by line, starting with loading the necessary libraries:

   ```
   import numpy as np
   import matplotlib.pyplot as plt
   ```

   The `numpy` library is a Python library that contains numerous functions that can help us manage multidimensional matrices. Furthermore, it contains a large collection of high-level mathematical functions we can use on these matrices. We will use two functions: `random.seed()` and `random.choose()`.

   The `matplotlib` library is a Python library for printing high-quality graphics. With `matplotlib`, it is possible to generate graphs, histograms, bar graphs, power spectra, error graphs, scatter graphs, and so on with a few commands. It is a collection of command-line functions like those provided by the MATLAB software. Let's move on and illustrate the code:

   ```
   np.random.seed(3)
   ```

   The `random.seed()` function initializes the seed of the random number generator. In this way, the simulation that uses random numbers will be reproducible. The reproducibility of the experiment will be guaranteed by the fact that the random numbers that will be generated will always be the same.

2. Now, let's define the possible states of the weather conditions:

   ```
   StatesData = ["Sunny","Rainy"]
   ```

   Two states are provided: sunny and rainy. The transitions matrix representing the transition between the weather conditions will be set as follows:

   ```
   TransitionStates = [["SuSu","SuRa"],["RaRa","RaSu"]]
   TransitionMatrix = [[0.80,0.20],[0.25,0.75]]
   ```

   The transition matrix returns the conditional probabilities P (A | B), which indicate the probability that event A occurs after event B has occurred. All the numbers in a row must be non-negative and the sum must be equal to 1. Let's move on and set the variable that will contain the list of state transitions:

   ```
   WeatherForecasting = list()
   ```

   The `WeatherForecasting` variable will contain the results of the weather forecast. This variable will be of the `list` type.

> **Important note**
> A list is an ordered collection of values and can be of various types. It is an editable container and allows us to add, delete, and modify existing values.

For our purpose, which is to continuously update our values through the subsequent steps of the path, the list represents the most suitable solution. The `list()` function accepts a sequence of values and converts them into lists.

3. Now, we decide on the number of days for which we will predict the weather conditions:

```
NumDays = 365
```

For now, we have decided to simulate the weather forecast for a 1-year time horizon; that is, 365 days. Let's fix a variable that will contain the forecast for the current day:

```
TodayPrediction = StatesData[0]
```

Furthermore, we also initialized it with the first vector value containing the possible states. This value corresponds to the `Sunny` condition. We print this value on the screen by doing the following:

```
print("Weather initial condition =",TodayPrediction)
```

At this point, we can predict the weather conditions for each of the days set by the `NumDays` variable. To do this, we will use a `for` loop that will execute the same piece of code several times equal to the number of days that we have set in advance:

```
for i in range(1, NumDays):
```

Now, we will analyze the main part of the entire program. Within the `for` loop, the forecast of the time for each consecutive day occurs through an additional conditional structure: the `if` statement. Starting from a meteorological condition contained in the `TodayPrediction` variable, we must predict that of the next day. We have two conditions: sunny and rainy. In fact, there are two control conditions, as shown in the following code:

```
if TodayPrediction == "Sunny":
        TransCondition = np.random.choice(TransitionStates[0],replace=True,
p=TransitionMatrix[0])
        if TransCondition == "SuSu":
            pass
        else:
            TodayPrediction = "Rainy"

  elif TodayPrediction == "Rainy":
```

Markov chain applications

```
            TransCondition = np.random.choice
    (TransitionStates[1],replace=True,p=TransitionMatrix[1])
        if TransCondition == "RaRa":
            pass
        else:
            TodayPrediction = "Sunny"
```

- If the current state is Sunny, we use the `numpy random.choice()` function to forecast the weather condition for the next state. A common use for random number generators is to select a random element from a sequence of enumerated values, even if these values are not numbers. The `random.choice()` function returns a random element of the non-empty sequence passed as an argument. Three arguments are passed:

- `TransitionStates[0]`: The first row of the transition states

- `replace=True`: The sample is with replacement

- `p=TransitionMatrix[0]`: The probabilities associated with each entry in the state passed

The `random.choice()` function returns random samples of the SuSu, SuRa, RaRa, and RaSu types, according to the values contained in the `TransitionStates` matrix. The first two will be returned starting from sunny conditions and the remaining two starting from rainy conditions. These values will be stored in the `TransCondition` variable.

Within each `if` statement, there is an additional `if` statement. This is used to determine whether to update the current value of the weather forecast or to leave it unchanged. Let's see how:

```
    if TransCondition == "SuSu":
            pass
        else:
            TodayPrediction = "Rainy"
```

If the `TransCondition` variable contains the SuSu value, the weather conditions of the current day remain unchanged. Otherwise, it is replaced by the Rainy value. The `elif` clause performs a similar procedure, starting from the rainy condition. At the end of each iteration of the `for` loop, the list of weather forecasts is updated, and the current forecast is printed:

```
WeatherForecasting.append(TodayPrediction)
print(TodayPrediction)
```

Now, we need to predict the weather forecast for the next 365 days.

4. Let's draw a graph with the sequence of forecasts for the next 365 days:

```
plt.plot(WeatherForecasting)
plt.show()
```

The following graph is printed:

Figure 5.7 – Plot of the weather forecast

Here, we can see that the forecast of sunny days prevails over the rainy ones.

> **Important note**
> The flat points at the top represent all the sunny days, while the dips in-between are the rainy days.

To quantify this prevalence, we can draw a histogram. In this way, we will be able to count the occurrences of each condition:

```
plt.figure()
plt.hist(WeatherForecasting)
plt.show()
```

The following is a histogram of the weather condition for the next 365 days:

Figure 5.8 – Histogram of the weather forecast

With this, we can confirm the prevalence of sunny days. The result we've obtained derives from the transition matrix. In fact, we can see that the probability of the persistence of a solar condition is greater than that of rain. In addition, the initial condition has been set to the sunny condition. We can also try to see what happens when the initial condition is set to the rainy condition.

Bellman equation explained

In 1953, Richard Bellman introduced the principles of dynamic programming in order to efficiently solve sequential decision problems. In such problems, decisions are periodically implemented and influence the size of the model. In turn, these influence future decisions. The principle of optimality, enunciated by Bellman, allows you, through an intelligent application, to efficiently deal with the complexity of the interaction between the decisions and the sizes of the model. Dynamic programming techniques were also applied from the outset to problems in which there is no temporal or sequential aspect.

> **Important note**
> Although dynamic programming can be applied to a wide range of problems by providing a common abstract model, from a practical point of view, many problems require models of such dimensions to preclude, then as now, any computational approach. This inconvenience was then called the curse of dimensionality and was an anticipation, still in informal terms, of concepts of computational complexity.

The greatest successes of dynamic programming have been obtained in the context of sequential decision models, especially of the stochastic type, such as the Markovian decision processes, but also in some combinatorial models.

Dynamic programming concepts

Dynamic programming (**DP**) is a programming technique designed to calculate an optimal policy based on a perfect model of the environment in the form of a **Markov decision-making process** (**MDP**). The basis of dynamic programming is the use of state values and action values in order to identify good policies.

DP methods are applied to MDP using two processes called policy evaluation and policy improvement, which interact with each other:

- **Policy evaluation**: Is done through an iterative process that seeks to solve Bellman's equation. The convergence of the process for $k \to \infty$ imposes approximation rules, thus introducing a stop condition.
- **Policy improvement**: Improves the policy based on current values.

In the policy iteration technique, the two phases just described alternate, and each concludes before the other begins.

> **Important note**
> The iterative process, when evaluating policies, obliges us to evaluate a policy at each step through an iterative process whose convergence is not known a priori and depends on the starting policy. To address this problem, we can stop evaluating the policy at some point while still ensuring that we converge to an optimal value.

Principle of optimality

The validity of the dynamic optimization procedure is ensured by Bellman's principle of optimality: *An optimal policy has the property that whatever the initial state and initial decision are, the remaining decisions must constitute an optimal policy with regard to the state resulting from the first decision.*

Based on this principle, it is possible to divide the problem into stages and solve the stages in sequence, using the dynamically determined values of the objective function, regardless of the decisions that led to them. This allows us to optimize one stage at a time, reducing the initial problem to a sequence of smaller subproblems, therefore, easier to solve.

Bellman equation

Bellman's equation helps us solve MDP by finding the optimal policy and value functions. The optimal value function V * (S) is the one that returns the maximum value of a state. This maximum value is the one that corresponds to the action that maximizes the reward value of the optimal action in each state. It then adds a discount factor multiplied by the value of the next state by the Bellman equation through a recursive procedure. The following is an example of a Bellman equation:

$$V(s) = max_a(R(s,a) + \gamma * V(s'))$$

In the previous equation, we have the following:

- $V(s)$ is the function value at state s
- $R(s,a)$ is the reward we get after acting a in state s
- γ is the discount factor
- $V(s')$ is the function value at the next state

For a stochastic system, when we take an action, it is not said that we will end up in a later state, but that we can only indicate the probability of ending up in that state.

In everyday life, we know that unity makes strength; we can think that even an intelligent system can benefit from the joint action of several intelligent agents. Making multiple agents operate in a single environment is not an easy thing, so let's see how to model this interaction and see what benefits we can derive from it.

Multi-agent simulation

An agent can be defined as anything that is able to perceive an environment through sensors and act in it through actuators. Artificial intelligence focuses on the concept of a rational agent, or an agent who always tries to optimize an appropriate performance measure. A rational agent can be a human agent, a robotic agent, or a software agent. In the following diagram, we can see the interaction between the agent and the environment:

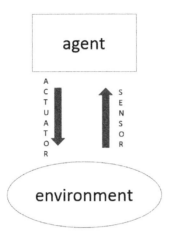

Figure 5.9 – Interaction between the agent and the environment

An agent is considered autonomous when it can flexibly and independently choose the actions to be taken to achieve its goals without constantly resorting to the intervention of an external decision system. Note that, in most complex domains, an agent can only partially obtain information and have control in the environment that it has been inserted into, thus exerting, at most, a certain influence on it.

An agent can be considered autonomous and intelligent if it has the following characteristics:

- **Reactivity**: It must perceive the environment, managing to adapt in good time to the changes that take place
- **Proactivity**: It must show behavior oriented toward achieving the objectives set with initiative
- **Social skills**: It must be able to interact with other agents for the pursuit of its goals

There are numerous situations where multiple agents coexist in the same environment and interact with each other in different ways. In fact, it is quite rare for an agent to represent an isolated system. We can define a **multi-agent system** (**MAS**) as a group of agents that can potentially interact with each other. An MAS can be competitive, which is where each agent tries to exclusively maximize their own interests, even at the expense of those of others, rather than cooperative, which is where agents are willing to give up part of their objectives in an attempt to maximize the global utility of the system.

The types of possible interactions are as follows:

- **Negotiation**: This occurs when agents must seek an agreement on the value to be assigned to some variables
- **Cooperation**: This occurs when there are common goals for which agents try to align and coordinate their actions
- **Coordination**: This is a type of interaction aimed at avoiding situations of conflict between agents

The use of MAS systems introduces a series of advantages:

- **Efficiency and speed**: Thanks to the possibility of performing computations in parallel
- **Robustness**: The system can overcome single-agent failures
- **Flexibility**: Adding new agents to the system is extremely easy
- **Modularity**: Extremely useful in the software design phase due to the possibility of reusing the code
- **Cost**: The single-unit agent has a very low cost compared to the overall system

The growing attention that's being paid to multi-agent systems for the treatment of problems based on decision-making processes is linked to some characteristics that distinguish them, such as flexibility and the possibility of representing independent entities through distinct computational units that interact with each other. The various stakeholders of a decision-making system can, in fact, be modeled as autonomous agents. Several practical applications of the real world have recently adopted an approach based on problems such as satisfying and optimizing distributed constraints and identifying regulations that have an intermediate efficiency between the centralized (optimal) and the non-coordinated (bad) ones.

Schelling's model of segregation

Schelling's model, which earned Schelling the Nobel Prize in 1979, models the social segregation of two populations of individuals who can move within a city by obeying a few simple rules:

- Each agent is satisfied with their position if there is at least a consistent percentage of neighbors like them around them
- Otherwise, the agents move

The neighbors are, for example, those of the Moore neighborhood, which is its eight closest neighbors. By letting the system evolve according to these rules, it is shown that starting from a mixed population, we obtain a city segregated into neighborhoods in which there are individuals of a single population. This phenomenon has been interpreted as an emergent property as it is typical of the macro scale of the system. It is not due to any centralized control or to explicit decisions by the two groups as a whole, and there is no direct causal link between the rules on individual agents (micro) and the aggregate result of the evolution of the system (macro).

Schelling's model of segregation, thanks to a model with cellular automata, studies how territorial segregation evolves given two types of agents who prefer to live in areas where they are not a minority. In the segregation model, the formation of segregated neighborhoods is analyzed, giving a territorial notion to the concept of a social group. Schelling considers a population of agents randomly distributed on a plane: the characterization of the agents is general, and the analysis can concern any dichotomous distinction of the population. The agents want to occupy an area where at least 50% of the neighbors are the same color as them. Neighbors are agents who occupy the squares adjacent to the agent's area under consideration.

Each zone can be occupied by only one agent. The agent's utility function can be formalized as follows:

$$U(q) = \begin{cases} 0 \text{ if } q < 0.5 \\ 1 \text{ if } q > 0.5 \end{cases}$$

$q \in 0.1$ is the share of agents like the agent among its neighbors. Agents can move from area to area. This happens if both of the following conditions apply:

- The current area is mainly inhabited by agents of the same type > 0.5
- There is an adjacent free area

A third condition can be added:

- There is an adjacent free zone that is more satisfactory

This condition could represent the hypothesis that the move is expensive and is carried out only if it brings an improvement. The third condition, by limiting movements, reduces segregation without, however, eliminating it.

Schelling's result is that the population segregates into areas of homogeneous color, although the preference does not require segregation, such as agents are happy to live in areas where the population is divided exactly in half between the two types or in general in areas in which they are not a minority. Each agent that moves influences the utility of the agents it leaves and the agents it encounters, creating a chain reaction.

Python Schelling model

Now let's see a practical example of how to implement Schelling's model in Python. Suppose we have two types of agents: A and B. The two types of agents can represent any two groups in any context. To begin, we place the two populations of agents in random positions in a delimited area. After placing all the agents in the grid, each cell is either occupied by an agent of the two groups or is empty.

Our task will then be to check whether each agent is satisfied with their current position. An agent is satisfied if they are surrounded by at least a percentage of agents of the same type: this threshold will apply to all agents in the model. The higher the threshold, the greater the likelihood that agents are dissatisfied with their current position.

If an agent is not satisfied, they can be moved to any free position on the grid. The choice of the new position can be made by adopting several strategies. For example, a random position can be chosen, or the agent can move to the nearest available position.

You can find the Python code on GitHub as `schelling_model.py`.

As always, we analyze the code line by line:

1. Let's start by importing the libraries:

    ```python
    import matplotlib.pyplot as plt
    from random import random
    ```

    The `matplotlib` library is a Python library for printing high-quality graphics. Then, we imported the `random()` function from the `random` module. The random module implements PRNGs for various distributions (see the *Random number generation using Python* section from *Chapter 2, Understanding Randomness and Random Numbers*).

2. Let's now define the class that contains the methods for managing agents:

    ```python
    class SchAgent:
        def __init__(self, type):
            self.type = type
            self.ag_location()
    ```

    The first method (`__init__`) allows us to initialize the attributes of the class and will be called automatically when an object of the `SchAgent` class is instantiated:

    ```python
    def ag_location(self):
        self.location = random(), random()
    ```

    The second method (`ag_location`) defines the location of the agents. In this case, we have chosen to place the agents in a random position; in fact, we have used the `random()` function. The `random()` function returns the next nearest floating-point value from the generated sequence. All return values are enclosed between 0 and 1.0. This means that we are going to place our agents on a 1.0 X 1.0 grid:

    ```python
    def euclidean_distance(self, new):
        eu_dist = ((self.location[0] - new.location[0])**2 \
            + (self.location[1] - new.location[1])**2)**(1/2)
        return eu_dist
    ```

The third method (`euclidean_distance`) simply calculates the distance between the agents. To do this, we used the Euclidean distance defined as follows:

$$\text{Euclidean distance} = \sqrt{\sum_i (x_i - y_i)^2}$$

```
def satisfaction(self, agents):
    eu_dist = []
    for agent in agents:
        if self != agent:
            eu_distance = 
                self.euclidean_distance(agent)
            eu_dist.append((eu_distance, agent))
    eu_dist.sort()
    neigh_agent = [agent for k, agent in 
                              eu_dist[:neigh_num]]
    neigh_itself = sum(self.type == agent.type 
                       for agent in neigh_agent)
    return neigh_itself >= neigh_threshold
```

The fourth method (`satisfaction`) evaluates the degree of satisfaction of the agent. As we anticipated, an agent is satisfied if they are surrounded by at least a percentage of agents of the same type: this threshold will apply to all agents in the model. This threshold will then be set when we are going to define the simulation parameters of the model:

```
def update(self, agents):
    while not self.satisfaction(agents):
        self.ag_location()
```

The last method (`update`) moves the agent's position in case they are not satisfied.

3. After defining the methods for the `SchAgent` class, we now need to define the function that will allow us to view the evolution of the model. In this way, we will be able to verify how the agents move on the grid:

```
def grid_plot(agents, step):
    x_A, y_A = [], []
    x_B, y_B = [], []
    for agent in agents:
        x, y = agent.location
        if agent.type == 0:
```

```
            x_A.append(x)
            y_A.append(y)
        else:
            x_B.append(x)
            y_B.append(y)
    fig, ax = plt.subplots(figsize=(10, 10))
    ax.plot(x_A, y_A, '^', markerfacecolor='b',markersize= 10)
    ax.plot(x_B, y_B, 'o', markerfacecolor='r',markersize= 10)
    ax.set_title(f'Step number = {step}')
    plt.show()
```

The function creates vectors x and y for each type of agent and inserts the current position. Next, create a chart with the location of each agent. To better appreciate the position of the two types of agents, we have denoted the two types of agents with blue triangles for type A and with red circles for type B.

4. Now we can define the simulation parameters:

```
num_agents_A = 500
num_agents_B = 500
neigh_num = 8
neigh_threshold = 4
```

We have defined the number of agents of the two types (num_agents_A, num_agents_B). The other two parameters require a more detailed description. The neigh_num variable defines the next neighbors; to set them, we used Moore's neighborhood definition, such as its eight closest neighbors. The neigh_threshold variable defines the satisfaction threshold; in this case, we requested that satisfaction is obtained when the agent is surrounded by at least 50% of agents of the same type.

5. Now we can instantiate objects:

```
agents = [SchAgent(0) for i in range(num_agents_A)]
agents.extend(SchAgent(1) for i in range(num_agents_B))
```

6. We then move on to run the simulation:

```
step = 0
k=0
while (k<(num_agents_A + num_agents_B)):
```

```
        print('Step number = ', step)
        grid_plot(agents, step)
        step += 1
        k=0
        for agent in agents:
            old_location = agent.location
            agent.update(agents)
            if agent.location == old_location:
                k=k+1
    else:
        print(f'Satisfied agents with
                        {neigh_threshold/neigh_num*100}\
                        % of similar neighbors')
```

To do this, we used a `while` loop that checks a `k` counter. The variable `k` simply counts the number of satisfied agents; therefore, if the number of satisfied agents is less than the total number of agents, a new update of the agent positions is performed. The cycle, on the other hand, stops when the positions of all the agents remain unchanged.

The following output is printed on the screen:

```
Step number =   0
Step number =   1
Step number =   2
Step number =   3
Step number =   4
Satisfied agents with 50.0 % of similar neighbors
```

The evolution of the model is appreciable in the diagrams that are returned. To show the evolution, we compared the initial grid in which the random positions of all agents are shown with the final grid in which all people are satisfied (*Figure 5.10*).

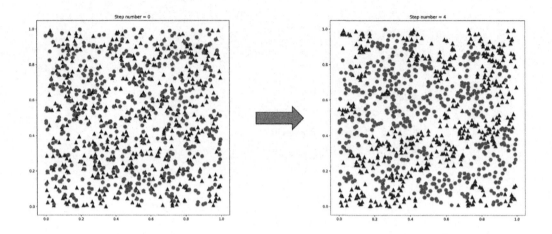

Figure 5.10 – Comparison between the initial and final agent's position

We can see that in the grid on the left, the agents belonging to the two groups are positioned randomly. On the contrary, in the grid on the right, the agents are grouped into groups of the same type.

Summary

In this chapter, we learned the basic concepts of the Markov process. This is where the future evolution of the process depends only on the instant of observation of the system and in no way depends on the past. We have seen how an agent and the surrounding environment interact and the elements that characterize its actions. We now understand the reward and policy concepts behind decision-making. We then went on to explore Markov chains by analyzing the matrices and transition diagrams that govern their evolution.

Then, we addressed some applications to put the concepts we'd learned about into practice. We dealt with a casual walk and a forecast model of weather conditions by adopting an approach based on Markov chains. Next, we studied Bellman equations as coherence conditions for optimal value functions to determine optimal policy. We also introduced multi-agent systems, which allow us to consider different stakeholders in a decision-making process.

Finally, we studied Schelling's segregative models with an example of Python code in which two types of agents randomly positioned in a grid are repositioned, creating groups of the same type.

In the next chapter, we will understand how to obtain robust estimates of confidence intervals and standard errors of population parameters, as well as how to estimate the distortion and standard error of a statistic. We will then discover how to perform a test for statistical significance and how to validate a forecast model.

6
Resampling Methods

Resampling methods are a set of techniques used to repeat data sampling – they simply rearrange the data to estimate the accuracy of a statistic. If we are developing a simulation model and we get unsatisfactory results, we can try to reorganize the starting data to remove any wrong correlations and re-check the capabilities of the model. Resampling methods are one of the most interesting inferential applications of stochastic simulations and random numbers. They are particularly useful in the nonparametric field, where the traditional inference methods cannot be correctly applied. They generate random numbers to be assigned to random variables or random samples. They require machine time related to the growth of repeated operations. They are very simple to implement and once implemented, they are automatic. The required elements must be placed in a sample that is, or at least can be, representative of the population. To achieve this, all the characteristics of the population must be included in the sample. In this chapter, we will try to extrapolate the results obtained from the representative sample of the entire population. Given the possibility of making mistakes in this extrapolation, it will be necessary to evaluate the degree of accuracy of the sample and the risk of arriving at incorrect predictions. In this chapter, we will learn how to apply resampling methods to approximate some characteristics of the distribution of a sample to validate a statistical model. We will analyze the basics of the most common resampling methods and learn how to use them by solving some practical cases.

In this chapter, we're going to cover the following main topics:

- Introducing resampling methods
- Exploring the Jackknife technique
- Demystifying bootstrapping
- Applying bootstrapping regression
- Explaining permutation tests
- Performing a permutation test
- Approaching cross-validation techniques

Technical requirements

In this chapter, we will address resampling method technologies. To deal with the topics in this chapter, you must have a basic knowledge of algebra and mathematical modeling. To work with the Python code in this chapter, you'll need the following files (available on GitHub at the following URL: https://github.com/PacktPublishing/Hands-On-Simulation-Modeling-with-Python-Second-Edition):

- jakknife_estimator.py
- bootstrap_estimator.py
- bootstrap_regression.py
- permutation_test.py
- kfold_cross_validation.py

Introducing resampling methods

Resampling methods are a set of techniques based on the use of subsets of data, which can be extracted either randomly or according to a systematic procedure. The purpose of this technology is to approximate some characteristics of the sample distribution – a statistic, a test, or an estimator – to validate a statistical model.

Resampling methods are one of the most interesting inferential applications of stochastic simulations and the generation of random numbers. These methods became widespread during the 1960s, originating from the basic concepts of Monte Carlo methods. The development of Monte Carlo methods took place mainly in the 1980s, following the progress of information technology and the increase in the power of computers. Their usefulness is linked to the development of non-parametric methods, in situations where the methods of classical inference cannot be correctly applied.

The following details can be observed from resampling methods:

- They repeat simple operations many times
- They generate random numbers to be assigned to random variables or random samples
- They require more machine time as the number of repeated operations grows
- They are very simple to implement and once implemented, they are automatic

Over time, various resampling methods have been developed and can be classified based on some characteristics.

> **Important note**
> A first classification can be made between methods based on randomly extracting subsets of sample data and methods in which resampling occurs according to a non-randomized procedure.

Further classification can be performed as follows:

- The bootstrap method and its variants, such as subsampling, belong to the random extraction category
- Procedures such as Jackknife and cross-validation fall into the non-randomized category
- Statistical tests, called permutation or exact tests, are also included in the family of resampling methods

Sampling concepts overview

Sampling is one of the fundamental topics of all statistical research. Sampling generates a group of elementary units, that is, a subset of a population, with the same properties as the entire population, at least with a defined risk of error.

By population, we mean the set, finite or unlimited, of all the elementary units to which a certain characteristic is attributed, which identifies them as homogeneous.

> Important note
> For example, this could be the population of temperature values in each place, in a period that can be daily, monthly, or yearly.

Sampling theory is an integral and preparatory part of statistical inference, along with the resulting sampling techniques, and allows us to identify the units whose variables are to be analyzed.

Statistical sampling is a method used to randomly select items so that every item in a population has a known, non-zero probability of being included in the sample. Random selection is considered a powerful means of building a representative sample that, in its structure and diversity, reflects the population under consideration. Statistical sampling allows us to obtain an objective sample: the selection of an element does not depend on the criteria defined for reasons of research convenience or availability and does not systematically exclude or favor any group of elements within a population.

In random sampling, associated with the calculation of probabilities, the following actions are performed:

- The results are extrapolated through mathematical formulas and an estimate of the associated error is provided
- Control is given over the risk of reaching an opposite conclusion to reality
- The minimum sample size necessary is calculated – through a formula – to obtain a given level of accuracy and precision

Reasoning about sampling

Now, let's learn why it may be preferable to analyze the data of a sample rather than that of the entire population:

- Consider a case in which the statistical units do not present variability. A statistical unit represents a single element of a set of elements studied, so it represents the smallest element to which a methodology is applied. Here, it is useless to make many measurements because the population parameters are determined with few measurements. A statistical population is a set of items that have been collected to perform a specific experiment. For example, if we wanted to determine the average of 1,000 identical statistical units, this value will be equal to that obtained if we only considered 10 units.

> **Important note**
> Sampling is used if not all the elements of the population are available. For example, investigations into the past can only be done on available historical data, which is often incomplete.

- Sampling is indicated when a considerable amount of time is saved when achieving results. This is because even if electronic computers are used, the data-entry phase is significantly reduced if the investigation is limited to a few elements of the overall population.

Pros and cons of sampling

When information is collected, a survey is performed on all the units that make up the population under study. When an analysis is carried out on the information collected, it is only possible to use it on part of the units that make up the population.

The pros of sampling are as follows:

- Cost reduction
- Reduction of time
- Reduction of the organizational load

The disadvantages of sampling are as follows:

- The sampling base is not always available or easy to know

Sampling can be performed by forced choice in cases where the reference population is partially unknown in terms of composition or size. Sampling cannot always replace a complete investigation, such as in the case of surveys regarding the movement of marital status, births, and deaths: all individual cases must be known.

Probability sampling

In probability sampling, the probability that each unit of the population must be extracted is known. In contrast, in non-probability sampling, the probability that each unit of the population must be extracted is not known.

Let's take a look at an example. If we extract a sample of university students by drawing lots from those present on any day in university, we do not get a probabilistic sample for the following reasons: non-attending students have no chance of entering, and the students who attend the most are more likely to be extracted than the other students of the following years.

How sampling works

The sampling procedure involves a series of steps that need to be followed appropriately to extract data that can adequately represent the population. Sampling is carried out as follows:

1. Define the objective population in the detection statistics.
2. Define the sampling units.
3. Establish the size of the sample.
4. Choose the sample or samples on which the *la* will be statistical detection according to a method of sampling.
5. Finally, formulate a judgment on the goodness of the sample.

Now that we've adequately introduced the various sampling techniques, let's look at a practical case.

Exploring the Jackknife technique

This method is used to estimate characteristics such as the distortion and the standard deviation of a statistic. This technique allows us to obtain the desired estimates without necessarily resorting to parametric assumptions. A statistical parameter is a value that defines an essential characteristic of a population, so it is essential for its description. Jackknife is based on calculating the statistics of interest for the sub-samples we've obtained, leaving out one sample observation at a time. The Jackknife estimate is consistent for various sample statistics, such as mean, variance, correlation coefficient, maximum likelihood estimator, and others.

Defining the Jackknife method

The Jackknife method was proposed in 1949 by M. H. Quenouille who, due to the low computational power of the time, created an algorithm that requires a fixed number of operations.

> **Important note**
>
> The main idea behind this method is to cut a different observation from the original sample each time and to re-evaluate the parameter of interest. The estimate will be compared with the same one that was calculated on the original sample.

Since the distribution of the variable is not known, the distribution of the estimator is not known either.

Jackknife samples are constructed by leaving an observation, x_i, out of the original sample each time, as shown in the following equation:

$$x_i = (x_1, x_2, \ldots, x_{i-1}, x_{i+1}, \ldots, x_n)$$

Then, n samples of size $m = n-1$ are obtained. Let's take a look at an example. Consider a sample of size $n = 5$ that produces five Jackknife samples of size $m = 4$, as follows:

$$x_{(1)} = (x_2, x_3, x_4, x_5)$$
$$x_{(2)} = (x_1, x_3, x_4, x_5)$$
$$x_{(3)} = (x_1, x_2, x_4, x_5)$$
$$x_{(4)} = (x_1, x_2, x_3, x_5)$$
$$x_{(5)} = (x_1, x_2, x_3, x_4)$$

The pseudo-value, $\hat{\theta}$, is recalculated on the generic i-th sample Jackknife. The procedure is iterated n times on each of the available Jackknife samples:

$$x_{(1)} = (x_2, x_3, x_4, x_5) \rightarrow \hat{\theta}_{(1)}$$
$$x_{(2)} = (x_1, x_3, x_4, x_5) \rightarrow \hat{\theta}_{(2)}$$
$$x_{(3)} = (x_1, x_2, x_4, x_5) \rightarrow \hat{\theta}_{(3)}$$
$$x_{(4)} = (x_1, x_2, x_3, x_5) \rightarrow \hat{\theta}_{(4)}$$
$$x_{(5)} = (x_1, x_2, x_3, x_4) \rightarrow \hat{\theta}_{(5)}$$

The following diagram shows this preliminary procedure:

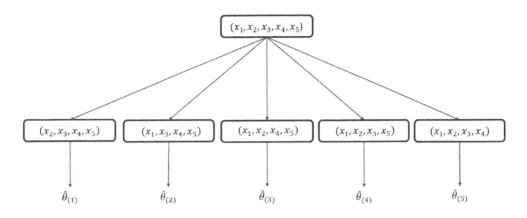

Figure 6.1 – Representation of the Jackknife method

To calculate the variance of the Jackknife estimate, the following equation must be used:

$$Variance_{Jakknife} = \sqrt{\frac{n-1}{n}\sum_{i=1}^{n}(\hat{\theta}_{(i)} - \hat{\theta}_{(.)})^2}$$

In the previous equation, the term $\hat{\theta}_{(.)}$ is defined as follows:

$$\hat{\theta}_{(.)} = \frac{1}{n}\sum_{i=1}^{n}\hat{\theta}_{(i)}$$

The calculated standard deviation will be used to build confidence intervals for the parameters.

To evaluate and possibly reduce the estimator distortion, the Jackknife estimate of the distortion is calculated as follows:

$$\hat{\theta}_i^* = n * \hat{\theta} - (n-1) * \hat{\theta}_i$$

Essentially, the Jackknife method reduces bias and evaluates variance for an estimator.

Estimating the coefficient of variation

To make comparisons regarding variability between different distributions, we can use the **coefficient of variation** (**CV**) since it considers the average of the distribution. The variation coefficient is a relative measure of dispersion and is a dimensionless magnitude. It allows us to evaluate the dispersion of the values around the average, regardless of the unit of measurement.

> **Important note**
> For example, the standard deviation of a sample of income expressed in dollars is completely different from the standard deviation of the same income expressed in euros, while the dispersion coefficient is the same in both cases.

The coefficient of variation is calculated using the following equation:

$$CV = \frac{\sigma}{|\mu|} * 100$$

In the previous equation, we use the following parameters:

- σ is the standard deviation of the distribution
- $|\mu|$ is the absolute value of the mean of the distribution

The variance is the average of the differences squared between each of the observations in a group of data and the arithmetic mean of the data:

$$\sigma^2 = \frac{1}{N}\sum_{i=1}^{N}(x_i - \mu)^2$$

So, it represents the squared error that we commit, on average, replacing a generic observation, x_i, with the average, µ. The standard deviation is the square root of the variance and therefore represents the square root of the mean squared error:

$$\sigma = \sqrt{\frac{1}{N}\sum_{i=1}^{N}(x_i - \mu)^2}$$

The coefficient of variation, which can be defined starting from the average and standard deviation, is the appropriate index for comparing the variability of two characters. CV is particularly useful when you want to compare the dispersion of data with different units of measurement or with different ranges of variation.

If the mean of the distribution approaches zero, the coefficient of variation will approach infinity. In this case, it is sensitive to small variations in the mean.

Applying Jackknife resampling using Python

Now, let's look at some Python code that compares the CV of a distribution and the one obtained with resampling according to the Jackknife method:

1. Let's see the code step by step, starting with loading the necessary libraries:

```
import random
import statistics
import matplotlib.pyplot as plt
```

The `random` module implements pseudo-random number generators for various distributions. The `random` module is based on the Mersenne Twister algorithm. Mersenne Twister is a pseudorandom number generator. Originally developed to produce inputs for Monte Carlo simulations, almost uniform numbers are generated via Mersenne Twister, making them suitable for a wide range of applications.

The `statistics` module contains numerous functions for calculating mathematical statistics from numerical data. With the tools available in this module, it will be possible to calculate the averages and make measurements of the central position and diffusion measures.

The `matplotlib` library is a Python library for printing high-quality graphics. With `matplotlib`, it is possible to generate graphs, histograms, bar graphs, power spectra, error graphs, scatter graphs, and so on with a few commands. This is a collection of command-line functions like those provided by the MATLAB software.

2. Now, we must generate a distribution that represents our data population. We will use this data to extract samples using the sampling methods we are studying. To do this, we will create an empty list that will contain such data:

```
PopData = list()
```

A list is an ordered collection of values and can be of various types. It is an editable container – it allows us to add, delete, and modify existing values. For our purpose, which is to continuously update our values, the list represents the most suitable solution. The `list()` function accepts a sequence of values and converts them into lists. With this command, we simply initialized the list, which is currently empty.

3. The list will be populated through the generation of random numbers. Then, to make the experiment reproducible, we will fix the seed in advance:

```
random.seed(5)
```

The `random.seed()` function is useful if we want to have the same set of data available to be processed in different ways as this makes the simulation reproducible.

> **Important note**
> This function initializes the basic random number generator. If you use the same seed in two successive simulations, you always get the same sequence of pairs of numbers.

4. Now, we can populate the list with 100 randomly generated values:

```
for i in range(100):
    DataElem = 10 * random.random()
    PopData.append(DataElem)
```

In the previous piece of code, we generated 100 random numbers between 0 and 1 using the `random()` function. Then, for each step of the for loop, this number was multiplied by 10 to obtain a distribution of numbers comprised between 0 and 10.

5. Now, let's define a function that calculates the coefficient of variation, as follows:

```
def CVCalc(Dat):
    CVCalc = statistics.stdev(Dat)/statistics.mean(Dat)
    return CVCalc
```

As indicated in the *Estimating the coefficient of variation* section, this coefficient is simply the ratio between the standard deviation and the mean. To calculate the standard deviation, we used the `statistics.stdev()` function. This function calculates the sample standard deviation, which represents the square root of the sample variance. To calculate the mean of the data, we used the `statistics.mean` function. This function calculates the sample arithmetic mean of the data. We can immediately use the newly created function to calculate the variation coefficient of the distribution that we have created:

```
CVPopData = CVCalc(PopData)
print(CVPopData)
```

The following result is returned:

```
0.6569398125747403
```

For now, we will leave this result out, but we will use it later to compare the results we obtained by resampling.

6. Now, we can move on and resample according to the Jackknife method. To begin, we must fix the variables that we will need in the following calculations:

```
N = len(PopData)
JackVal = list()
PseudoVal = list()
```

N represents the number of samples present in the starting distribution. The `JackVal` list will contain the `Jackknife` sample, while the `PseudoVal` list will contain the `Jackknife` pseudo values.

7. The two newly created lists must be initialized to zero to avoid problems in subsequent calculations:

   ```
   for i in range(N-1):
       JackVal.append(0)
   for i in range(N):
       PseudoVal.append(0)
   ```

   The JackVal list has a length of N-1 and relates to what we discussed in the *Defining the Jackknife method* section.

8. At this point, we have all the tools necessary to apply the Jackknife method. We will use two for loops to extract the samples from the initial distribution by calculating the pseudo value at each step of the external loop:

   ```
   for i in range(N):
       for j in range(N):
           if(j < i):
               JackVal[j] = PopData[j]
           else:
               if(j > i):
                   JackVal[j-1]= PopData[j]
       PseudoVal[i] = N*CVCalc(PopData) -
                       (N-1)*CVCalc(JackVal)
   ```

   Jackknife samples (JackVal) are constructed by leaving an observation, x_i, out of the original sample at each step of the external loop (for i in range(N)). At the end of each step of the external cycle, the pseudo value is evaluated using the following equation:

   $$\hat{\theta}_i^* = n * \hat{\theta} - (n-1) * \hat{\theta}_i$$

9. To analyze the distribution of pseudo values, we can draw a histogram:

   ```
   plt.hist(PseudoVal)
   plt.show()
   ```

The following graph will be printed:

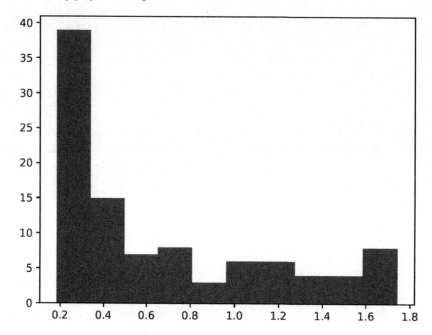

Figure 6.2 – Distribution of pseudo values

10. Now, let's calculate the average of the pseudo values that we have obtained:

```
MeanPseudoVal=statistics.mean(PseudoVal)
print(MeanPseudoVal)
```

The following result is returned:

```
0.6545985339842991
```

As we can see, the value we've obtained is comparable with what we obtained from the starting distribution. Now, we will calculate the variance of the pseudo values:

```
VariancePseudoVal=statistics.variance(PseudoVal)
print(VariancePseudoVal)
```

The following result is returned:

```
0.2435929299444099
```

Finally, let's evaluate the variance of the Jackknife estimator:

```
VarJack = statistics.variance(PseudoVal)/N
print(VarJack)
```

The following result is returned:

```
0.002435929299444099
```

We can use these results to compare the different resampling methods.

The resampling techniques are different and each approaches the problem from a different point of view. Now, let's learn how to resample the data through bootstrapping.

Demystifying bootstrapping

The most well-known resampling technique is the one defined as bootstrapping, as introduced by B. Efron in 1993. The logic of the bootstrap method is to build samples that are not observed but statistically like those observed. This is achieved by resampling the observed series through an extraction procedure where we reinsert the observations.

Introducing bootstrapping

This procedure is like extracting a number from an urn, with subsequent reinsertion of the number before the next extraction. Once a statistical test has been chosen, it is calculated both on the observed sample and on a large number of samples of the same size as that observed and obtained by resampling. The N values of the test statistic then allow us to define the sample distribution – that is, the empirical distribution of the chosen statistic.

> **Important note**
> A statistical test is a rule for discriminating samples that, if observed, lead to the rejection of an initial hypothesis, from those which, if observed, lead to accepting the same hypothesis until proven otherwise.

Since the bootstrapped samples derive from a random extraction process with reintegration from the original series, any temporal correlation structure of the observed series is not preserved. It follows that bootstrapped samples have properties such as the observed sample, but respect, at least approximately, the hypothesis of independence. This makes them suitable for calculating test statistics distributions, assuming there's a null hypothesis for the absence of trends, change points, or a generic systematic temporal trend.

Once the sample distribution of the generic test statistic under the null hypothesis is known, it is possible to compare the value of the statistic itself, as calculated on the observed sample with the quantiles, deduced from the sample distribution, and check whether the value falls into critical regions with a significance level of 5% and 10%, respectively. Alternatively, you can define the percentage of times that the value of the statistic calculated on the observed sample is exceeded by the values coming from the N samples. This value is the statistic p-value for the observed sample and checks if this percentage is far from the commonly adopted meaning of 5% and 10%.

Bootstrap definition problem

Bootstrapping is a statistical resampling technique with reentry so that we can approximate the sample distribution of a statistic. Therefore, it allows us to approximate the mean and variance of an estimator so that we can build confidence intervals and calculate test p-values when the distribution of the statistics of interest is not known.

> **Important note**
> Bootstrap is based on the fact that the only available sample is used to generate many more samples and to build the theoretical reference distribution. You use the data from the original sample to calculate a statistic and estimate its sample distribution without making any assumptions about the distribution model.

So, the original sample is used to generate the distribution; that is, the estimate of θ is constructed by substituting the empirical equivalent of the unknown distribution function of the population. The distribution function of the sample is obtained by constructing a distribution of frequencies of all the values it can assume in that experimental situation.

In the simple case of simple random sampling, the operation is as follows. Consider an observed sample with n elements, as described by the following equation:

$$x = (x_1, \ldots, x_n)$$

From this distribution, m other samples of a constant number equal to n, say x * 1, ..., x * m are resampled. In each bootstrap extraction, the data from the first element of the sample can be extracted more than once. Each one that's provided has a probability equal to 1 / n of being extracted.

Let E be the estimator of θ that interests us to study, say, E(x) = θ. Here, θ is a parameter of the static distribution of essential interest for its description. This quantity is calculated for each bootstrap sample, E(x * 1),..., E(x * m). In this way, m estimates of θ are available, from which it is possible to calculate the bootstrap mean, the bootstrap variance, the bootstrap percentiles, and so on. These values are approximations of the corresponding unknown values and carry information on the distribution of E(x). Therefore, starting from these estimated quantities, it is possible to calculate confidence intervals, test hypotheses, and so on.

Bootstrap resampling using Python

We will proceed in a similar way to what we did for Jackknife resampling: we will generate a random distribution, carry out a resampling according to the bootstrap method, and then compare the results. Let's see the code step by step to understand the procedure:

1. Let's start by importing the necessary libraries:

   ```
   import random
   import numpy as np
   import matplotlib.pyplot as plt
   ```

   The random module implements pseudo-random number generators for various distributions. The random module is based on the Mersenne Twister algorithm. Mersenne Twister is a pseudorandom number generator. Originally developed to produce inputs for Monte Carlo simulations, almost uniform numbers are generated via Mersenne Twister, making them suitable for a wide range of applications.

   numpy is a Python library that contains numerous functions that help us manage multidimensional matrices. Furthermore, it contains a large collection of high-level mathematical functions that we can use on these matrices.

   matplotlib is a Python library for printing high-quality graphics. With matplotlib, it is possible to generate graphs, histograms, bar graphs, power spectra, error graphs, scatter graphs, and so on with a few commands. It is a collection of command-line functions like those provided by the MATLAB software.

2. Now, we will generate a distribution that represents our data population. We will use this data to start extracting samples using the sampling methods we have studied. To do this, we will create an empty list that will contain such data:

   ```
   PopData = list()
   ```

   A list is an ordered collection of values and can be of various types. It is an editable container – it allows us to add, delete, and modify existing values. For our purpose, which is to provide continuous updates for our values, the list represents the most suitable solution. The list() function accepts a sequence of values and converts them into lists. With this command, we simply initialized the list that is currently empty. This list will be populated by generating random numbers.

3. To make the experiment reproducible, we will fix the seed in advance:

   ```
   random.seed(7)
   ```

   The random.seed() function is useful if we want to have the same set of data available to be processed in different ways as it makes the simulation reproducible.

4. Now, we can populate the list with 1,000 randomly generated values:

   ```
   for i in range(1000):
       DataElem = 50 * random.random()
       PopData.append(DataElem)
   ```

In the previous piece of code, we generated 1,000 random numbers between 0 and 1 using the `random()` function. Then, for each step of the for loop, this number was multiplied by 50 to obtain a distribution of numbers comprised between 0 and 50.

5. At this point, we can start extracting a sample of the initial population. The first sample can be extracted using the `random.choices()` function, as follows:

```
PopSample = random.choices(PopData, k=100)
```

This function extracts a sample of size k elements chosen from the population with substitution. We extracted a sample of 100 elements from the original population of 1,000 elements.

6. Now, we can apply the bootstrap method, as follows:

```
PopSampleMean = list()
for i in range(10000):
    SampleI = random.choices(PopData, k=100)
    PopSampleMean.append(np.mean(SampleI))
```

In this piece of code, we created a new list that will contain the sample. Here, we used a for loop with 10,000 steps. At each step, a sample of 100 elements was extracted using the `random.choices()` function from the initial population. Then, we obtained the average of this sample. This value was then added to the end of the list.

> **Important note**
> We resampled the data with the replacement, thereby keeping the resampling size equal to the size of the original dataset.

7. Now, we can print a histogram of the sample we obtained to visualize its distribution:

```
plt.hist(PopSampleMean)
plt.show()
```

The following graph will be printed:

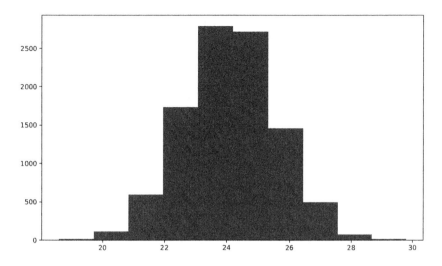

Figure 6.3 – Histogram of the sample distribution

Here, we can see that the sample has a normal distribution.

8. Now, let's calculate the mean of the three distributions that we have generated. Let's start with the bootstrap estimator:

```
MeanPopSampleMean = np.mean(PopSampleMean)
print("The mean of the Bootstrap estimator is ",MeanPopSampleMean)
```

The following result is returned:

```
The mean of the Bootstrap estimator
is  24.105354873028915
```

Then, we can calculate the mean of the initial population:

```
MeanPopData = np.mean(PopData)
print("The mean of the population is ",MeanPopData)
```

The following result is returned:

```
The mean of the population is  24.087053989747968
```

Finally, we can calculate the mean of the simple sample that was extracted from the initial population:

```
MeanPopSample = np.mean(PopSample)
print("The mean of the simple random sample is ",MeanPopSample)
```

The following result is returned:

```
The mean of the simple random sample
is   23.140472976536497
```

We can now compare the results. Here, the population and bootstrap sample means are practically identical, while the generic sample mean deviates from these values. This tells us that the bootstrap sample is more representative of the initial population than a generic sample that was extracted from it.

Comparing Jackknife and bootstrap

In this section, we will compare the two sampling methods that we have studied by highlighting their strengths and weaknesses:

- Bootstrap requires approximately 10 times more computational effort. Jackknife can, at least theoretically, be done by hand.
- Bootstrap is conceptually simpler than Jackknife. Jackknife requires n repetitions for a sample of n, while bootstrap requires a certain number of repetitions. This leads to choosing a number to use, which is not always an easy task. A general rule of thumb is that this number is 1,000 unless you have access to a great deal of computing power.
- Bootstrap introduces errors due to having additional sources of variation due to the finished resampling. Note that this error is reduced for large sizes or where only specific bootstrap sample sets are used.
- Jackknife is more conservative than bootstrap as it produces slightly larger estimated standard errors.
- Jackknife always provides the same results due to the small differences between the replicas. Bootstrap, on the other hand, provides different results each time it is run.
- Jackknife tends to work best for estimating the confidence interval for pair agreement measures.
- Bootstrap performs better for distorted distributions.
- Jackknife is best suited for small samples of original data.

Now, let's learn how to use bootstrapping in the case of a regression problem.

Applying bootstrapping regression

Linear regression analysis is used to determine a linear relationship between two variables, x and y. If x is the independent variable, we try to verify if there is a linear relationship with the dependent variable, y: we try to identify the line capable of representing the distribution of points on a two-dimensional plane. If the points corresponding to the observations are close to the line, the model will effectively describe the link between x and y. The lines that can approximate the observations are infinite, but only one of them optimizes the representation of the data. In the case of a linear mathematical relationship, the observations of y can be obtained from a linear function of the observations of x:

$$y = \alpha * x + \beta + \epsilon$$

In the previous equation, the terms are defined as follows:

- x is the explanatory variable
- α is the slope of the line
- β is the intercept with the y axis
- ϵ is a random error variable with zero mean
- y is the response variable

The α and β parameters must be estimated starting from the observations collected for the two variables, x and y.

The slope, α, represents the variation of the mean response for every single increment of the explanatory variable:

- if the slope is positive, the regression line increases from left to right
- if the slope is negative, the line decreases from left to right
- if the slope is zero, the x variable does not affect the value of y

Therefore, regression analysis aims to identify the α and β parameters that minimize the difference between the observed values of y and the estimated ones.

In the *Demystifying bootstrapping* section, we adequately introduced the bootstrapping technique: we can use this methodology to evaluate, for example, the uncertainty returned by an estimation model. In the example we will analyze, we will apply this methodology to study how possible outliers cause a significant deviation in the variance of a dependent variable that is explained by an independent variable in a regression model.

Here is the Python code (`bootstrap_regression.py`). As always, we will analyze the code line by line:

1. Let's start by importing the libraries:

    ```
    import numpy as np
    from sklearn.linear_model import LinearRegression
    import matplotlib.pyplot as plt
    import pandas as pd
    import seaborn as sns
    ```

    The numpy library is a Python library that contains numerous functions that can help us manage multidimensional matrices. Furthermore, it contains a large collection of high-level mathematical functions we can use on these matrices. Here, we imported the `LinearRegression()` function from `sklearn.linear_model`: this function performs an ordinary least squares linear regression.

The `matplotlib` library is a Python library for printing high-quality graphics.

The `pandas` library is an open source BSD-licensed library that contains data structures and operations to manipulate high-performance numeric values for the Python programming language.

Finally, we imported the `seaborn` library. It is a Python library that enhances the data visualization tools of the `matplotlib` module. In the `seaborn` module, there are several features we can use to graphically represent our data. Some methods facilitate the construction of statistical graphs with `matplotlib`.

2. Now, let's generate the distributions:

```
x = np.linspace(0, 1, 100)
y = x + (np.random.rand(len(x)))
for i in range(30):
    x=np.append(x, np.random.choice(x))
    y=np.append(y, np.random.choice(y))
x=x.reshape(-1, 1)
y=y.reshape(-1, 1)
```

To start, we generated 100 values for the independent variable, x, using the `linspace()` function: the `linspace()` function of the `numpy` library allows us to define an array composed of a series of *N* numerical elements equally distributed between two extremes (0, 1). Then, we generated the dependent variable, y, by adding a random term to the value of x using the `random.rand()` function: this function computes random values in a given shape. It creates an array of the given shape and populates it with random samples from a uniform distribution over [0, 1]. Next, to add outliers artificially, we added 30 more observations using the `random.choice()` function: this function returns a random element of the non-empty sequence passed as an argument (see the *The random.choice() function* section in *Chapter 2, Understanding Randomness and Random Numbers*). Finally, we used the `reshape()` function to format the data as required by the linear regression model we will be using. In this case, we have passed the parameters (-1, 1) to indicate that we do not indicate how many rows there will be -1 because those already defined will remain, while we set several columns equal to 1.

3. Now, we can fit the linear regression model using the `sklearn` library:

```
reg_model = LinearRegression().fit(x, y)
r_sq = reg_model.score(x, y)
print(f"R squared = {r_sq}")
```

The `LinearRegression()` function minimizes the residual sum of squares between the observed targets in the dataset and the predicted targets from the linear approximation. To do this, we simply passed the x and y variable Subsequently, to evaluate the performance of the model, we evaluated the coefficient of determination, R-squared. R-squared measures how well a model can predict the data and falls between zero and one. The higher the value of the coefficient of determination, the better the model is at predicting the data. The following value is printed on the screen:

```
R squared = : 0.286581509708418
```

The value of the coefficient tells us that only about 28% of the variance is specified by the model. We must retrieve the values of the model parameters – slope and intercept:

```
alpha=float(reg_model.coef_[0])
print(f"slope: {reg_model.coef_}")
beta=float(reg_model.intercept_[0])
print(f"intercept: {reg_model.intercept_}")
```

The following values are printed on the screen:

```
slope: [[0.79742372]]
intercept: [0.5632016]
```

Then, we must draw a graph in which we report all the observations and the regression line. To do this, we must use the model to obtain the values of *y* starting from the values of *x*:

```
y_pred = reg_model.predict(x)
plt.scatter(x, y)
plt.plot(x, y_pred, linewidth=2)
plt.xlabel('x')
plt.ylabel('y')
plt.show()
```

The following graph will be returned:

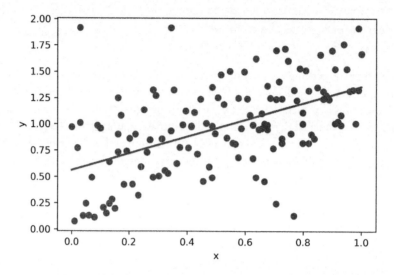

Figure 6.4 – Regression line on a scatter plot

4. Now, let's try to improve the linear regression model by approaching the problem from a data point of view. Then, we will apply resampling, looking for which combination of data gives us the model with the highest value of R-squared. To start, we must initialize the variables that we will use in the bootstrapping procedure:

```
boot_slopes = []
boot_interc = []
r_sqs= []
n_boots = 500
num_sample = len(x)
data = pd.DataFrame({'x': x[:,0],'y': y[:,0]})
```

We will need the first three lists to collect the slope, intercept, and R-squared values obtained in each boot. Successively, we set the number of boots and the number of samples to be extracted from the initial observations. Finally, we have created a DataFrame with the initial data (x, y): this will be used for resampling.

Now, let's create a frame for the diagram we're going to draw and then set up a `for` loop that repeats the instructions several times equal to the boot number:

```
plt.figure()
for k in range(n_boots):
  sample = data.sample(n=num_sample, replace=True)
  x_temp=sample['x'].values.reshape(-1, 1)
  y_temp=sample['y'].values.reshape(-1, 1)
  reg_model = LinearRegression().fit(x_temp, y_temp)
  r_sqs_temp = reg_model.score(x_temp, y_temp)
  r_sqs.append(r_sqs_temp)
  boot_interc.append(float(reg_model.intercept_[0]))
  boot_slopes.append(float(reg_model.coef_[0]))
  y_pred_temp = reg_model.predict(x_temp)
  plt.plot(x_temp, y_pred_temp, color='grey', alpha=0.2)
```

To start, we extracted the sample of the initial distribution using the `sample()` function. This function returns a random sample of elements; two parameters were passed: n = num_sample, replace = True. The first sets the number of elements that is set equal to the number of initial observations. The second parameter (replace = True) allows the same observation to be sampled more than once, in the sense that the same observation can be resampled as many times as it can be omitted.

After extracting the new sample, we must extract the values to be used to adapt the model. After evaluating the parameters of the regression line and the coefficient of determination r_squared, these values are added to the previously initialized lists. Finally, the regression line for the current model is evaluated and it is added to the graph. This procedure is repeated several times equal to the expected number of boots.

5. Now, we can proceed to a first visual evaluation of the results:

```
plt.scatter(x, y)
plt.plot(x, y_pred, linewidth=2)
plt.xlabel('x')
plt.ylabel('y')
plt.show()
```

First, we add all the regression lines we have evaluated to the scatter plot of the observations. The result is shown in the following figure:

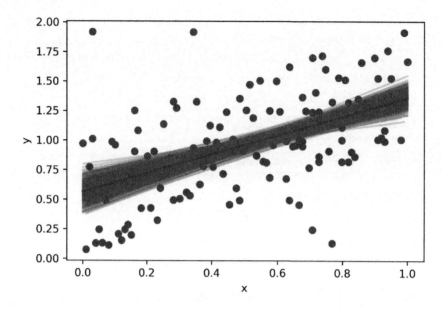

Figure 6.5 – Regression lines evaluation

6. Next, we must plot two histograms with the density curve of the parameters (slope and intercept) evaluated in the bootstrapping procedure:

```
sns.histplot(data=boot_slopes, kde=True)
plt.show()
sns.histplot(data=boot_interc, kde=True)
plt.show()
```

The following graph shows the slope distribution of all 500 models fitted to the samples extracted from the initial distributions:

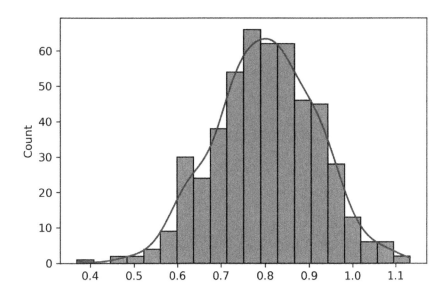

Figure 6.6 – The model's slope distribution

The following graph shows the distribution of the intercept of all 500 models fitted to the samples extracted from the initial distributions:

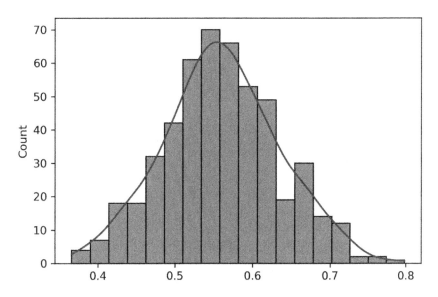

Figure 6.7 – The model's intercept distribution

7. In the final part of this script, we try to evaluate the performance of the 500 models we have worked out in more detail. To start, we will draw a graph with the values of the coefficient of determination r_squared:

```
plt.plot(r_sqs)
```

The following graph is returned:

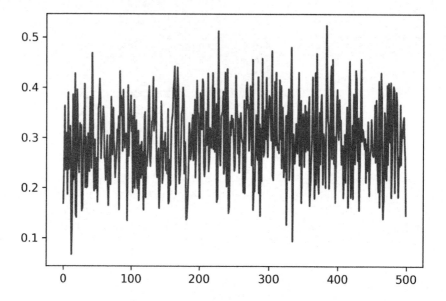

Figure 6.8 – The model's performance evaluation

We can see that the values oscillate between 0.1 and 0.5, so we try to extract the maximum value:

```
max_r_sq=max(r_sqs)
print(f"Max R squared = {max_r_sq}")
```

The following value is displayed:

```
Max R squared = 0.5245632432772953
```

If we compare this value with what we obtained with the initial distribution of data (R-squared = 0.286581509708418), the improvement that we've obtained is evident. We went from about 28% of the explained variance to 52%. So, we can say that bootstrapping returned a very good result.

Now, let's try to extract the values of the parameters of the regression line that return these results:

```
pos_max_r_sq=r_sqs.index(max(r_sqs))
print(f"Boot of the best Regression model = {pos_max_r_
sq}")
```

```
max_slope=boot_slopes[pos_max_r_sq]
print(f"Slope of the best Regression model = {max_
slope}")
max_interc=boot_interc[pos_max_r_sq]
print(f"Intercept of the best Regression model = {max_
interc}")
```

The following results are shown:

```
Boot of the best Regression model = 383
Slope of the best Regression model = 1.1086506372800053
Intercept of the best Regression model =
0.3752482619581162
```

In this way, we will be able to draw the regression line that best approximates the resampled data.

Now that we've analyzed the resampling techniques in detail, let's learn how to perform permutation tests.

Explaining permutation tests

When observing a phenomenon belonging to a set of possible results, we ask ourselves what the law of probability is that we can assign to this set. Statistical tests provide a rule that allows us to decide whether to reject a hypothesis based on the sample observations.

Parametric approaches are very uncertain about the experiment plan and the population model. When these assumptions are not respected, particularly when the data law does not conform to the needs of the test, the parametric results are less reliable. When the hypothesis is not based on knowledge of the data distribution and assumptions have not been verified, nonparametric tests are used. Nonparametric tests offer a very important alternative since they need fewer hypotheses.

Permutation tests are a special case of randomization tests that use a series of random numbers formulated from statistical inferences. The computing power of modern computers has made their widespread application possible. These methods do not require that their assumptions about data distribution are met.

A permutation test is performed through the following steps:

1. A statistic is defined whose value is proportional to the intensity of the process or relationship being studied.
2. A null hypothesis, H_0, is defined.
3. A dataset is created based on the scrambling of those observed. The mixing mode is defined according to the null hypothesis.
4. The reference statistics are recalculated, and the value is compared with the one that was observed.

5. The last two steps are repeated many times.
6. If the observed statistic is greater than the limit obtained in 95% of the cases based on shuffling, H_0 is rejected.

Two experiments use values in the same sample space under the respective distributions, P_1 and P_2, both of which are members of an unknown population distribution. Given the same dataset, x, if the inference conditional on x, which is obtained using the same test statistic, is the same, assuming that the exchangeability for each group is satisfied in the null hypothesis. The importance of permutation tests lies in their robustness and flexibility. The idea of using these methods is to generate a reference distribution from the data and recalculate the test statistics for each permutation of the data concerning the resulting discrete law.

Now, let's look at a practical case of a permutation test in a Python environment.

Performing a permutation test

A permutation test is a powerful non-parametric test for comparing the central trends of two independent samples. No verification is needed on the variability of the population groups or the shape of the distribution. The limitation of this test is its application to small samples.

Therefore, the permutation test allows us to evaluate the correlation between the data by returning the distribution of the test statistic under the null hypothesis: it is obtained by calculating all the possible values of the test statistic using an adequate number of resamplings of the observed data. In a dataset, the data labels are associated with those features; if the labels are swapped under the null hypothesis, the resulting tests produce exact significance levels. The confidence intervals can then be derived from the tests.

As mentioned in the *Demystifying bootstrapping* section, statistical significance tests initially assume the so-called null hypothesis. When comparing two or more groups of data, the null hypothesis always states that there is no difference between the groups regarding the parameter considered: the null hypothesis specifies that the groups are equal to each other and any observed differences must be attributed to chance alone.

We proceed by applying a statistical significance test, the result of which must be compared with a critical value: if the test result exceeds the critical value, then the difference between the groups is declared statistically significant and, therefore, the null hypothesis is rejected; otherwise, the null hypothesis is accepted.

The results of a statistical test do not have a value of absolute and mathematical certainty, but only of probability. Therefore, a decision to reject the null hypothesis is probably right, but it could be wrong. The measure of this risk of falling into error is called the significance level of the test. The significance level of a test can be chosen at will by the researcher, but usually, a probability level of 0.05 or 0.01 is chosen. This probability (called the p-value) represents a quantitative estimate of the probability that the observed differences are due to chance.

The p-value represents the probability of obtaining a more extreme result than the one observed if the diversity is entirely due to the sampling variability alone, thus assuming that the initial null hypothesis is true. p is a probability and therefore can only assume values between 0 and 1. A p-value approaching 0 means a low probability that the observed difference may be due to chance.

Statistically significant does not mean relevant but simply means that what has been observed is hardly due to chance. Numerous statistical tests are used to determine with a certain degree of probability the existence or otherwise of significant differences in the data under examination.

In the example we will analyze, we will compare the data that's been collected on two observations: the first dataset is the well-known Iris flower dataset. In the second case, we will artificially generate a dataset. Our goal is to verify that, in the first dataset, there is a strong correlation between features and labels, something that is non-existent in the artificially generated dataset.

Here is the Python code (permutation_tests.py). As always, we will analyze the code line by line:

1. Let's start by importing the necessary libraries:

    ```
    from sklearn.datasets import load_iris
    import numpy as np
    from sklearn import tree
    from sklearn.model_selection import permutation_test_score
    import matplotlib.pyplot as plt
    import seaborn as sns
    ```

   The sklearn.datasets library contains the most widely used datasets in data analysis. Among, these we import the well-known Iris dataset. The numpy library is a Python library that contains numerous functions that can help us manage multidimensional matrices. Furthermore, it contains a large collection of high-level mathematical functions we can use on these matrices. Here, we imported the tree classification model from the sklearn library. Then, we imported the permutation_test_score() function from sklearn.model_selection, which calculates the significance of a cross-validated score with permutations. The matplotlib library is a Python library for printing high-quality graphics. Finally, we imported the seaborn library. It is a Python library that enhances the data visualization tools of the matplotlib module. In the seaborn module, there are several features we can use to graphically represent our data. Some methods facilitate the construction of statistical graphs with matplotlib.

2. Let's import the data:

    ```
    data=data = load_iris()
    X = data.data
    y = data.target
    ```

The Iris dataset is a multivariate dataset introduced by the British statistician and biologist Ronald Fisher in 1936 as an example of linear discriminant analysis. The dataset contains 50 samples from each of the three species of Iris (Iris setosa, Iris virginica, and Iris versicolor). Four features were measured from each sample: the length and the width of the sepals and petals, in centimeters.

The following variables are contained:

- Sepal.Length in centimeters
- Sepal.Width in centimeters
- Petal.Length in centimeters
- Petal.Width in centimeters
- Class: setosa, versicolour, virginica

3. Now, let's create a completely artificial dataset with several features equal to those of the Iris dataset, but which do not have any correlation with the labels of the Iris dataset:

```
np.random.seed(0)
X_nc_data = np.random.normal(size=(len(X), 4))
```

To do this, we used numpy's `random.normal()` function. This function generates a normal distribution by default and uses a mean equal to 0 and a standard deviation equal to 1. We added the seed for reproducibility.

4. Our goal is to evaluate the correlation in the data by performing a classification of the labels (y) based on the features contained in the matrix, X. Let's choose the classification model:

```
clf = tree.DecisionTreeClassifier(random_state=1)
```

Here, `DecisionTreeClassifier` was chosen. A decision tree algorithm is based on a non-parametric supervised learning method used for classification and regression. The aim is to build a model that predicts the value of a target variable using decision rules inferred from the data features.

5. Now, let's perform a permutation test on the data:

```
p_test_iris = permutation_test_score(
    clf, X, y, scoring="accuracy", n_permutations=1000
)
print(f"Score of iris flower classification = {p_test_iris[0]}")
print(f"P_value of permutation test for iris dataset = {p_test_iris[2]}")
```

```
p_test_nc_data = permutation_test_score(
    clf, X_nc_data, y, scoring="accuracy", n_
permutations=1000
)
print(f"Score of no-correletd data classification = {p_
test_nc_data[0]}")
print(f"P_value of permutation test for no-correletd
dataset = {p_test_nc_data[2]}")
```

For the test, we used the `permutation_test_score()` function, which evaluates the significance of a cross-validated score with permutations. This function performs a permutation of the target to resample the data and calculate the empirical p-value concerning the null hypothesis, which assumes the characteristics and objectives are independent. Three results are returned:

- `score`: The score of the real classification without trade-in on the targets
- `permutation_scores`: Scores obtained for each permutation
- `pvalue`: Approximate probability that the score is obtained by chance

The p-value represents the fraction of randomized datasets in which the classifier performed well: a small p-value tells us that there is a real dependence between characteristics and targets. A high p-value may be due to a lack of real dependency between characteristics and targets.

6. First, we performed the test on the data from the Iris dataset and obtained the following results:

```
Score of iris flower classification = 0.9666666666666668
P_value of permutation test for iris dataset =
0.000999000999000999
```

1,000 permutations have been made: the accuracy of the decision tree-based classifier is very high, which tells us that the model can predict the class of the type of Iris with excellent performance. However, the p-value is very low, which confirms the real dependence of the target, y, on the features contained in the matrix, X.

7. Then, we ran the same test on the artificially generated data. The following results were returned:

```
Score of no-correletd data classification =
0.2866666666666667
P_value of permutation test for no-correletd dataset =
0.8711288711288712
```

The classification score is low, indicating that the forecasts do not have good accuracy. However, the p-value is high, which indicates that a correlation between the features and the target has not been found.

8. Finally, we performed a visual analysis of the results of the permutation test:

```
pbox1=sns.histplot(data=p_test_iris[1], kde=True)
plt.axvline(p_test_iris[0],linestyle="-", color='r')
plt.axvline(p_test_iris[2],linestyle="--", color='b')
pbox1.set(xlim=(0,1))
plt.show()
```

The following graph was displayed:

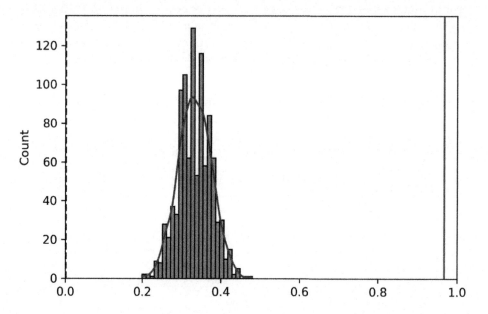

Figure 6.9 – Histogram with a density curve for the permutation test

This graph shows a histogram with the density curve of the distribution of the results of the permutation tests performed 100 times. A blue dashed vertical line is drawn to indicate the p-value, and a continuous red vertical line is drawn to indicate the accuracy of the classifier. Note that since these are the original data (Iris dataset), characterized by a strong correlation between features and targets, the score of the classifier is much higher than those obtained by permuting the targets: the permutations of the targets cause the correlation between the features and targets to be lost. Furthermore, the p-value is located on the far left of the graph to indicate a low value and therefore a strong correlation between the features and targets.

Let's see what happens to the randomly generated data:

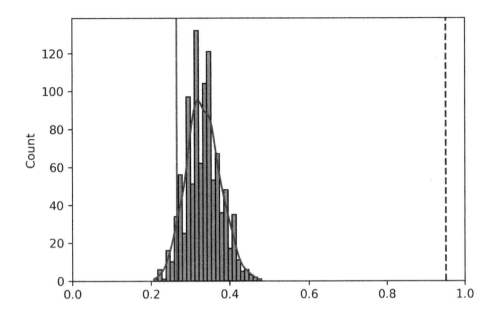

Figure 6.10 – Histogram with a density curve for the randomly generated data

In this case, being random data, characterized by no correlation between features and targets, the score of the classifier is low and in line with those obtained by permuting the targets. Furthermore, the p-value is on the far right of the graph to indicate a high value and therefore a low correlation between the features and targets.

Now that we've analyzed a practical case of permutation testing, let's learn how to perform data resampling by applying cross-validation.

Approaching cross-validation techniques

Cross-validation is a method used in model selection procedures based on the principle of predictive accuracy. A sample is divided into two subsets, of which the first (training set) is used for construction and estimation, while the second (validation set) is used to verify the accuracy of the predictions of the estimated model. Through a synthesis of repeated predictions, a measure of the accuracy of the model is obtained. A cross-validation method is like Jackknife in that it leaves one observation out at a time. In another method, known as k-fold validation, the sample is divided into k subsets and, in turn, each of them is left out as a validation set.

> **Important note**
> Cross-validation can be used to estimate the **mean squared error** (**MSE**) (or, in general, any measure of precision) of a statistical learning technique to evaluate its performance or select its level of flexibility.

Cross-validation can be used for both regression and classification problems. The three main validation techniques of a simulation model are the validation set approach, **leave-one-out cross-validation (LOOCV)**, and k-fold cross-validation. In the following sections, we will learn about these concepts in more detail.

Validation set approach

This technique consists of randomly dividing the available dataset into two parts:

- A training set
- A validation set, called the holdout set

A statistical learning model is adapted to the training data and subsequently used for predicting the data of the validation set.

The measurement of the resulting test error, which is typically the MSE in the case of regression, provides an estimate of the real test error. The validation set is the result of a sampling procedure and therefore different samplings result in different estimates of the test error.

This validation technique has various pros and cons. Let's take a look at a few:

- The method tends to have high variability; that is, the results can change substantially as the selected test set changes.
- Only a part of the available units is used for function estimates. This can lead to less precision in function estimating and overestimating the test error.

The LOOCV and k-fold cross-validation techniques try to overcome these problems.

Leave-one-out cross-validation

LOOCV also divides the observation set into two parts. However, instead of creating two subsets of comparable size, we do the following:

1. A single observation (x_1, y_1) is used for validation; the remaining observations make up the training set.
2. The function is estimated based on the n-1 observations of the training set.
3. The prediction, $\hat{y}_1$, is made using x_1. Since (x_1, y_1) was not used in the function estimate, an estimate of the test error is as follows:

$$MSE_1 = (y_1 - \hat{y}_1)^2$$

But even if MSE_1 is impartial to the test error, it is a poor estimate because it is very variable. This is because it is based on a single observation (x_1, y_1).

4. The procedure is repeated by selecting for validation (x_2, y_2), where a new estimate of the function is made based on the remaining n-1 observations, and calculating the test error again, as follows:

$$MSE_2 = (y_2 - \hat{y}_2)^2$$

5. Repeating this approach *n* times produces *n* test errors.
6. The LOOCV estimate for the MSE test is the average of the *n* MSEs available, as follows:

$$CV_n = \frac{1}{n} \sum_{i=1}^{n} MSE_i$$

LOOCV has some advantages over the validation set approach:

- Using an n-1 unit to estimate the function has less bias. Consequently, the LOOCV approach does not tend to overestimate the test error.
- As there is no randomness in the choice of the test set, there is no variability in the results for the same initial dataset.

LOOCV can be computationally intensive, so for large datasets, it takes a long time to calculate. In the case of linear regression, however, there are direct computational formulas with low computational intensity.

k-fold cross-validation

In **k-fold cross-validation (k-fold CV)**, the set of observations is randomly divided into *k* groups, or folders, of approximately equal size. The first folder is considered a validation set and the function is estimated on the remaining k-1 folders. The mean square error, MSE_1, is then calculated on the observations of the folder that's kept out. This procedure is repeated *k* times, each time choosing a different folder for validation, thus obtaining *k* estimates of the test error. The k-fold CV estimate is calculated by averaging these values, as follows:

$$CV_{(k)} = \frac{1}{k} \sum_{i=1}^{k} MSE_i$$

This method has the advantage of being less computationally intensive if k << n. Furthermore, the k-fold CV tends to have less variability than the LOOCV on different-sized datasets, *n*.

Choosing *k* is crucial in k-fold CV. What happens when i changes in cross-validation? Let's see what an extreme choice of *k* entails:

- A high *k* value results in larger training sets and therefore less bias. This implies small validation sets and therefore greater variance.
- A low *k* value results in smaller training sets and therefore greater bias. This implies larger validation sets and therefore low variance.

Cross-validation using Python

In this section, we will look at an example of the application of cross-validation. First, we will create an example dataset that contains simple data to identify to verify the procedure being performed by the algorithm. Then, we will apply k-fold CV and analyze the results:

1. As always, we will start by importing the necessary libraries:

    ```
    import numpy as np
    from sklearn.model_selection import KFold
    ```

    numpy is a Python library that contains numerous functions that help us manage multidimensional matrices. Furthermore, it contains a large collection of high-level mathematical functions we can use on these matrices.

    scikit-learn is an open source Python library that provides multiple tools for machine learning. In particular, it contains numerous classification, regression, and clustering algorithms; this includes support vector machines, logistic regression, and much more. Since it was released in 2007, scikit-learn has become one of the most widely used libraries in the field of machine learning, both supervised and unsupervised, thanks to the wide range of tools it offers, but also thanks to its API, which is documented, easy to use, and versatile.

    > **Important note**
    >
    > **Application programming interfaces** (**APIs**) are sets of definitions and protocols that application software is created and integrated with. They allow products or services to communicate with other products or services without knowing how they are implemented, thus simplifying app development, and allowing a net saving of time and money. When creating new tools and products or managing existing ones, APIs offer flexibility, simplify design, administration, and use, and provide opportunities for innovation.

    The scikit-learn API combines a functional user interface with an optimized implementation of numerous classification and meta-classification algorithms. It also provides a wide variety of data pre-processing, cross-validation, optimization, and model evaluation functions. scikit-learn is particularly popular for academic research since developers can use the tool to experiment with different algorithms by changing only a few lines of code.

2. Now, let's generate the starting dataset:

    ```
    StartedData=np.arange(10,110,10)
    print(StartedData)
    ```

    Here, we generated a vector containing 10 integers, starting from the value 10 up to 100 with a step equal to 10. To do this, we used the numpy `arange()` function. This function generates equidistant values within a certain range. Three arguments have been passed, as follows:

- `10`: Start of the interval. This value is included. If this value is omitted, the default value of 0 is used.
- `110`: End of range. This value is not included in the range except in cases of floating-point numbers.
- `10`: Spacing between values. This is the distance between two adjacent values. By default, this value is equal to 1.

The following array was returned:

```
[ 10  20  30  40  50  60  70  80  90 100]
```

3. Now, we can set the function that will allow us to perform k-fold CV:

   ```
   kfold = KFold(5, True, 1)
   ```

   scikit-learn's `KFold()` function performs k-fold CV by dividing the dataset into *k* consecutive folds without shuffling by default. Each fold is then used once as validation, while the remaining k - 1 folds form the training set. Three arguments were passed, as follows:

   - `5`: Number of folds required. This number must be at least 2.
   - `True`: Optional Boolean value. If it is equal to `True`, the data is mixed before it's divided into batches.
   - `1`: Seed used by the random number generator.

4. Finally, we can resample the data by using k-fold CV:

   ```
   for TrainData, TestData in kfold.split(StartedData):
       print("Train Data :", StartedData[TrainData],"Test
   Data :", StartedData[TestData])
   ```

   To do this, we used a loop for the elements generated by the `kfold.split()` method, which returns the indexes that the dataset is divided into. Then, for each step, which is equal to the number of folds, the elements of the subsets that were drawn are printed.

   The following results are returned:

   ```
   Train Data : [ 10  20  40  50  60  70  80  90]
   Test Data : [ 30 100]
   Train Data : [ 10  20  30  40  60  80  90 100]
   Test Data : [50 70]
   Train Data : [ 20  30  50  60  70  80  90 100]
   Test Data : [10 40]
   Train Data : [ 10  30  40  50  60  70  90 100]
   Test Data : [20 80]
   ```

```
Train Data : [ 10  20  30  40  50  70  80 100]
Test Data : [60 90]
```

These pairs of data (`Train Data`, `Test Data`) will be used in succession to train the model and validate it. This way, you can avoid overfitting and bias problems. Every time you evaluate the model, the extracted part of the dataset is used, and the remaining part of the dataset is used for training.

Summary

In this chapter, we learned how to resample a dataset. We analyzed several techniques that approach this problem through different techniques. First, we analyzed the basic concepts of sampling and learned about the reasons that push us to use a sample extracted from a population. Then, we examined the pros and cons of this choice. We also analyzed how a resampling algorithm works.

Then, we tackled the first resampling method: the Jackknife method. First, we defined the concepts behind the method and then moved on to the procedure, which allows us to obtain samples from the original population. To put the concepts we learned into practice, we applied Jackknife resampling to a practical case.

Next, we explored the bootstrap method, which builds unobserved but statistically, like the observed samples. This is accomplished by resampling the observed series through an extraction procedure where we reinsert the observations. After defining the method, we worked through an example to highlight the characteristics of the procedure. Furthermore, a comparison between Jackknife and bootstrap was made. Then, we analyzed a practical case of bootstrapping applied to a regression problem.

After analyzing the concepts underlying permutation tests and exploring an example of this test, we concluded this chapter by looking at various cross-validation methods. Our knowledge of the k-fold CV method was deepened through an example.

In the next chapter, we will learn about the basic concepts of various optimization techniques and how to implement them. We will understand the difference between numerical and stochastic optimization techniques, and we will learn how to implement stochastic gradient descent. Then, we will discover how to estimate missing or latent variables and optimize model parameters. Finally, we will discover how to use optimization methods in real-life applications.

7
Using Simulation to Improve and Optimize Systems

Simulation models allow us to obtain a lot of information using few resources. As often happens in life, simulation models are also subject to improvements to increase their performance. Through optimization techniques, we try to modify the performance of a model to obtain improvements both in terms of the results and when trying to exploit resources. Optimization problems are usually so complex that a solution cannot be determined analytically. Complexity is determined first by the number of variables and constraints, which define the size of the problem, and then by the possible presence of nonlinear functions. To solve an optimization problem, it is necessary to use an iterative algorithm that, given a current approximation of the solution, determines—with an appropriate sequence of operations—and updates this approximation. Starting from an initial approximation, a sequence of approximations that progressively improve the solution is determined.

In this chapter, we will learn how to use the main optimization techniques to improve the performance of our simulation models. We will learn how to use the gradient descent technique, the Newton-Raphson method, and stochastic gradient descent. We will also learn how to apply these techniques with practical examples.

In this chapter, we're going to cover the following main topics:

- Introducing numerical optimization techniques
- Exploring the gradient descent technique
- Understanding the Newton-Raphson method
- Deepening our knowledge of stochastic gradient descent
- Approaching the **Expectation-Maximization** (**EM**) algorithm
- Understanding **Simulated Annealing** (**SA**)
- Discovering multivariate optimization methods in Python

Technical requirements

In this chapter, we will learn how to use simulation models to improve and optimize systems. To deal with the topics in this chapter, it is necessary that you have a basic knowledge of algebra and mathematical modeling. To work with the Python code in this chapter, you'll need the following files (available on GitHub at the following URL: https://github.com/PacktPublishing/Hands-On-Simulation-Modeling-with-Python-Second-Edition):

- gradient_descent.py
- newton_raphson.py
- gaussian_mixtures.py
- simulated_annealing.py
- scipy_optimize.py

Introducing numerical optimization techniques

In real life, optimizing means choosing the best option among several available alternatives. Each of us optimizes an itinerary to reach a destination, organize our day, determine how we use savings, and so on. In mathematics, optimizing means determining the value of the variables of a function so that it assumes its minimum or maximum. Optimization is a discipline that deals with the formulation of useful models in applications, thereby using efficient methods to identify an optimal solution.

Optimization models have great practical interest for the many applications offered. In fact, there are numerous decision-making processes that require you to determine a choice that minimizes the cost or maximizes the gain and is, therefore, attributable to optimization models. In optimization theory, a relevant position is occupied by mathematical optimization models, for which the evaluation function and the constraints that characterize the permissible alternatives are expressed through equations and inequalities. Mathematical optimization models come in different forms:

- Linear optimization
- Integer optimization
- Nonlinear optimization

Defining an optimization problem

An optimization problem consists of trying to determine the points that belong to a set F in which a function f takes values that are as low as possible. This problem is represented in the following form:

$$min\, f(x) \quad \forall\, x \in F$$

Here, we have the following:

- *f* is called the objective function
- *F* is called the feasible set and contains all the admissible choices for *x*

> **Important note**
> If you have a maximization problem—that is, if you have to find a point where the function f takes on the highest possible value, you can always go back to the minimal problem, thus changing the sign of the objective function.

The elements that minimize the function *f* by satisfying the previous relationship are called global optimal solutions, also known as optimal solutions or minimum solutions. The corresponding value of the objective function *f* is called the global optimum value, also known as the optimal or minimum.

The complexity of the optimization problem—that is, its difficulty of resolution—obviously depends on the characteristics of the objective function and the structure of the flexible set. Usually, an optimization problem is characterized by whether there is complete freedom in the choice of the vector *x*. We can therefore state that there are two types of problems, as follows:

- Unconstrained minimization problem, if $F = Rn$; that is, if the flexible set *F* coincides with the whole set *Rn*
- Constrained minimization problem, if $F \subset Rn$; that is, if the flexible set *F* is only a part of the set *Rn*

When trying to solve an optimization problem, the first difficulty we face is understanding if it is well placed, in the sense that there may not be a point of *F* where the function *f (x)* takes the value pi as small. In fact, one of the following conditions could occur:

- The flexible set *F* may be empty
- The flexible set *F* may not be empty but the objective function could have a lower limit equal to $-\infty$
- The flexible set F may not be empty and the objective function could have a lower limit equal to $-\infty$ but also, in this case, there could be no global minimum points of *f* on *F*

A sufficient but not necessary condition for the existence of a global minimum point in an optimization problem is that expressed by the Weierstrass theorem through the following proposition: let $F \subset Rn$ be a non-empty and compact set. Let *f* be a continuous function on *F*. If so, a global minimum point of *f* exists in *F*.

The previous proposition applies only to the class of constrained problems in which the flexible set is compact. To establish existence results for problems with non-compact flexible sets—that is, in the case where $F = Rn$—it is necessary to try to characterize some subset of *F* containing optimal solutions to the problem.

> **Important note**
> A compact space is a topological space where every open covering of it contains a finished sub-covering.

In general, there isn't always an optimal solution for the problem at hand, and, where it exists, it isn't always unique.

Explaining local optimality

Unfortunately, all global optimality conditions have limited application interest. In fact, they are linked to the overall behavior of the objective function on the admissible set and, therefore, are necessarily described by complex conditions from a computational point of view. Next to the notion of global optimality, as introduced by defining the optimization model, it is appropriate to define the concept of local optimality.

We can define a local optimum as the best solution to a problem in a small neighborhood of possible solutions. In the following diagram, we can identify four local minimum conditions for the function $f(x)$, which therefore represent the local optimum:

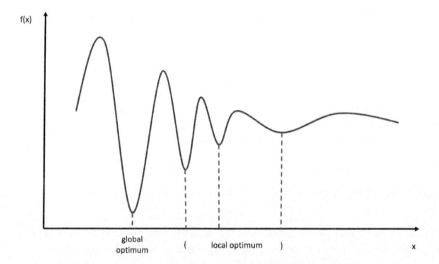

Figure 7.1 – Local minimum conditions for the f(x) function

However, only one of these is a global optimum, while the other three remain local optimums. Local optimality conditions are more useful from an application point of view. These are nothing but necessary conditions, but in general, they are not sufficient. This is because an assigned point is a local minimum point of a minimization problem. Therefore, from a theoretical point of view, they do not give a satisfactory characterization of the local minima of the optimization problem, but they do play an important role in the definition of minimization algorithms.

Many problems that are faced in real life can be represented as nonlinear optimization problems. This motivates the increasing interest, from a technical and scientific point of view, toward the study and development of methods that can face and solve this class of difficult mathematical problems.

Let's now see the most widely used optimization technique.

Exploring the gradient descent technique

The goal of any simulation algorithm is to reduce the difference between the values predicted by the model and the actual values returned by the data. This is because a lower error between the actual and expected values indicates that the algorithm has done a good simulation job. Reducing this difference simply means minimizing the objective function that the model being built is based on.

Defining descent methods

Descent methods are iterative methods that, starting from an initial point $x_0 \in R^n$, generate a sequence of points $\{x_n\}\ n \in N$ defined by the following equation:

$$x_{n+1} = x_n + \gamma_n * g_n$$

Here, the g_n vector is a search direction and the γ_n scalar is a positive parameter called step length, which indicates the distance by which we move in the g_n direction.

In a descent method, the g_n vector and the γ_n parameter are chosen to guarantee the decrease of the objective f function at each iteration, as follows:

$$f_{x_{n+1}} < f_{x_n} \ \forall\ n \geq 0$$

Using the g_n vector, we take a direction of descent, which is such that the line $x = x_n + \gamma_n * g_n$ forms an obtuse angle with the gradient vector $\nabla f(x_n)$. In this way, it is possible to guarantee the decrease of f, provided that γ_n is sufficiently small.

Depending on the choice of g_n, there are different descent methods. The most common are these methods:

- Gradient descent method
- Newton-Raphson method

Let's start by analyzing the gradient descent algorithm.

Approaching the gradient descent algorithm

A gradient is a vector-valued function that represents the slope of the tangent of the function graph, indicating the direction of the maximum rate of increase of the function. Let's consider the convex function represented in the following diagram:

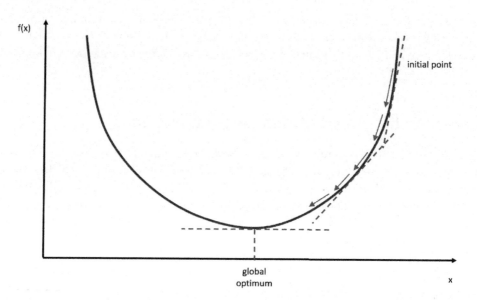

Figure 7.2 – The convex function

The goal of the gradient descent algorithm is to reach the lowest point of this function. In more technical terms, the gradient represents a derivative that indicates the slope or inclination of the objective function.

To understand this better, let's assume we got lost in the mountains at night with poor visibility. We can only feel the slope of the ground under our feet. Our goal is to reach the lowest point of the mountain. To do this, we will have to walk a few steps and move toward the direction of the highest slope. We will do this iteratively, moving one step at a time until we finally reach the mountain valley.

In mathematics, the derivative is the rate of change or slope of a function at a given point. So, the value of the derivative is the slope of the slope at a specific point. The gradient represents the same thing, with the addition that it is a vector value function that stores partial derivatives. This means that the gradient is a vector and that each of its components is a partial derivative with respect to a specific variable.

Let's analyze a function, $f(x, y)$—that is, a two-variable function, x and y. Its gradient is a vector containing the partial derivatives of f: the first with respect to x and the second with respect to y. If we calculate the partial derivatives of f, we get the following:

$$\frac{\delta f}{\delta x}, \frac{\delta f}{\delta y}$$

The first of these two expressions is called a partial derivative with respect to x, while the second partial derivative is with respect to y. The gradient is the following vector:

$$\nabla f(x, y) = \begin{bmatrix} \dfrac{\delta f}{\delta x} \\ \dfrac{\delta f}{\delta y} \end{bmatrix}$$

The preceding equation is a function that represents a point in a two-dimensional space, or a two-dimensional vector. Each component indicates the steepest climbing direction for each of the function variables. Hence, the gradient points in the direction where the function increases the most in.

Similarly, if we have a function with five variables, we would get a gradient vector with five partial derivatives. Generally, a function with n variables results in an n-dimensional gradient vector, as follows:

$$\nabla f(x, y, \ldots z) = \begin{bmatrix} \dfrac{\delta f}{\delta x} \\ \dfrac{\delta f}{\delta y} \\ \ldots \\ \ldots \\ \dfrac{\delta f}{\delta z} \end{bmatrix}$$

For gradient descent, however, we don't want to maximize f as fast as possible. Instead, we want to minimize it—that is, find the smallest point that minimizes the function.

Suppose we have a function $y = f(x)$. Gradient descent is based on the observation that if the function f is defined and differentiable in a neighborhood of x, then this function decreases faster if we move in the direction of the negative gradient. Starting from a value of x, we can write the following:

$$x_{n+1} = x_n - \gamma * \nabla f(x_n)$$

Here, we have the following:

- γ is the learning rate
- ∇ is the gradient

For sufficiently small γ values, the algorithm converges to the minimum value of the function f in a finite number of iterations.

> **Important note**
> Basically, if the gradient is negative, the objective function at that point is decreasing, which means that the parameter must move toward larger values to reach a minimum point. On the contrary, if the gradient is positive, the parameters move toward smaller values to reach the lower values of the objective function.

Understanding the learning rate

The gradient descent algorithm searches for the minimum of the objective function through an iterative process. At each step, an estimate of the gradient is performed to direct the descent in the direction that minimizes the objective function. In this procedure, the choice of **learning rate** parameter becomes crucial. This parameter determines how quickly or slowly we will move to the optimal values of the objective function:

- If it is too small, we will need too many iterations to converge to the best values
- If the learning rate is very high, we will skip the optimal solution

In the following diagram, you can see the two possible scenarios imposed by the value of the learning rate:

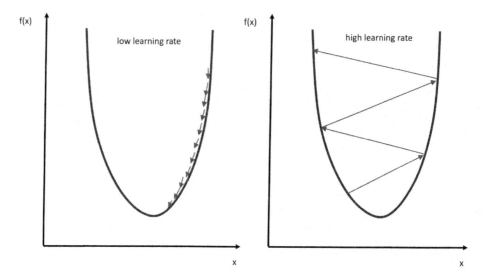

Figure 7.3 – Scenarios for the learning rate

Due to this, it is essential to use a good learning rate. The best way to identify the optimal learning rate is through trial and error.

Explaining the trial and error method

The term *trial and error* defines a heuristic method that aims to find a solution to a problem by attempting it and checking if it has produced the desired effect. If so, the attempt constitutes a solution to the problem; otherwise, we continue with a different attempt.

Let's analyze the essential characteristics of the method:

- **It is oriented toward the solution**: It does not aim to find out why an attempt works, but simply seeks it.
- **It is specific to the problem in question**: It has no claim to generalize to other problems.
- **It is not optimal**: It usually limits itself to finding a single solution that will usually not be the best possible one.
- **It does not require having a thorough knowledge of it**: It proposes to find a solution to a problem of which little or nothing is known.

The trial and error method can be used to find all solutions to the problem or the best solution among them if there is more than one. In this case, instead of stopping at the first attempt that provided the desired result, we take note of it and continue in the attempts until all the solutions are found. In the end, these are compared based on a given criterion, which will determine which of them is to be considered the best.

Implementing gradient descent in Python

In this section, we will apply what we have learned so far on gradient descent by completing a practical example. We will define a function and then use that method to find the minimum point of the function. As always, we will analyze the code line by line:

1. Let's start by importing the necessary libraries:

    ```
    import numpy as np
    import matplotlib.pyplot as plt
    ```

    numpy is a Python library that contains numerous functions that help us manage multidimensional matrices. Furthermore, it contains a large collection of high-level mathematical functions we can use on these matrices.

    matplotlib is a Python library for printing high-quality graphics. With matplotlib, it is possible to generate graphs, histograms, bar graphs, power spectra, error graphs, scatter graphs, and so on with a few commands. It is a collection of command-line functions, like those provided by MATLAB.

2. Now, let's define the function:

    ```
    x = np.linspace(-1,3,100)
    y=x**2-2*x+1
    ```

    First, we defined an interval for the dependent variable x. We will only need this to visualize the function and draw a graph. To do this, we used the linspace() function. This function creates numerical sequences. Then, we passed three arguments: the starting point, the ending point, and the number of points to be generated. Next, we defined a parabolic function.

3. Now, we can draw a graph and display it:

```
fig = plt.figure()
axdef = fig.add_subplot(1, 1, 1)
axdef.spines['left'].set_position('center')
axdef.spines['bottom'].set_position('zero')
axdef.spines['right'].set_color('none')
axdef.spines['top'].set_color('none')
axdef.xaxis.set_ticks_position('bottom')
axdef.yaxis.set_ticks_position('left')
plt.plot(x,y, 'r')
plt.show()
```

First, we defined a new figure and then set the axes so that the *x* axis coincides with the minimum value of the function and the *y* axis coincides with the center of the parabola. This will make it easier to visually locate the minimum point of the function. The following diagram is printed:

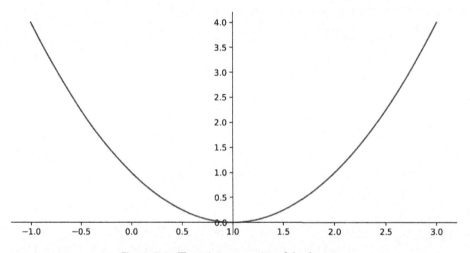

Figure 7.4 – The minimum point of the function

As we can see, the minimum point of the function corresponding to y = 0 occurs for a value of x equal to 1. This will be the value that we will have to determine through the gradient descent method.

4. Now, let's define the gradient function:

```
Gradf = lambda x: 2*x-2
```

Recall that the gradient of a function is its derivative. In this case, doing this is easy since it is a single-variable function.

5. Before applying the iterative procedure, it is necessary to initialize a series of variables:

   ```
   actual_X = 3
   learning_rate = 0.01
   precision_value = 0.000001
   previous_step_size = 1
   max_iteration = 10000
   iteration_counter = 0
   ```

   So, let's describe in detail the meaning of these variables:

   - The `actual_X` variable will contain the current value of the independent variable x. To start, we initialize it at x = 3, which represents the far-right value of the display range of the function in the graph.

   - The `learning_rate` variable contains the learning rate. As explained in the *Understanding the learning rate* section, it is set to 0.01. We can try to see what happens if we change that value.

   - The `precision_value` variable will contain the value that defines the degree of precision of our algorithm. Being an iterative procedure, the solution is refined at each iteration and tends to converge. But this convergence may come after a very large number of iterations, so to save resources, it is advisable to stop the iterative procedure once adequate precision has been reached.

   - The `previous_step_size` variable will contain the calculation of this precision and will be initialized to 1.

   - The `max_iteration` variable contains the maximum number of iterations that we have provided for our algorithm. This value will be used to stop the procedure if it does not converge.

   - Finally, the `iteration_counter` variable will be the iteration counter.

6. Now, we are ready for the iteration procedure:

   ```
   while previous_step_size > precision_value and iteration_counter < max_iteration :
       PreviousX = actual_X
       actual_X = actual_X - learning_rate * Gradf(PreviousX)
       previous_step_size = abs(actual_X - PreviousX)
       iteration_counter = iteration_counter +1
       print("Number of iterations = ",iteration_counter ,"\
   nActual value of x is = ",actual_X )
       print("X value of f(x) minimum = ", actual_X )
   ```

The iterative procedure uses a `while` loop, which will repeat itself until both conditions are verified (TRUE). When at least one of the two assumes a FALSE value, the cycle will be stopped. The two conditions provide a check on the precision and the number of iterations.

This procedure, as anticipated in the *Defining descent methods* section, requires that we update the current value of x in the direction of the gradient's descent. We do this using the following equation:

$$x_{n+1} = x_n - \gamma * \nabla f(x_n)$$

At each step of the cycle, the previous value of x is stored so that we can calculate the precision of the previous step as the absolute value of the difference between the two x values. In addition, the step counter is increased at each step. At the end of each step, the number of iterations and the current value of the x are printed, as follows:

```
Number of iterations =  520
Actual value of x  is =  1.0000547758790321
Number of iterations =  521
Actual value of x  is =  1.0000536803614515
Number of iterations =  522
Actual value of x  is =  1.0000526067542224
Number of iterations =  523
Actual value of x  is =  1.000051554619138
Number of iterations =  524
Actual value of x  is =  1.0000505235267552
Number of iterations =  525
Actual value of x  is =  1.0000495130562201
Number of iterations =  526
Actual value of x  is =  1.000048522795095 7
```

As we can see at each step, the value of x progressively approaches the exact value. Here, 526 iterations were executed.

7. At the end of the procedure, we can print the result:

```
print("X value of f(x) minimum = ", actual_X )
```

The following result is returned:

```
X value of f(x) minimum =  1.0000485227950957
```

As we can verify, the returned value is very close to the exact value, which is equal to 1. It differs precisely from the value of the precision that we imposed as the term for the iterative procedure.

Understanding the Newton-Raphson method

After analyzing in detail a practical case of application of the optimization technique based on the gradient descent method, we will now turn to the Newton-Raphson method, an iterative method to calculate the zeros of a function.

Understanding the Newton-Raphson method

Newton's method is the main numerical method for the approximation of roots of nonlinear equations. The function is linearly approximated at each iteration to obtain a better estimate of the zero point.

Using the Newton-Raphson algorithm for root finding

Given a nonlinear function f and an initial approximation x_0, Newton's method generates a sequence of approximations $\{x_k\}$ k> 0 by constructing, for each k, a linear model of the function f in a neighborhood of x_k and approximating the function with the model itself. This model can be constructed starting from Taylor's development of the function f at a point x belonging to a neighborhood of the iterated current point x_k, as follows:

$$f(x) = f(x_k) + (x - x_k) * f'(x_k) + (x - x_k)^2 * \frac{f''(x_k)}{2!} + \cdots$$

Truncating Taylor's first-order development gives us the following linear model:

$$f(x) = f(x_k) + (x - x_k) * f'(x_k)$$

The previous equation remains valid in a sufficiently small neighborhood of x_k.

Given x_0 as the initial data, the first iteration consists of calculating x_1 as the zero value of the previous linear model with k = 0—that is, to solve the following scalar linear equation:

$$f(x) = 0$$

The previous equation leads to the next iterated x_1 in the following form:

$$x_1 = x_0 - \frac{f(x_0)}{f'(x_0)}$$

Similarly, the subsequent equation iterates x_2, where x_3 is constructed so that we can elaborate on a general validity equation, as follows:

$$x_{n+1} = x_n - \frac{f(x_n)}{f'(x_n)}$$

The form of the update equation is like the generic formula of descent methods. From a geometric point of view, the previous update equation represents the line tangent to the function f at the point $(x_k, f(x_k))$. It is for this reason that the method is also called the tangent method.

Geometrically, we can define this procedure through the following steps:

- The tangent of the function is plotted at the starting point x_0.
- The intercept of this line is identified with the x axis. This point represents the new value x_1.
- This procedure is repeated until convergence.

The following diagram shows this procedure:

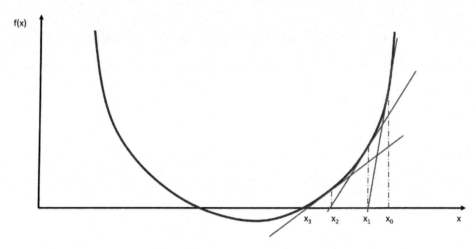

Figure 7.5 – The procedure of finding a tangent

This algorithm is well-defined if f'(x_k) = 0 for every k. With regard to the computational cost, it can be noted that, at each iteration, the evaluation of the function f and its derivative before in point x_k is required.

Approaching Newton-Raphson for numerical optimization

The Newton-Raphson method is also used for solving numerical optimization problems. In this case, the method takes the form of Newton's method for finding the zeros of a function but is applied to the derivative of the function f. This is because determining the minimum point of the function f is equivalent to determining the root of the first derivative f'. The root of the first derivative is an optimality condition.

In this case, the update formula takes the following form:

$$x_{n+1} = x_n - \frac{f'(x_n)}{f''(x_n)}$$

In the previous equation, we have the following:

- $f'(x_n)$ is the first derivative of the function f
- $f''(x_n)$ is the second derivative of the function f

> **Important note**
> The Newton-Raphson method is usually preferred over the descending gradient method due to its speed. However, it requires knowledge of the analytical expression of the first and second derivatives and converges indiscriminately to the minima and maxima.

There are variants that bring this method to global convergence and that lower the computational cost by avoiding having to determine the direction of the research with direct methods.

Applying the Newton-Raphson technique

In this section, we will apply what we have learned so far about the Newton-Raphson method by completing a practical exercise. We'll define a function and then use that method to find the minimum point of the function. As always, we will analyze the code line by line:

1. Let's start by importing the necessary libraries:

    ```
    import numpy as np
    import matplotlib.pyplot as plt
    ```

2. Now, let's define the function:

    ```
    x = np.linspace(0,3,100)
    y=x**3 -2*x**2 -x + 2
    ```

    First, we defined an interval for the dependent variable x. We will only need this to visualize the function in order to draw a graph. To do this, we used the `linspace()` function. This function creates numerical sequences. Then, we passed three arguments: the starting point, the ending point, and the number of points to be generated. Next, we defined a cubic function.

3. Now, we can draw a graph and display it:

    ```
    fig = plt.figure()
    axdef = fig.add_subplot(1, 1, 1)
    axdef.spines['left'].set_position('center')
    axdef.spines['bottom'].set_position('zero')
    axdef.spines['right'].set_color('none')
    axdef.spines['top'].set_color('none')
    ```

```
axdef.xaxis.set_ticks_position('bottom')
axdef.yaxis.set_ticks_position('left')
plt.plot(x,y, 'r')
plt.show()
```

First, we defined a new figure and then we set the axes so that the *x* axis coincides with the minimum value of the function and the *y* axis coincides with the center of the parabola. This will make it easier to visually locate the minimum point of the function. The following diagram is printed:

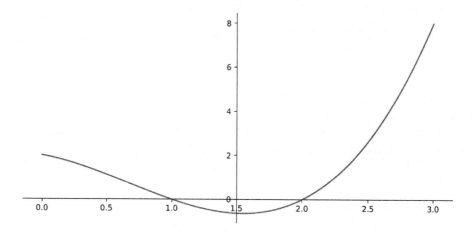

Figure 7.6 – The minimum point of the function

Here, we can see that the minimum of the function occurs for a value of x roughly equal to 1.5. This will be the value that we will have to determine through the gradient descent method. But to have the precise value so that we can compare it with what we will get later, we need to extract this value:

```
print('Value of x at the minimum of the function', x[np.
argmin(y)])
```

To determine this value, we used `numpy`'s `argmin()` function. This function returns the position index of the minimum element in a vector. The following result is returned:

```
Value of x at the minimum of the function
1.5454545454545454
```

4. Now, let's define the first and second derivative functions:

```
FirstDerivative = lambda x: 3*x**2-4*x -1
SecondDerivative = lambda x: 6*x-4
```

5. Now, we will initialize some parameters:

   ```
   actual_X           = 3
   precision_value    = 0.000001
   previous_step_size = 1
   max_iteration      = 10000
   iteration_counter  = 0
   ```

   These parameters have the following meaning:

   - The `actual_X` variable will contain the current value of the independent variable x. To start, we initialize it at x = 3, which represents the far-right value of the display range of the function in the graph.

   - The `precision_value` variable will contain the value that defines the degree of precision of our algorithm. Being an iterative procedure, the solution is refined at each iteration and tends to converge. But this convergence may come after a very large number of iterations, so to save resources, it is advisable to stop the iterative procedure once adequate precision has been reached.

   - The `previous_step_size` variable will contain the calculation of this precision and will be initialized to 1.

   - The `max_iteration` variable contains the maximum number of iterations that we have provided for our algorithm. This value will be used to stop the procedure if it does not converge.

   - Finally, the `iteration_counter` variable will be the iteration counter.

6. Now, we can apply the Newton-Raphson method, as follows:

   ```
   while previous_step_size > precision_value  and 
   iteration_counter  < max_iteration :
       PreviousX = actual_X
       actual_X  = actual_X  - FirstDerivative(PreviousX)/ 
   SecondDerivative(PreviousX)
       previous_step_size = abs(actual_X  - PreviousX)
       iteration_counter  = iteration_counter +1
       print("Number of iterations = ",iteration_counter ,"\
   nActual value of x  is = ",actual_X )
   ```

   This procedure is similar to what we adopted to solve our problem in the *Implementing gradient descent in Python* section. A `while` loop, which will repeat itself until both conditions are verified (TRUE), was used here. When at least one of the two assumes a FALSE value, the cycle will be stopped. These two conditions provide a check on the precision and the number of iterations.

The Newton-Raphson method updates the current value of x, as follows:

$$x_{n+1} = x_n - \frac{f'(x_n)}{f''(x_n)}$$

At each step of the cycle, the previous value of x is stored in order to calculate the precision of the previous step as the absolute value of the difference between the two x values. In addition, the step counter is increased at each step. At the end of each step, the number of the iteration and the current value of x are printed, as follows:

```
Number of iterations = 1
Actual value of x is = 2.0
Number of iterations = 2
Actual value of x is = 1.625
Number of iterations = 3
Actual value of x is = 1.5516304347826086
Number of iterations = 4
Actual value of x is = 1.5485890147300967
Number of iterations = 5
Actual value of x is = 1.5485837703704566
Number of iterations = 6
Actual value of x is = 1.5485837703548635
```

As we mentioned in the *Approaching Newton-Raphson for numerical optimization* section, the number of iterations necessary to reach the solution is drastically skewed. In fact, we went from the 526 iterations necessary to bring the method based on the gradient's descent to convergence to only 6 iterations for the Newton-Raphson method.

7. Finally, we will print the result:

```
print("X value of f(x) minimum = ", actual_X )
```

The following result is returned:

```
X value of f(x) minimum = 1.5485837703548635
```

As we can verify, the returned value is very close to the exact value, which is equal to 1.5454545454545454. It differs precisely in terms of the value of the precision that we imposed as the term for the iterative procedure.

The secant method

A variant of Newton's method is the secant method in which, at the step (n + 1), instead of considering the tangent to the curve of equation y = f (x) at the point of abscissa x_n, the secant to the curve is constructed at the points of abscissa x_n and x_{n-1} respectively.

In other words, x_{n+1} is calculated as the intersection of this secant with the abscissa axis. As with Newton's method, the secant method also proceeds in an iterative way:

1. Given two initial values x_0 and x_1, x_2 is calculated.
2. Then, with x_1 and x_2 known, x_3 is calculated.

    And so on.

Compared to Newton's method, the secant method offers the advantage of not requiring the evaluation of the derivative of the function. Therefore, unlike the latter, it is also applicable when the function has no *a priori* known derivative. However, precisely in the vicinity of the solution, some computational problems may arise from the calculations.

Let's now introduce stochastic gradient descent, an iterative method for the optimization of differentiable functions.

Deepening our knowledge of stochastic gradient descent

As we mentioned in the *Exploring the gradient descent technique* section, the implementation of the gradient descent method consists of initially evaluating both the function and its gradient, starting from a configuration chosen randomly in the space of dimensions.

From here, we try to move in the direction indicated by the gradient. This establishes a direction of descent in which the function tends to a minimum and examines whether the function actually takes on a value lower than that calculated in the previous configuration. If so, the procedure continues iteratively, recalculating the new gradient. This can be totally different from the previous one. After this, it starts again in search of a new minimum.

This iterative procedure requires that, at each step, the entire system status is updated. This means that all the parameters of the system must be recalculated. From a computational point of view, this equates to an extremely expensive operating cost and greatly slows down the estimation procedure. With respect to the standard gradient descent method, in which the weights are updated after calculating the gradient for the entire dataset, in the stochastic method, the system parameters are updated after a certain number of examples. These are chosen randomly in order to speed up the process and to try to avoid any local minimum situations.

Consider a dataset that contains *n* observations of a phenomenon. Here, let *f* be an objective function that we want to minimize with respect to a series of parameters *x*. Here, we can write the following equation:

$$f(x) = \frac{1}{n}\sum_{i=1}^{n} f_i(x)$$

From the analysis of the previous equation, we can deduce that the evaluation of the objective function f requires n evaluations of the function f, one for each value contained in the dataset.

In the classic gradient descent method, at each step, the function gradient is calculated in correspondence with all the values of the dataset through the following equation:

$$x_{n+1} = x_n - \gamma * \frac{1}{n}\sum_{i=1}^{n} \nabla f_i(x_n)$$

In some cases, the evaluation of the sum present in the previous equation can be particularly expensive, such as when the dataset is particularly large and there is no elementary expression for the objective function. The stochastic descent of the gradient solves this problem by introducing an approximation of the gradient function. At each step, instead of the sum of the gradients being evaluated in correspondence to the data contained in the dataset, the evaluation of the gradient is used only in a random subset of the dataset.

So, the previous equation replaces the following:

$$x_{n+1} = x_n - \gamma * \nabla f_i(x_n)$$

In the previous equation, $\nabla f_i(x_n)$ is the gradient of one of the observations in the dataset, chosen randomly.

The pros of this technique are set out here:

- Based only on a part of the observations, the algorithm allows a wider exploration of the parametric space, with the greater possibility of finding new and potentially better points of the minimum
- Taking a step of the algorithm is computationally much faster, which ensures faster convergence toward the minimum point
- The parameter estimates can also be calculated by loading only a part of the dataset into memory at a time, allowing this method to be applied to large datasets

Now, let's see how an iterative algorithm widely used to calculate maximum likelihood estimates works in the case of incomplete data.

Approaching the EM algorithm

The EM method is based on the notion of **Maximum Likelihood Estimation (MLE)** of θ, an unknown parameter of the data distribution. Given a sample space X, let x ∈ X be an observation extracted from the density f (x | θ), which depends on the parameter θ; we define the likelihood function of θ, given the single observation x in the following function:

$$L(\theta|X) = f(x|\theta)$$

The likelihood function is a conditional probability function, taken as a function of its second argument, keeping the first argument fixed.

When the sample consists of *n* independent observations, then the likelihood becomes this:

$$L(\theta|X) = \prod_{i=1}^{n} f(\theta|x_i)$$

Since the values of the likelihood are very close to 0 and to simplify the calculation of the derivative for the maximum likelihood estimates of the parameters, it is appropriate to transform the function with a logarithmic transformation and therefore to study what is called log likelihood:

$$l(\theta|x) = \sum_{i=1}^{n} \log(L(\theta|x_i))$$

The goal of maximum MLE is to choose the parameters θ * that maximize the likelihood:

$$\theta^* = \arg\max_{\theta}\{l(\theta)\}$$

The EM algorithm is an iterative algorithm for MLE: the simplicity of the method is the reason why it has become very popular in the study of these types of problems. The idea of the algorithm alternates an **Expectation** phase, called **E-step**, which calculates the expected value of the log-likelihood of the parameter conditioned by the complete data and previous estimates of the parameter, and a maximization phase (**Maximization**), called **M-step**, which exploits the data just updated by the E-step to find the new estimated maximum likelihood value of the parameter. The procedure is repeated until the difference between the last iterations does not reach the predetermined tolerance threshold, demonstrating that the algorithm has reached convergence.

The E-step uses the elements of the mean vector and the covariance matrix to construct a set of regression equations that predict incomplete variable values from the observed variables. The purpose of this step is to predict the values of the parameters in a way that resembles the imputation of stochastic regression.

The next M-step applies the standard formulas for complete data to the newly created data to update the vector estimates of means and the variance and covariance matrix. The new parameter estimates are moved on to the next E-step, where a new set of regression equations are built to predict the unknown parameters again.

The EM algorithm repeats these two steps until the mean and the covariance matrix do not vary for several consecutive steps, and at that point, the algorithm converges to the MLEs. The concept behind EM is to use observable samples of latent variables to predict the values of samples that are not observable.

The EM algorithm offers the tools to solve a large number of problems; among these, we can list the following:

- Estimation of the values of latent variables
- Imputation of missing data in a dataset
- Estimation of the parameters of **Finite Mixture Models** (**FMMs**)
- Estimation of the parameters of **Hidden Markov Models** (**HMMs**)
- Unsupervised cluster learning

So, let's see how to use EM to deal with Gaussian mixture problems.

EM algorithm for Gaussian mixture

The use of latent variables is a widely practiced solution for modeling complex distributions: properly selected, latent variables can greatly simplify the structure of the model. Latent variables are variables that are not directly observable and therefore not measurable, which therefore are hypothesized and analyzed through their effects. The links, the relationships, and the influences that a latent variable has on the other measurable variables become a way to go back to this hidden variable. Although larger, the joint distribution of observed and latent variables is often easier to manage than the marginal distribution of observed variables only. Models that make use of latent variables are referred to as latent variable models.

A popular class of latent variable models is mixed models: the methodology is based on dividing the responsibility of data modeling among several usually relatively simple components. The components are then combined using a blend distribution, giving each component a weight. Mixture models have several interesting properties: they can model arbitrary complex distributions while being easy to work with. For many common operations, the calculations can be performed independently for the individual components and for the distribution of the mixture. In addition, adapting the number of components provides an intuitive way to control the complexity of the model.

A mixture model is used as a probabilistic model to represent the presence of subpopulations within a population. It can also be defined as a mixed distribution that represents the probability distribution of some observations in the general population. Mixture models are used to create inferences, approximations, and predictions on the properties of subpopulations from observations or data acquired from the population under analysis.

Among the simple distributions, the Gaussian one is the most widely used. In the *Normal distribution* section of *Chapter 3*, *Probability and Data Generation Processes*, we defined this distribution. A normal distribution—also called a Gaussian distribution—has some important characteristics, is symmetrical and bell-shaped; its central position measures—the expected value and the median—coincide; its interquartile range is 1.33 times the mean square deviation; the random variable in a normal distribution takes values between -∞ and + ∞.

The goal of the **Gaussian mixture model** (**GMM**) is to find an approximation or estimate of its components, finding accommodation of the data that contains these components. It is not enough to explain the distributions of some data through a single statistical distribution if the data can be grouped into subpopulations or associated with different generation processes. It is necessary to use a composition of distributions, the same ones that are usually described by mixture models, which are defined by the parameters of each component and the proportions in which each of them contributes to the general distribution.

Gaussian mixtures are probabilistic models based on the assumption that the data belong to a mixture of different Gaussian distributions with unknown parameters. These models provide information on the center (mean) and on the variability (variance) of each cluster and provide the posterior probabilities. For example, if we have two distributions, if the data labels are known, we can go back to the parameters of the distributions (mean and variance); if the parameters are known, we can identify the labels. If, on the other hand, we do not have any information (parameters and labels), we can apply probabilistic algorithms to estimate the parameters.

The set of parameters that define these models can be estimated using many techniques: the EM algorithm is an iterative tool for estimating the maximum likelihood of mixed distributions. The principle of this tool is to introduce a multinomial indicator variable that identifies belonging to a specific cluster of each observation present in the dataset. In these models, the starting point is a random variable that is assumed to be extracted from a population that is an additive mixture of several subpopulations. The GMM generates a natural and intuitive representation of heterogeneity in a finite, usually small, number of latent classes, each considered as a type or group.

So, let's see a practical application of the EM algorithm for estimating the parameters of a Gaussian mixture distribution. In this example, we will create two Gaussian distributions by setting the mean and standard deviation of each. Next, we will merge the newly generated data and try to estimate the parameters of this new distribution by modeling it as a Gaussian mixture. Our goal is to recover the parameters of the two starting Gaussian distributions.

As always, we will analyze the code line by line:

1. Let's start by importing the libraries:

    ```
    import numpy as np
    import seaborn as sns
    from matplotlib import pyplot as plt
    from sklearn.mixture import GaussianMixture
    import pandas as pd
    ```

The numpy library is a Python library that contains numerous functions that can help us manage multidimensional matrices. Furthermore, it contains a large collection of high-level mathematical functions we can use on these matrices. Then, we imported the `seaborn` library, which is a Python library that enhances the data visualization tools of the `matplotlib` module. In the `seaborn` module, there are several features we can use to graphically represent our data. There are methods that facilitate the construction of statistical graphs with `matplotlib`. The `matplotlib` library is a Python library for printing high-quality graphics. Then, we imported the `GaussianMixture()` function from the `sklearn.mixture` module. This function estimates the parameters of a Gaussian mixture distribution using the EM algorithm. Finally, we imported the `pandas` library, which is an open source BSD-licensed library that contains data structures and operations to manipulate high-performance numeric values for the Python programming language.

2. Now, let's set the Gaussian distribution parameters:

```
mean_1=25
st_1=9
mean_2=50
st_2=5
```

We then set the means and standard deviations of the two starting Gaussian distributions.

3. Now, let's create two distributions:

```
n_dist_1 = np.random.normal(loc=mean_1, scale=st_1, size=3000)
n_dist_2 = np.random.normal(loc=mean_2, scale=st_2, size=7000)
```

To do this, we used the `random.normal()` function of the `numpy` library. Three parameters are passed: the mean, the standard deviation, and the number of samples to generate.

4. Now, we can merge the two distributions creating a single population:

```
dist_merged = np.hstack((n_dist_1, n_dist_2))
```

The `hstack()` function of the `numpy` library stacks arrays in sequence horizontally (column-wise).

So, let's see the population we created by drawing a histogram:

```
sns.set_style("white")
sns.histplot(data=dist_merged, kde=True)
plt.show()
```

We used the `histplot()` function of the `seaborn` library to return a univariate or bivariate histogram to show distributions of datasets. The following diagram is plotted:

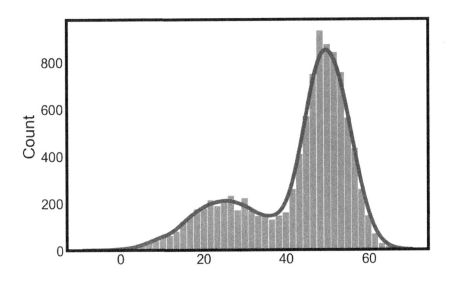

Figure 7.7 – Bivariate histograms of the datasets' distributions

We can easily understand that the data contains two distributions—we just need to evaluate their parameters.

5. To do this, we will use the GMM:

    ```
    dist_merged_res = dist_merged.reshape((len(dist_merged),
    1))
    gm_model = GaussianMixture(n_components=2, init_
    params='kmeans')
    gm_model.fit(dist_merged_res)
    ```

    We first reshaped the population data into a format compatible with the GaussianMixture() function. K-means clustering is a partitioning method, and as anticipated, this method decomposes a dataset into a set of disjoint clusters. Given a dataset, a partitioning method constructs several partitions of this data, with each partition representing a cluster. These methods relocate instances by moving them from one cluster to another, starting from an initial partitioning. Finally, we applied the fit() function to fit the model using the merged initial Gaussian distribution.

6. Now, we can compare the parameters obtained from the simulation with the GMM with those of the initial distributions:

    ```
    print(f"Initial distribution means = {mean_1,mean_2}")
    print(f"Initial distribution standard deviation =
    {st_1,st_2}")
    ```

```
print(f"GM_model distribution means = {gm_model.means_}")
print(f"GM_model distribution standard deviation = {np.
sqrt(gm_model.covariances_)}")
```

The following data is displayed:

```
Initial distribution means = (25, 50)
Initial distribution standard deviation = (9, 5)
GM_model distribution means = [[24.12193283]
                                [49.87502388]]
GM_model distribution standard deviation =
              [[[8.36021272]] [[5.15620167]]]
```

Against initial means of 25 and 50, we obtained an estimate of 24.12 and 49.87—very close to the initial values. For the standard deviations, we estimated values equal to 8.3 and 5.15 compared to initial values equal to 9 and 5.

7. Using these parameters, we can now predict the class of each value of the distribution obtained by joining the two initial distributions:

```
dist_labels = gm_model.predict(dist_merged_res)
```

In the previous line of code, we used the `predict()` function of the GMM. This function predicts the labels for the data samples in `dist_merged_res` using a trained model.

Let's now see a representation of the distribution of the two classes:

```
sns.set_style("white")
data_pred=pd.DataFrame({'data':dist_merged, 'label':dist_
labels})
sns.histplot(data = data_pred, x = "data", kde = True,
hue = "label")
plt.show()
```

The following diagram is returned:

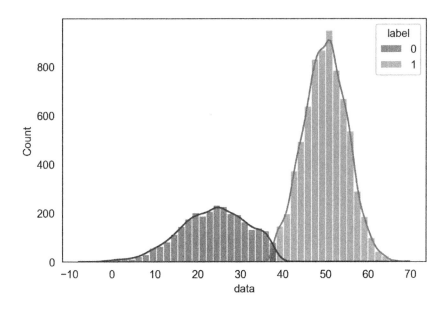

Figure 7.8 – Representation of the distribution of the two classes

8. To make a comparison between the initial distributions and those estimated with the GMM, we also represent the initial distributions:

```
label_0 = np.zeros(3000, dtype=int)
label_1 = np.ones(7000, dtype=int)
labels_merged = np.hstack((label_1, label_0))
data_init=pd.DataFrame({'data':dist_merged,
'label':labels_merged})
```

We first created two vectors containing the labels of the two classes with the same number of elements as the initial distributions. We then stacked them in a single vector and created a pandas DataFrame of two columns: in the first, we inserted the merged distribution, while in the second column, we added the labels.

All that's required is to display the result:

```
sns.set_style("white")
sns.histplot(data = data_init, x = "data", kde = True,
hue = "label")
plt.show()
```

The following diagram is displayed:

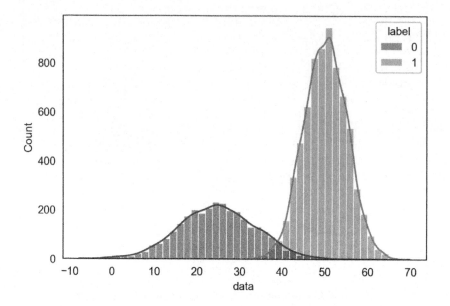

Figure 7.9 – Initial distributions of the two classes

From the comparison of *Figures 7.8* and *7.9*, we can appreciate the quality of the estimate of the parameters of the two distributions: it should be noted that this estimate was carried out without knowing any information on the starting distributions.

After analyzing in detail, a practical case of application of the EM method, we will now study another optimization method.

Understanding Simulated Annealing (SA)

The SA algorithm is a general optimization technique for solving combinatorial optimization problems. The algorithm is based on stochastic techniques and iterative improvement algorithms.

Iterative improvement algorithms

Search algorithms and iterative improvement of solutions are known as local search algorithms. The operation of these algorithms involves the following steps:

1. Starting from configuration A, a sequence of iterations is performed, each of which consists of a transition from the current configuration to another that belongs to the neighborhood of A.

2. If the transition gives rise to an improvement of the cost function, the current configuration is replaced by its neighbor; otherwise, a new configuration among the neighboring ones is selected for a new comparison.

3. The algorithm ends when a configuration is obtained that has a cost no worse than any other neighboring configuration.

This class of algorithms has some disadvantages:

- By definition, the local search algorithm ends in a local minimum, and there is no information on how much this minimum differs from the global minimum
- The quality of the obtained local minimum depends on the initial configuration chosen and no general criterion establishes a way to select a starting point from which to obtain good solutions
- In general, no limits can be given to the completion time of the algorithm

However, the local search has the advantage of being generally applicable with ease: the definition of the configurations, the cost function, and the generation mechanism do not generally create problems.

To remedy the aforementioned disadvantages, some variations have been introduced in the algorithm:

- Execution of the algorithm on a high number of initial configurations uniformly distributed over the set of possible configurations
- Use of the information obtained from the previous executions to improve the choice of the initial configuration for the subsequent computation
- Introduction of a more complex generation mechanism to make the algorithm capable of surpassing the local optimum

In addition, configurations are accepted that provide a worsening of the value from the cost function; in general, however, the local search algorithm accepts only improvements in the cost of the solutions. This variant of improvement is provided for by the SA algorithm: the solutions obtained with the approximation algorithm based on the SA algorithm do not depend on the initial configuration and are generally able to approximate the overall optimum well. SA finds the global optimum asymptotically, does not show the disadvantages of the local search algorithm, and is equally an algorithm applicable to the generality of problems.

SA in action

The SA algorithm is based on the analogy between the solidification process of metals and the problem of solving complex combinatorial problems. In physics, the term *annealing* denotes a physical process in which a solid is progressively heated up to a temperature at which it reaches the liquid state, and then slowly cooled. In this way, if the temperature reached is high enough and the cooling process is slow enough, all the molecules are arranged in a configuration of minimum energy.

During this phase, changes of state occur in the direction of decreasing energies: transitions from an i state to a j state of the solid involving energy increases can still occur with a probability defined by the following equation:

$$p = e^{\left(\frac{E_i - E_j}{kT}\right)}$$

In the previous equation:

- E_i, E_j are the energies associated with the two states
- k is Boltzmann's constant
- T is the temperature of the metal

This model is exploited in SA to solve optimization problems. In this case, an analogy is proposed between the objective function of the problem and energy, while the temperature becomes the control variable of the algorithm. During the search for the optimal solution, therefore, in addition to solutions that lead to improvements in the objective function, those that lead to worsening are also accepted, with a temperature-dependent acceptance probability according to the previous equation. This feature translates into the main advantage of SA, which is the ability to expand the search space for solutions and overcome the local minima of the objective function.

The SA algorithm foresees the following steps: initially, a high value is given to the control parameter; then, a sequence of configurations is generated, up to the minimum point of the objective function. The configurations are chosen among those present in the neighborhood of the current solution, according to the following criterion: if the difference between the values of the objective function <0, then the new configuration replaces the old one; otherwise, if the difference between the values of the objective function ≥ 0, the probability of replacing the old configuration with the new one is calculated according to the previous equation. So, there is a non-negative probability of accepting solutions with higher objective functions. The process continues until the probability distribution of the configurations approaches the Boltzmann distribution.

The value of the control parameter is progressively lowered in discrete steps, and the system is allowed to reach equilibrium at each step. In the initial part of the algorithm, when the temperature value is high, the acceptability criterion will also be high: in this first phase, it will therefore be possible to examine many permutations and then explore the entire search domain. As the temperature drops and the model approaches its minimum energy state, only the smallest energy deviations are accepted: this is reminiscent of gradient descent as it only considers the local search space and focuses only on improving the solution.

The temperature parameter plays an essential role in finding the global optimum of an optimization problem: it is set to regulate the slow decrease in the probability of accepting worse solutions while the entire solution space is explored.

The algorithm ends with a control value, beyond which no configuration that worsens the value of the objective function is accepted. The solution thus obtained is the solution to the problem treated. The method, therefore, leverages a global optimizer that explores the search space early on and a local optimizer that leverages only what is critical to achieving good results. It should be noted that this method does not protect us from possible stagnation in a position of local minimum. The method is very sensitive to parameter values, which makes tuning difficult: parameter adjustment difficulty is the greatest weakness of SA.

So, let's see a practical application of this algorithm (`simulated_annealing.py`). As always, we will analyze the code line by line:

1. Let's start by importing the necessary libraries:

    ```
    import numpy as np
    import matplotlib.pyplot as plt
    ```

    We first imported the `numpy` library, which is a Python library that contains numerous functions that can help us manage multidimensional matrices. Furthermore, it contains a large collection of high-level mathematical functions we can use on these matrices. Then, we imported the `matplotlib` library, a Python library for printing high-quality graphics.

2. Let's now create an exploration domain and a function we want to minimize:

    ```
    x= np.linspace(0,10,1000)
    def cost_function(x):
        return x*np.sin(2.1*x+1)
    ```

    We first created an array of 1000 elements equally spaced between the extreme values 0 and 10. This represents the domain of existence of our function—that is, the space in which we are going to look for the solution. Next, we created a sinusoidal function that represents our objective function; the form of the function was chosen because it has different minima. To confirm this, let's see a representation of it:

    ```
    plt.plot(x,cost_function(x))
    plt.xlabel('X')
    plt.ylabel('Cost Function')
    plt.show()
    ```

The following diagram is displayed:

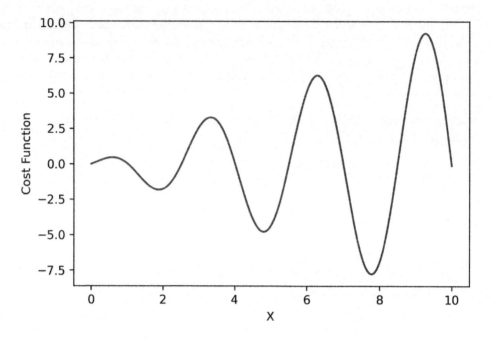

Figure 7.10 – Cost function representation

In this way, we confirm the presence of three lows in the 0-10 range.

3. Now, let's fix the parameters of the method:

```
temp = 2000
iter = 2000
step_size = 0.1
```

4. Now, let's move on to initializing the system variables:

```
np.random.seed(15)
xi = np.random.uniform(min(x), max(x))
E_xi = cost_function(xi)
xit, E_xit = xi, E_xi
cost_func_eval = []
acc_prob = 1
```

To start, we set the seed of the random number generation using the np.random.seed() numpy function: this function initializes the basic random number generator. Next, we set the first value of x in which to evaluate the cost function. To do this, we used the random.

`uniform()` numpy function: this function generates numbers within a defined numeric range. In this case, the range is between the extreme values of the x variable setting range. Next, we made an evaluation of the cost function for that value of x. After doing this, we initialized some temporary variables that we will need in the iterative procedure. Our aim is to set a temporary x variable, evaluate the cost function, and check if this value is acceptable. Acceptable values are stored in an array created ad hoc (`cost_func_eval`). Finally, we initialized the `acc_prob` variable, which will contain the acceptance probability.

5. Now, we can start the iterative procedure:

```
for i in range(iter):
    xstep = xit + np.random.randn() * step_size
    E_step = cost_function(xstep)
```

At each cycle, we set a value for x by adding to the previous value a term obtained by multiplying the step size by a random number. In correspondence with this new value of x, we then carry out an evaluation of the function.

We can therefore carry out our first check:

```
if E_step < E_xi:
    xi, E_xi = xstep, E_step
    cost_func_eval.append(E_xi)
    print('Iteration = ',i,
        'x_min = ',xi,'Global Minimum =', E_xi,
        'Acceptance Probability =', acc_prob)
```

If the estimated energy is less than what has been accepted so far, then we update the improvement values. This value is first added to the `cost_func_eval` array, and the information on the improvement obtained is printed on the screen:

```
diff_energy = E_step - E_xit
t = temp /(i + 1)
acc_prob = np.exp(-diff_energy/ t)
```

In this block of code, we evaluate the acceptance probability according to what is indicated in the equation seen at the beginning of this section. We first evaluate the difference between the energies, then update the temperature value, and finally evaluate the acceptance probability.

Finally, we decide whether to accept the new point:

```
if diff_energy < 0 or
        np.random.randn() < acc_prob:
    xit, E_xit = xstep, E_step
```

In conclusion, we draw a graph with the cost function improvement procedure during the iterative procedure:

```
plt.plot(cost_func_eval, 'bs--')
plt.xlabel('Improvement Step')
plt.ylabel('Cost Function improvement')
plt.show()
```

The following diagram is displayed:

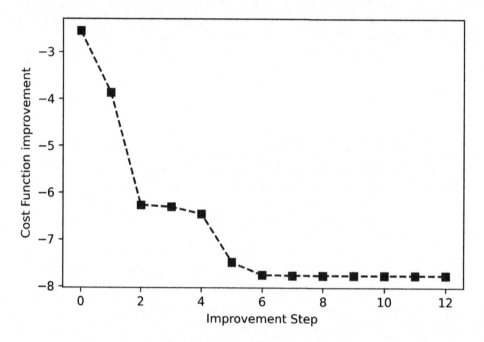

Figure 7.11 – Cost function improvement

To conclude, we report an extract of the information regarding some accepted improvement steps:

```
Iteration =   0 x_min =   8.352060345432603 Global Minimum
= -2.549691087509061 Acceptance Probability = 1
Iteration =   1 x_min =   8.268146731695177 Global
Minimum = -3.865265080181498 Acceptance Probability =
1.0011720530797679
Iteration =   2 x_min =   8.077074224781931 Global
Minimum = -6.264757086553908 Acceptance Probability =
1.0013164397397474
Iteration =   4 x_min =   8.072960126990173 Global
Minimum = -6.3053561226675905 Acceptance Probability =
```

```
0.9993874938212183
Iteration =   14 x_min =   8.057426141282635 Global
Minimum = -6.45398523520768 Acceptance Probability =
1.0011249991583007
Iteration =   17 x_min =   7.908942504770391 Global
Minimum = -7.482144101692382 Acceptance Probability =
0.9960707945643333
Iteration =   18 x_min =   7.805571271239743 Global
Minimum = -7.755842336539771 Acceptance Probability =
1.0092963751415767
Iteration =   21 x_min =   7.767739570791674 Global
Minimum = -7.763382995462104 Acceptance Probability =
1.0072493114830527
Iteration =   27 x_min =   7.793961859355448 Global
Minimum = -7.763418087146118 Acceptance Probability =
0.9975392161808777
Iteration =   36 x_min =   7.775446728598834 Global
Minimum = -7.765853968129766 Acceptance Probability =
1.007367947545534
Iteration =   41 x_min =   7.778624797903834 Global
Minimum = -7.766277157406623 Acceptance Probability =
0.9981491502724658
Iteration =  138 x_min =   7.781197205345049 Global
Minimum = -7.766364644616736 Acceptance Probability =
1.036566292077183
Iteration = 1174 x_min =   7.78075358444397 Global
Minimum = -7.7663658475723265 Acceptance Probability =
1.3178155041756248
```

From the comparison between the graph of the trend of the cost function and the final value obtained, we can conclude that the algorithm has brought us in correspondence with the global minimum.

After analyzing some optimization procedures with practical examples, let's now see other multivariate optimization methods available in the SciPy library.

Discovering multivariate optimization methods in Python

In this section, we will analyze some numerical optimization methods contained in the Python SciPy library. SciPy is a collection of mathematical algorithms and functions based on numpy. It contains a series of commands and high-level classes that can be used to manipulate and display data. With SciPy, functionality is added to Python, making it a data processing and system prototyping environment, similar to commercial systems such as MATLAB.

Scientific applications that use SciPy benefit from the development of add-on modules in numerous fields of numerical computing made by developers around the world. Numerical optimization problems are also covered among the available modules.

The SciPy `optimize` module contains numerous functions for the minimization/maximization of objective functions, both constrained and unconstrained. It treats nonlinear problems with support for both local and global optimization algorithms. In addition, problems regarding linear programming, constrained and nonlinear least squares, search for roots, and the adaptation of curves are treated. In the following sections, we will analyze some of them.

The Nelder-Mead method

Most of the well-known optimization algorithms are based on the concept of derivatives and on the information that can be deduced from the gradient. However, many optimization problems deriving from real applications are characterized by the fact that the analytical expression of the objective function is not known, which makes it impossible to calculate its derivatives, or because is particularly complex, so coding the derivatives may take too long. To solve this type of problem, several algorithms have been developed that do not attempt to approximate the gradient but rather use the values of the function in a set of sampling points to determine a new iteration by other means.

The Nelder-Mead method tries to minimize a nonlinear function by evaluating test points that constitute a geometric form called a simplex.

> **Important note**
> A simplex is defined as a set of closed and convex points of a Euclidean space that allow us to find the solution to the typical optimization problem of linear programming.

The choice of geometric figure for the simplex is mainly due to two reasons: the ability of the simplex to adapt its shape to the trend in the space of the objective function deforming itself, and the fact that it requires the memorization of only n + 1 points. Each iteration of a direct search method based on the simplex begins with a simplex, specified by its n + 1 vertices and the values of the associated functions. One or more test points and the respective values of the function are calculated, and the iteration ends with a new simplex so that the values of the function in its vertices satisfy some form of descent condition with respect to the previous simplex.

The Nelder-Mead algorithm is particularly sparing in terms of its evaluation of the function at each iteration, given that, in practice, it typically requires only one or two evaluations of the function to build a new simplex. However, since it does not use any gradient assessment, it may take longer to find the minimum.

This method is easily implemented in Python using the `minimize` routine of the SciPy `optimize` module. Let's look at a simple example of using this method:

1. Let's start by loading the necessary libraries:

   ```
   import numpy as np
   from scipy.optimize import minimize
   import matplotlib.pyplot as plt
   from matplotlib import cm
   from matplotlib.ticker import LinearLocator, FormatStrFormatter
   from mpl_toolkits.mplot3d import Axes3D
   ```

   The library that's needed to generate 3D graphics is imported (`Axes3D`).

2. Now, let's define the function:

   ```
   def matyas(x):
       return 0.26*(x[0]**2+x[1]**2)-0.48*x[0]*x[1]
   ```

   The `matyas` function is continuous, convex, unimodal, differentiable, and non-separable, and is defined on two-dimensional space. The `matyas` function is defined as follows:

   $$f(x,y) = 0.26 * (x^2 + y^2) - 0.48 * x * y$$

   This function is defined on a x,y ∈ [-10,10]. This function has one global minimum in f(0,0) = 0.

3. Let's visualize the `matyas` function:

   ```
   x = np.linspace(-10,10,100)
   y = np.linspace(-10,10,100)
   x, y = np.meshgrid(x, y)
   z = matyas([x,y])

   fig = plt.figure()
   ax = fig.gca(projection='3d')
   surf = ax.plot_surface(x, y, z, rstride=1, cstride=1,
                          cmap=cm.RdBu,linewidth=0, antialiased=False)

   ax.zaxis.set_major_locator(LinearLocator(10))
   ax.zaxis.set_major_formatter(FormatStrFormatter('%.02f'))

   fig.colorbar(surf, shrink=0.5, aspect=10)

   plt.show()
   ```

To start, we defined independent variables, x and y, in the range that we have already specified [-10.10]. So, we created a grid using the `numpy meshgrid()` function. This function creates an array in which the rows and columns correspond to the values of x and y. We will use this matrix to plot the corresponding points of the z variable, which corresponds to the `matyas` function. After defining the x, y, and z variables, we traced a three-dimensional graph to represent the function. The following diagram is plotted:

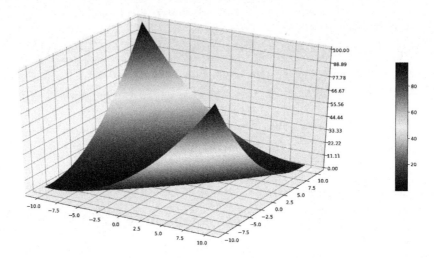

Figure 7.12 – Meshgrid plot to represent the function

4. As we already mentioned, the Nelder-Mead method does not require us to calculate a derivative as it is limited to evaluating the function. This means that we can directly apply the method, like so:

```
x0 = np.array([-10, 10])
NelderMeadOptimizeResults = minimize(matyas, x0,
        method='nelder-mead',
        options={'xatol': 1e-8, 'disp': True})
print(NelderMeadOptimizeResults.x)
```

To do this, we first defined an initial point to start the search procedure from for the minimum of the function. So, we used the `minimize()` function of the SciPy `optimize` module. This function finds the minimum of the scalar functions of one or more variables. The following parameters have been passed:

- `matyas`: The function you want to minimize
- `x0`: The initial vector
- `method = 'nelder-mead'`: The method used for the minimization procedure

Additionally, the following two options have been added:

- `'xatol': 1e-8`: Defines the absolute error acceptable for convergence
- `'disp': True`: Set to True to print convergence messages

5. Finally, we printed the results of the optimization method, as follows:

```
Optimization terminated successfully.
        Current function value: 0.000000
        Iterations: 77
        Function evaluations: 147
[3.17941614e-09 3.64600127e-09]
```

The minimum was identified in the value 0, as already anticipated. Furthermore, this value was identified in correspondence with the following values:

```
X = 3.17941614e-09
Y = 3.64600127e-09
```

These are values that are very close to zero, as we expected. The deviation from this value is consistent with the error that we set for the method.

Powell's conjugate direction algorithm

Conjugate direction methods were originally introduced as iterative methods for solving linear systems with a symmetric and positive definite coefficient matrix, and for minimizing strictly convex quadratic functions.

The main feature of conjugated direction methods for minimizing quadratic functions is that of generating, in a simple way, a set of directions that, in addition to being linearly independent, enjoy the further important property of being mutually conjugated.

The idea of Powell's method is that if the minimum of a quadratic function is found along each of the p (p <n) directions in a stage of the research, then when taking a step along each direction, the final displacement from the beginning up to the p-th step is conjugated with respect to all the p subdirections of research.

For example, if points 1 and 2 are obtained from one-dimensional searches in the same direction but from different starting points, then the line formed by 1 and 2 will be directed toward the maximum. The directions represented by these lines are called conjugate directions.

Let's analyze a practical case of applying the Powell method. We will use the `matyas` function, which we defined in the *The Nelder-Mead method* section:

1. Let's start by loading the necessary libraries:

```
import numpy as np
from scipy.optimize import minimize
```

2. Now, let's define the function:

```
def matyas (x):
    return 0.26 * (x [0] ** 2 + x [1] ** 2) -0.48 * x [0] * x [1]
```

3. Now, let's apply the method:

```
x0 = np.array([-10, 10])
PowellOptimizeResults = minimize(matyas, x0,
        method='Powell',
        options={'xtol': 1e-8, 'disp': True})
print(PowellOptimizeResults.x)
```

The `minimize()` function of the SciPy `optimize` module was used here. This function finds the minimum of the scalar functions of one or more variables. The following parameters were passed:

- `matyas`: The function we want to minimize
- `x0`: The initial vector
- `method = 'Powell'`: The method used for the minimization procedure

Additionally, the following two options have been added:

- `'xtol': 1e-8`: Defines the absolute error acceptable for convergence
- `'disp': True`: Set to `True` to print convergence messages

4. Finally, we printed the results of the optimization method. The following results are returned:

```
Optimization terminated successfully.
        Current function value: 0.000000
        Iterations: 3
        Function evaluations: 66
[-6.66133815e-14 -1.32338585e-13]
```

The minimum was identified in the value 0, as specified in the *The Nelder-Mead method* section. Furthermore, this value was identified in correspondence with the following values:

```
X = -6.66133815e-14
Y = -1.32338585e-13
```

These are values very close to zero, as we expected. We can now make a comparison between the two methods we applied to the same function. We can note that the number of iterations necessary to reach convergence is equal to 3 for the Powell method, while it is equal to 77 for the Nelder-Mead method. A drastic reduction in the number of evaluations of the function is also noted: 66 against 147. Finally, the difference between the calculated value and the expected value is reduced by the Powell method.

Summarizing other optimization methodologies

The `minimize()` routine of the SciPy `optimize` package contains numerous methods for unconstrained and constrained minimization. We analyzed some of them in detail in the previous sections. In the following list, we have summarized the most used methods provided by the package:

- **Newton-Broyden-Fletcher-Goldfarb-Shanno** (**BFGS**): This is an iterative unconstrained optimization method used to solve nonlinear problems. This method looks for the points where the first derivative is zero.

- **Conjugate Gradient** (**CG**): This method belongs to the family of conjugate gradient algorithms and performs a minimization of the scalar function of one or more variables. This method requires that the system matrix be symmetric and positive definite.

- **dog-leg trust-region** (**dogleg**): The method first defines a region around the current best solution, where the original objective function can be approximated. The algorithm thus takes a step forward within the region.

- **Newton-CG**: This method is also called **truncated Newton's method**. It is a method that identifies the direction of research by adopting a procedure based on the conjugate gradient, to roughly minimize the quadratic function.

- **Limited-memory BFGS** (**L-BFGS**): This is part of the family of quasi-Newton methods. It uses the BFGS method for systematically saving computer memory.

- **Constrained Optimization By Linear Approximation** (**COBYLA**): The operating mechanism is iterative and uses the principles of linear programming to refine the solution found in the previous step. Convergence is achieved by progressively reducing the pace.

Summary

In this chapter, we learned how to use different numerical optimization techniques to improve the solutions offered by a simulation model. We started by introducing the basic concepts of numerical optimization, defining a minimization problem, and learning to distinguish between local and global minimums. We then moved on and looked at the optimization techniques based on gradient descent. We defined the mathematical formulation of the technique and gave it a geometric representation. Furthermore, we deepened our knowledge of the concepts surrounding the learning rate and trial and error. By doing this, we addressed a practical case to reinforce the concepts we learned by solving the problem of searching for the minimum of a quadratic function.

Subsequently, we learned how to use the Newton-Raphson method to search for the roots of a function and then how to exploit the same methodology for numerical optimization. We also analyzed a practical case for this technology to immediately put the concepts we learned into practice. We did this by looking for the local minimum of a convex function.

We then went on to study the stochastic gradient descent algorithm, which allows us to considerably reduce the computational costs of a numerical optimization problem. This result is obtained by using a single estimate of the gradient at each step, which is chosen in a stochastic way among those available. Then, we analyzed the EM algorithm and the SA method to improve the optimization procedure.

Finally, we explored the multivariate numerical optimization algorithms contained in the Python SciPy package. For some of them, we defined the mathematical formulation and proposed a practical example of using the method. For the others, a summary was drawn up to list their characteristics.

In the next chapter, we will learn the basic concepts of soft computing and how to implement genetic programming. We will also understand genetic algorithm techniques, and we will learn how to implement symbolic regression and how to use the **Cellular Automation** (**CA**) model.

8
Introducing Evolutionary Systems

Evolutionary algorithms are a family of stochastic techniques for solving problems that are part of the broader category of natural metaphor models. They find their inspiration in biology and are based on the imitation of the mechanisms of so-called natural evolution. Over the last few years, these techniques have been applied to many problems of great practical importance.

In this chapter, we will learn the basic concepts of SC and how to implement genetic programming. We will also understand the genetic algorithm techniques, how to implement symbolic regression, and how to use the **cellular automata (CA)** model.

In this chapter, we're going to cover the following main topics:

- Introducing SC
- Understanding genetic programming
- Applying a genetic algorithm for search and optimization
- Performing symbolic regression
- Exploring the CA model

Technical requirements

In this chapter, we will learn how to use genetic programming to optimize systems. To deal with the topics in this chapter, it is necessary that you have a basic knowledge of algebra and mathematical modeling. To work with the Python code in this chapter, you'll need the following files (available on GitHub at the following URL: https://github.com/PacktPublishing/Hands-On-Simulation-Modeling-with-Python-Second-Edition):

- `genetic_algorithm.py`

- `symbolic_regression.py`
- `cellular_automata.py`

Introducing SC

Over the past few decades, many researchers have developed numerous methods and systems, many of which have been successfully used in real-world applications. In these applications, most of the methods are based on probabilistic paradigms, such as the well-known Bayesian inference, rules of thumb, and decision systems. Since the 1960s, several great and epoch-making theories have been proposed using fuzzy logic, genetic algorithms, evolutionary computation, and neural networks – all methods that are referred to as **SC**. When combined with well-established approaches to probability, these new SC methods become effective and powerful in real-world applications. The ability of these techniques to include inaccuracy and incompleteness of information and model very complex systems makes them useful tools in many sectors.

SC takes on the following characteristics:

- The ability to model and control uncertain and complex systems, as well as to represent knowledge efficiently through linguistic descriptions typical of fuzzy set theory
- The ability to optimize genetic algorithms whose computation is inspired by the laws of selection and mutation typical of living organisms
- The ability to learn complex functional relationships as in the case of neural networks, inspired by those in brain tissue

In the SC, among its characteristic features, we find, in fact, uncertain, ambiguous, or incomplete data, consistent parallelism, randomness, approximate solutions, and adaptive systems.

> **Important note**
> The constitutive methodologies of SC require considerable computing powers; all, in fact, presuppose a significant computational effort, which only modern computers have made it possible to sustain in a reasonable time.

From a historical point of view, we can consider that neural networks were born in 1959, fuzzy logic in 1965, probabilistic reasoning in 1967, and genetic algorithms in 1975. Originally, each algorithm had well-defined labels and usually could be identified with specific scientific communities. In recent years, by improving the understanding of the strengths and weaknesses of these algorithms, we have begun to make the most of their characteristics and the development of hybrid algorithms. These denominations indicate a new integration trend that reflects the current high degree of integration between scientific communities. These interactions have given birth to SC, a new field that combines the versatility of fuzzy logic to represent qualitative knowledge with efficient data from neural networks to provide adequate refinements through a local search, and the ability of genetic algorithms to efficiently perform a global search. The result is the development of hybrid algorithms that are superior to each underlying component of SC, providing us with the best real-world problem-solving tools.

At the beginning of this section, we said that under the name of SC, there are several theories: fuzzy logic, neural networks, evolutionary computation, and genetic algorithms. Let us now see a brief description of these methodologies.

Fuzzy logic (FL)

FL corresponds to a mathematical approach to translating the fuzziness of linguistic concepts into a representation that computers can understand and manipulate. Since FL can transform linguistic variables into numerical ones without losing the sense of truth along the way, it allows for the construction of improved models of human reasoning and in-depth knowledge. FL and fuzzy set theory in general provide an approximate indication, an effective and flexible tool, to describe the behavior of systems that are too complex or too ill-defined to admit a precise mathematical analysis with classical methods and tools. Since fuzzy set theory is a generalization of classical set theory, there is more flexibility to faithfully capture the various aspects of an information situation under conditions of incompleteness or ambiguity. In this way, in addition to operating only with linguistic variables, modern fuzzy set systems are designed to manage any type of information uncertainty.

Artificial neural network (ANN)

ANNs take inspiration from neural biological systems. ANNs have attributes such as universal approximation, the ability to learn and adapt to their environment, and the ability to invoke weak assumptions about the understanding of phenomena responsible for generating input data. ANNs are suitable for solving problems where no analytical model exists or where the analytical model is too complex to be applied. The basic components that make up an ANN are called artificial neurons, which roughly model the operating principles of their biological counterparts. Furthermore, ANNs model not only biological neurons but also their interconnected mechanisms and some global functional properties. A detailed discussion of ANNs will follow in *Chapter 10, Simulating Physical Phenomena by Neural Networks*.

Evolutionary computation

The mechanisms during natural reproduction are evolution, mutation, and the survival of the fittest. They allow the adaptation of lifeforms in an environment through subsequent generations. From a computational point of view, this can be seen as an optimization process. The application of evolution mechanisms to artificial computing systems is called evolutionary computing. From this, we can say that evolutionary algorithms use the power of selection to transform computers into automatic optimization tools. These methodologies are efficient, adaptable, and robust research processes, which produce solutions close to optimal ones and have a great capacity for implicit parallelism.

To better understand how the mechanisms of natural evolution are translated into SC, let's analyze in detail the functioning of genetic algorithms.

Understanding genetic programming

Since their introduction, electronic calculators have been used to speed up calculations, but not only to build models that explain and reproduce the biological characteristics of nature and its evolution. If the electronic reproduction of the brain's behavior and its way of learning gave rise to neural networks, the simulation of biological evolution gave rise to what is now called evolutionary computation. The first studies on computer evolutionary systems were carried out in the 1950s and 1960s, with the aim of identifying mechanisms of biological evolution that could be useful as optimization tools for engineering problems. The term *evolutionary strategies* was introduced by Rechenberg in 1965 to indicate the method he used to optimize some parameters of aerodynamic structures. Later, the field of evolutionary strategies caught the interest of other researchers and became a research area with specialized congresses.

Introducing the genetic algorithm (GA)

GAs were introduced and developed by Holland in the 1960s, with the aim of studying the phenomenon of the adaptation of natural systems and translating this feature into computer systems. Holland's GA was an abstraction of biological evolution, in which a population of chromosomes composed of strings of genes, valued at 0 or 1, was made to evolve into a new population using some genetic operators, such as selection, crossing, mutation, and inversion:

- The selection operator classifies the chromosomes so that the most suitable ones have more chances to reproduce
- The crossover has the task of exchanging parts of chromosomes
- The mutation randomly changes the value of genes in some positions of the chromosome
- Inversion changes the order of the arrangement of genes in a part of the chromosome

GAs are inherently very flexible and, at the same time, robust. These features have allowed their use in different fields; one of the main ones is, obviously, the optimization of complicated numerical functions.

> **Important note**
> In many cases, GAs have proved to be more effective than other techniques, such as the gradient one, because the continuous mixing of genes by crossover and mutation prevents us from stopping on a local maximum or minimum.

Due to their characteristics of flexibility and robustness, GAs have found use in many fields:

- **Combinatorial optimization**: They have been effectively used in problems in which it is necessary to find the optimal sequential arrangement of a series of objects.
- **Bin Packing**: They have been used in the search for the optimal allocation of limited resources for maximizing yield or production.

- **Design**: By implementing a mix of combinatorial optimization and function optimization, GAs have also been used in the field of design. Operating without preconceptions, they can often try and find things that a human designer would never have thought of.
- **Image processing**: They were used to align images of the same area taken at different times or to create an identikit of suspicious people, starting from the description of a witness.
- **Machine learning**: In the field of artificial intelligence, GAs are often used to train machines in certain problems.

The basics of GA

All living organisms are composed of cells containing one or more chromosomes. A chromosome is a DNA strand that acts as a blueprint for an organism. A chromosome can ideally be divided into genes, each of which encodes a particular protein (see *Figure 8.1*). In simple terms, we can imagine that each gene encodes a particular characteristic; in this case, the term *allele* specifies the different possible configurations of that characteristic. The totality of an individual's genetic material is called a genome, while a particular set of genes contained in a genome is called a genotype.

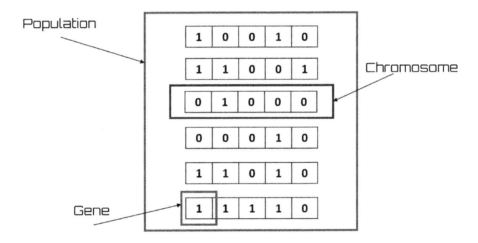

Figure 8.1 – The essential elements of a GA

During the development of an organism, the genotype gives rise to the phenotype that governs the characteristics of the organism, such as eye color, height, and brain size. In nature, most organisms that reproduce sexually are diploid, and during reproduction, a recombination of genes occurs between each pair of chromosomes. The descendants are subject to a possible mutation of genes, often resulting from an error in copying the genes. The fitness of an individual is typically defined as the probability that the organism lives long enough to reproduce or as a function of the number of descendants it generates. Typically, applications that use GAs use stateless individuals with only one chromosome.

The term *chromosome* identifies a solution to the problem, often encoded as a string of bits. Genes can be either single bits or small blocks of bits. In binary encoding, an allele can be 0 or 1.

In the literature, there is no rigorous definition of GA accepted by all researchers in the field of evolutionary computation or one that distinguishes GAs from other methods of evolutionary computation. However, it is possible to identify some characteristics common to all GAs: a population of chromosomes, a selection that acts based on fitness, a crossover to produce new descendants, and their random mutation. Chromosomes are generally represented by bit strings in which the alleles can take the value of 0 or 1 (binary coding). Each chromosome represents a point in the solution search space, and the GA modifies the population of chromosomes by creating descendants in each evolutionary era and replacing the population of the parents with them. Each chromosome must, therefore, be associated with a degree of fitness for reproduction in relation to its ability to solve the problem in question. The fitness of a chromosome is calculated based on an appropriate function defined by the user, depending on the problem considered.

Although GAs are relatively easy to program, their behavior can be complicated. Traditional theory assumes that GAs work by discovering and recombining good building blocks of solutions in parallel. The good solutions tend to be formed, compared to the bad ones, by combinations of bit values more suitable to represent the solution of the considered problem. Once the initial population is formed, recombination occurs through several genetic operators.

Genetic operators

Let's say we are given a population, P, with a set of m individuals belonging to the environment being considered. If an optimization problem is being considered, in which there are n variables to be optimized, a population will consist of m sets of n variables:

$$P = \{p_1, p_2, ... p_m\}$$

Each p_k element is an individual of the population, composed of the following n variables:

$$p_k = [p_1^k, p_2^k, ..., p_n^k], k = 1, ..., m$$

The vector represented in the previous equation must be encoded in the language used by the GA. The initial population of individuals can be generated randomly or based on heuristics.

The m size of the population influences the efficiency of the GA and must be chosen with care. If the population is made up of too small several individuals, the diversity of the population is not guaranteed, and the algorithm converges too quickly. However, if there are too many individuals in the population, the selection pressure is not adequate, and the waiting time to reach convergence becomes very long. An adequate estimate for the proper choice of the m number can be the following interval: $2n \leq m \leq 4n$.

A specific fitness function must be built for each problem to be solved. Given a particular chromosome, the fitness function returns a single numerical value called `fitness`, which is supposed to be proportional to the utility or ability of the individual that the chromosome represents. For many problems, especially for optimization functions, the fitness function measures the value of the function itself.

Understanding genetic programming 261

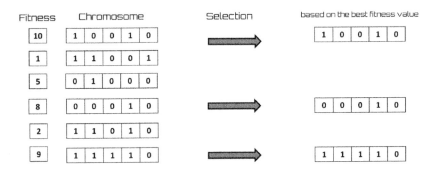

Figure 8.2 – A selection of the best parents in the population

During the reproduction phase of a GA, individuals undergo a selection process (*Figure 8.2*) and are recombined, producing the offspring that will give rise to the next generation. Parents are randomly selected using a scheme that favors the best individuals. These will likely be selected multiple times for reproduction, while the worst ones may never be chosen. The most popular selection method consists of directly associating the probability of selection with individuals based on their fitness in solving the given problem. The mechanism that is defined as roulette is then implemented and which statistically allows to ensure a selection strictly proportional to the fitness function of each individual. Alternatives are available: several random evaluations at the same time that should be avoided if the number of individuals is not particularly high, that an individual who does not deserve it is not favored. Another solution is to preserve a certain number of the best individuals unchanged, so that their genetic characteristics are not lost during the evolution of the population. In this case, however, there is the risk of a premature convergence of the algorithm.

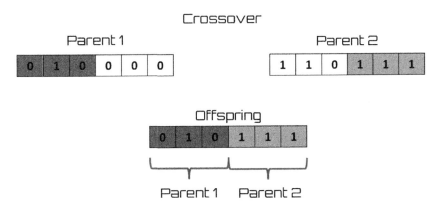

Figure 8.3 – A crossover operation

After selecting two individuals, their chromosomes are recombined using the crossover and mutation operators. The crossover considers two individuals and cuts a piece of the strings of their two chromosomes in a randomly chosen position or at a specific point (*Figure 8.3*) to produce two new chromosomes, in which the piece of string cut by the first is exchanged with the piece cut by the second. Each of the offsprings will inherit some genes from each parent but will not be identical to one of them. Crossover is not usually applied to all pairs of individuals selected for mating. A random choice is made, where the probability of the crossover operation occurring is typically between 0.6 and 1.0. If the crossover is not applied, the offspring is generated simply by duplicating the parents.

The mutation is applied to each child individually after crossover (*Figure 8.4*). The mutation operator randomly alters the value of a gene. The probability of occurrence of this operator is generally very low, typically 0.001. Traditional theory holds that crossover is more important than mutation as regards the speed of exploration of the search space. The mutation brings only a factor of randomness, which serves to ensure that no unexplored regions remain in this space.

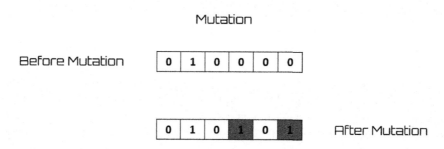

Figure 8.4 – A mutation operator in a GA

If the GA is correctly implemented, the population evolves over several generations so that the fitness of the best individual and the average in each generation grows toward the global optimum. Convergence is represented by the progression toward increasing uniformity. To formalize the concept of the building block, Holland introduced the notion of a schema. A pattern is a set of bit strings that can be described by a pattern consisting of ones, zeroes, and asterisks (wildcards). A pattern is a subset of the space of all possible individuals for which all genes match the pattern for the pattern itself.

Let's analyze the following schema:

$$A = [0 ***** 1]$$

It represents the set of all seven-bit strings, starting with zero and ending with one. Different schemes have different characteristics, but there are important properties common to all, such as the order and the definition length.

The order of scheme A, denoted by o (A), is the number of defined positions – that is, the number of 0s and 1s present – therefore, the number of non*. The order defines the specificity of a scheme and is useful for calculating the probability of a scheme's survival from mutation.

The defining length of scheme A, denoted by δ (H), is the distance between the first and last position fixed in the string. It defines the compactness of the information contained in a schema. Note that a pattern with only one fixed position has a defining length equal to zero. The scheme presented previously contains two defined bits (not asterisks) of order 2. Its definition length, the distance between the outermost defined bits, is 6.

A fundamental assumption of the traditional theory of GAs is that the schemes are, implicitly, the building blocks on which the algorithm acts by means of selection, mutation, and crossover operators. At a given generation, the algorithm evaluates the fitness of n individuals but implicitly estimates the average fitness of a much greater number of schemes.

> **Important note**
> Short, low-order schemes with above-average fitness receive an exponentially increasing number of individuals from one generation to the next.

The convergence of the GA strongly depends on the fitness function adopted. The competition between the schemes, during evolution, generally proceeds from the lower-order partitions toward the higher-order partitions. In this way, it would be difficult for a GA to find the maximum point of a certain fitness function if the lower-order partitions contained misleading indications about the higher-order ones.

After being properly introduced to the basic concept of genetic programming in the previous section, we will analyze a practical case of using such algorithms to optimize the parameters of a model.

Applying a GA for search and optimization

After introducing numerical optimization techniques in *Chapter 7, Using Simulation to Improve and Optimize Systems*, we should be adequately familiar with optimization methodologies.

In the search for a solution to a problem, one of the challenges that we find ourselves facing often is being able to coordinate a series of aspects that, although in contrast with each other, must be able to coexist in the best possible way. A researcher will be called upon to implement a series of compromises between the various possibilities that are presented to them to be able to optimize a characteristic, chosen based on the constraints imposed by the remaining variables of the problem considered. The optimization process seeks better performance by looking for some point, or points, of optimum. Often, when judging optimization procedures, the focus is solely on convergence, neglecting temporary performance entirely. This originates from the concept of optimization present in mathematical calculation. By analyzing most problems, we realize that convergence toward the best solution is not the main objective; rather, an attempt is made to improve and overcome previous experiences. It is thus clear that the main goal of optimization is improvement.

Basically, a search for an optimal solution through genetic programming consists of the following steps:

1. Random initialization of the population
2. Selection of parents by evaluating fitness
3. Parental crossing for reproduction

4. Mutation of an offspring
5. Evaluation of descendants
6. A union of descendants with the main population and order

So, let's see how to implement this procedure in a Python environment. To understand the algorithm as easily as possible, we will face a simple optimization problem, namely the search for coefficients that maximize a multi-variable equation (`genetic_algorithm.py`). Suppose we have the following equation:

$$= a * x_1 + b * x_2 + c * x_3 + d * x_4$$

In the previous equation, the parameters have the following meaning:

- y: The result of the equation that will become our fitness
- x_i: The four variables of our equation
- a, b, c, d: The coefficients of the equation to be searched

As always, we analyze the code line by line:

1. Let's start by importing the libraries:

```
import numpy as np
```

The `numpy` library is a Python library that contains numerous functions that can help us manage multidimensional matrices. Furthermore, it contains a large collection of high-level mathematical functions we can use on these matrices.

2. Let's initialize some parameters:

```
var_values = [1,-3,4.5,2]
num_coeff = 4
pop_chrom = 10
sel_rate = 5
```

Let's examine the meaning of each parameter:

- `var_values`: The values of the independent variables present in the starting equation. Our optimization procedure concerns a single sequence of variables; if you want to explore other combinations of values, just repeat the procedure by setting other values of x_i.
- `num_coeff`: The number of coefficients we are looking to optimize.
- `pop_chrom`: The number of population chromosomes, each chromosome representing a candidate solution for the optimization of the problem.
- `sel_rate`: The number of parents that will be selected for reproduction.

3. Now, we can define the size of the population that we are going to consider:

   ```
   pop_size = (pop_chrom,num_coeff)
   ```

   This is simply a matrix with a number of rows equal to the number of chromosomes (candidate solutions for the optimization of the problem) and a number of columns equal to the number of coefficients of the equation (values that we want to optimize).

   At this point, we can initialize the population:

   ```
   pop_new = np.random.uniform(low=-10.0, high=10.0,
   size=pop_size)
   print(pop_new)
   ```

   We initialized the population randomly using the NumPy `random.uniform` function. This function generates numbers within a defined numeric range. The following population was returned:

   ```
   [[ 5.62001356 -0.33698498  5.98386158  7.99349938]
    [-8.99933866  4.36708159  0.1752488   8.24435752]
    [-2.7168974  -0.62478605  3.96715342  0.95252127]
    [-2.49535006 -1.31442411  3.93149064 -3.00483574]
    [-6.78770529  0.33676806  8.36153357  4.16832494]
    [-2.5726188  -1.01028351  8.18507797 -2.48511823]
    [ 3.03714886 -4.21650291 -8.75350175 -4.97657289]
    [-2.50146921 -5.51742108 -1.09468644 -6.07296704]
    [-4.67969645  5.85248018 -5.2888357  -2.39211734]
    [-1.77859384 -7.64974142 -2.88171404 -6.88599093]]
   ```

   Now, we have all the initial parameters to proceed with the optimization of the coefficients.

4. To begin the search for the optimal solution, we first set the number of generations:

   ```
   num_gen = 100
   ```

   The evolutionary algorithm iteratively generates a new population of individuals by completely replacing the previous one and applying genetic operations to the current generation, until a predetermined termination criterion is met. In this case, we adopt as termination criterion precisely the maximum number of generations, but we could also have chosen the achievement of an acceptable fitness value for the given problem.

   Now, we use a `for` loop that iterates the procedure for a number equal to the number of generations we have set:

   ```
   for k in range(num_gen):
       fitness = np.sum(pop_new *var_values, axis=1)
       par_sel = np.empty((sel_rate, pop_new.shape[1]))
   ```

With each generation, the procedure evaluates the fitness of the population. In our case, this is simply equivalent to the evaluation of the dependent *y* variable, according to the equation set at the beginning of this section. To do this, simply add the products between coefficients and the x_i variable. Then, we initialize a matrix (`par_sel`), which will contain the population selection that will be selected for the genetic evolution. To do this, we used the NumPy `empty()` function, which returns a new array of a given shape and type, without initializing entries. This matrix will have rows equal to the `sel_rate` parameter that we defined as the number of solutions chosen for the evolution, and the same number of columns as the population matrix.

Now, let's try to print the values that will help us to follow the evolution of the system. We then print the current generation value and the best fitness value:

```
print("Current generation = ", k)
    print("Best fitness value : ", np.max(fitness))
```

5. Now, let's apply the genetic operations, starting with the selection:

```
for i in range(sel_rate):
    sel_id = np.where(fitness == np.max(fitness))
    sel_id = sel_id[0][0]
    par_sel[i, :] = pop_new[sel_id, :]
    fitness[sel_id] =np.min(fitness)
```

The selection operator behaves in a similar way to that of natural selection, its operation based on the following assumption: the poorest performing individuals are eliminated, and the best performing individuals are selected because they have an above-average chance to promote the information they contain within the next generation. Basically, we have simply identified the one with the maximum fitness value in the current population and selected it by moving it to the `par_sel` matrix. We eventually overwrote that fitness value with the minimum value. This is necessary; otherwise, in subsequent iterations, the same would always be selected. We iterated this operation several times equal to `sel_rate`, thus selecting the strongest chromosomes.

6. Let's now move on to perform the crossover operation, which consists of dividing the chromosomes of two individuals into one or more points and exchanging them, giving rise to two new individuals:

```
offspring_size=(pop_chrom-sel_rate, num_coeff)
offspring = np.empty(offspring_size)
crossover_lenght = int(offspring_size[1]/2)
for j in range(offspring_size[0]):
        par1_id = np.random.randint(0,par_sel.shape[0])
        par2_id = np.random.randint(0,par_sel.shape[0])
        offspring[j, 0:crossover_lenght] = par_sel[par1_id, 0:crossover_lenght]
```

```
            offspring[j, crossover_lenght:] = par_sel[par2_
    id, crossover_lenght:]
```

The crossover combines the genetic heritages of the two parents to build children, who have half the genes of one and half of the other. There are different types of crossover. In this example, we have adopted the one-point crossover. We have randomly chosen two parents, and then we have chosen the crossover length. Assuming that the string is of the `offspring_size[1]` length, the crossover length is chosen by dividing into two parts the length of the chromosomes, so the first child will have the genes from `1..crossover_lenght` of the first parent and from `crossover_lenght` to the end of the other.

The crossover is the driving force of the GA and is the operator that most influences convergence, even if the unscrupulous use of it could lead to premature convergence of research on some local maximum.

Basically, we first defined the size of the offsprings, which will be a matrix with rows equal to the number of chromosomes (`pop_chrom`) minus the number of parents that will be selected for reproduction (`sel_rate`). We then initialized an array of these dimensions (`offspring_size`) using the NumPy `empty()` function, which returns a new array of a given shape and type, without initializing entries. So, we fixed the length of the crossover by dividing the length of the chromosomes in half. After having set the essential parameters of the crossover, we performed the operation using a `for` loop that iterates the process a number of times equal to the row of the offspring matrix, each loop generating a new offspring. As mentioned previously, each offspring is achieved by joining the pieces of two parents.

7. Now, we operate the mutation that is used to randomly change the value of one or more genes:

```
for m in range(offspring.shape[0]):
        mut_val = np.random.uniform(-1.0, 1.0)
        mut_id = np.random.randint(0,par_sel.shape[1])
        offspring[m, mut_id] = offspring[m, mut_id] +
    mut_val
```

The mutation modifies a gene, randomly chosen with a uniform distribution, of the individual's genotype and thus inserts diversity into the population, leading research into new spaces or recovering an allele that had been previously lost. We iterated with a `for` loop that traverses all the rows of the offspring array. At each cycle, the value to add to a gene to add a mutation is first randomly generated. Then, the gene to be modified is randomly decided and the random value is added to the gene.

8. We have reached the end of the generative cycle, so we can update the population:

```
pop_new[0:par_sel.shape[0], :] = par_sel
pop_new[par_sel.shape[0]:, :] = offspring
```

We simply added the selected parents' chromosomes first and then queued the offspring generated with the crossover and modified with the mutation.

9. At the end of the iteration of generations, we just have to recover the results:

```
fitness = np.sum(pop_new *var_values, axis=1)
best_id = np.where(fitness == np.max(fitness))
print("Optimized coefficient values = ", pop_new[best_id,
:])
print("Maximum value of y = ", fitness[best_id])
```

To start, we reevaluated the fitness of the final population, and then we identified the position of the maximum value. We then printed on the screen both the best combination of the coefficients of the equation and the maximum value of the *y* value they determine. The following results are displayed:

```
Optimized coefficient values =   [[[ 10.58453911
-19.11676776

               23.11509817    2.38742604]]]
Maximum value of y =   [176.72763626]
```

The parameters that regulate the minor or major influence of one operator on another are the probability of crossover, mutation, and reproduction; their sum must be equal to one, even if there are variants of a GA that carry out the mutation in any case, even after the crossover to maintain greater diversity in the population. The choice of parameters and the operator used strongly depends on the domain of the problem; therefore, it is not possible, a priori, to establish the specifications of a GA.

After analyzing in detail a practical example of optimization with the use of GAs, we will now see how genetic programming can be useful to carry out a symbolic regression.

Performing symbolic regression (SR)

A mathematical model is a set of equations and parameters that return output from input. The mathematical model is always a compromise between precision and simplicity. In fact, it is useless to resort to sophisticated models when the values of the parameters that appear in them are known only approximately. The search for mathematical models can sometimes be very complicated – for example, when trying to symbolize a markedly nonlinear phenomenon. In these cases, the researcher can find valuable help from the process of extrapolating information and relationships present in the input data, in the form of a symbolic equation. In this case, it is a symbolic regression process, different from its classical counterpart in that it has the advantage of simultaneously modifying the underlying structure and the parameters of the mathematical model. A symbolic regression model can be represented by the following equation:

$$y = f(x_1, x_2, \dots x_n)$$

In the previous equation, the parameters have the following meaning:

- y = system output
- x_i = system input

The model will return a combination of the input data expressed by a function or combination of functions, f. The x_i variables may or may not be related to the desired answer, y; in fact, it will be the system itself that directs us on the quality of the solution found.

With SR, we can identify the optimal mathematical formula suitable for a multivariate dataset represented by input/output pairs. To do this, SR exploits the **evolutionary algorithms** (**EAs**) based on the Darwinian theory of evolution, with its ability to provide a class of possible real solutions. Each solution is called a population, while each member is called an individual. The main objective of the methodology is to discover the best individual, who represents the best solution in the population through an iterative process of generating new populations. Each individual represents a mathematical function in the form of a tree structure. In this tree, the nodes correspond to the mathematical operations, while the leaves represent the operands of the formula. The result of the formula is calculated for each data record and the evaluation is performed through the application of the fitness function, which returns an error between the current result and the expected one. The evolution of the formula toward an acceptable result takes place according to the principles already seen in the *Understanding genetic programming* section: selection, crossover, and mutation.

The selection makes the choice of the best formulas that returns a result closest to the real model, based on the indications provided by the fitness function. The selected formulas are then combined by operating the crossover, swapping some subtrees of two formulas. The new formula thus obtained contains the best genetic information of two parents – that is, the offspring is made up of pieces of functions belonging to the best solutions. Finally, the mutation operation guarantees genetic diversity by altering one or more genes, based on a user-defined probability of mutation. The set of new formulas represents the new population, which is used in the next cycle of evolution. The cycle is repeated until the fitness function reaches the desired value.

So let's see a practical case of symbolic regression (`symbolic_regression.py`). To describe how to use symbolic regression to derive a mathematical model from the data, we will artificially generate a distribution of data. To do this, we will use the following equation:

$$f(x, y) = x^2 + y^2$$

It is a simple equation with two variables, although we will later see from its graphic representation that it is anything but simple. Then, we will generate data from this equation and then try to trace the data back to its formula.

As always, we analyze the code line by line:

1. Let's start by importing the libraries:

   ```
   import numpy as np
   import matplotlib.pyplot as plt
   from mpl_toolkits.mplot3d import Axes3D
   from gplearn.genetic import SymbolicRegressor
   ```

   The numpy library is a Python library that contains numerous functions that can help us manage multidimensional matrices. Furthermore, it contains a large collection of high-level mathematical functions we can use on these matrices. The matplotlib library is a Python library for printing high-quality graphics. Furthermore, the library that's needed to generate 3D graphics is imported (Axes3D). Finally, we import from the gplearn.genetic toolbox the SymbolicRegressor() function.

2. Let's now generate the data to display a graphical representation of the function in two variables:

   ```
   x = np.arange(-1, 1, 1/10.)
   y = np.arange(-1, 1, 1/10.)
   x, y = np.meshgrid(x, y)
   f_values = x**2 + y**2
   ```

   To start, we define the independent variables, x and y, in the [-1,1] range. So, we create a grid using the NumPy meshgrid() function. This function creates an array in which the rows and columns correspond to the values of x and y. Then, we calculate the value of the function.

   Let's display this function:

   ```
   fig = plt.figure()
   ax = Axes3D(fig)
   ax.plot_surface(x, y, f_values)
   plt.xlabel('x')
   plt.ylabel('y')
   plt.show()
   ```

   The following graph is shown:

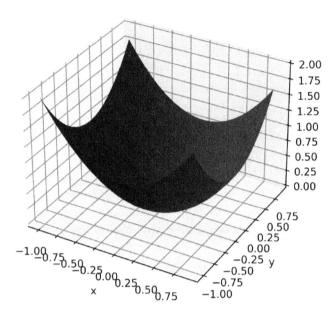

Figure 8.5 – A function of two variables obtained using symbolic regression

3. Now, let's generate the data we will use to derive the mathematical model:

   ```
   input_train = np.random.uniform(-1, 1, 200).reshape(100, 2)
   output_train = input_train[:, 0]**2 + input_train[:, 1]**2
   input_test = np.random.uniform(-1, 1, 200).reshape(100, 2)
   output_test = input_test[:, 0]**2 + input_test[:, 1]**2
   ```

   We have generated two datasets. The first set (the train set) we will use to train the algorithm. This data will be used as starting data to evolve individuals until the convergence criterion is satisfied. The second set of data (the test set) will be used to evaluate the performance of the algorithm. To generate the input data, we used the numpy.random.uniform() function, which generates random numbers from a uniform distribution within a defined numeric range. The output was instead generated by applying the function of two variables, which is obtained by adding the squares of the two variables.

4. Now, we can set the parameters needed to perform symbolic regression:

   ```
   function_set = ['add', 'sub', 'mul']
   sr_model = SymbolicRegressor(population_
   ```

```
size=1000,function_set=function_set,
                         generations=10, stopping_
criteria=0.001,
                         p_crossover=0.7, p_subtree_
mutation=0.1,
                         p_hoist_mutation=0.05, p_
point_mutation=0.1,
                         max_samples=0.9, verbose=1,
                         parsimony_coefficient=0.01,
random_state=1)
```

To start, we have defined the mathematical functions that will be exploited by the algorithm. We have limited the number of functions, since we already know what to look for; in practice, it is appropriate to foresee all the functions and then exclude those that appear inconsistent. The more functions that are provided, the larger the solution search domain will be. Next, we used the `SymbolicRegressor()` function from the `gplearn` library, which extends the scikit-learn library to perform genetic programming with symbolic regression.

The following parameters are passed:

- `population_size`: The number of programs in each generation
- `function_set`: The functions to use when building and evolving programs
- `generations`: The number of generations to evolve
- `stopping_criteria`: The value required to stop evolution early
- `p_crossover`: The probability of performing crossover on a tournament winner
- `p_subtree_mutation`: The probability of performing subtree mutation on a tournament winner
- `p_hoist_mutation`: The probability of performing hoist mutation on a tournament winner
- `p_point_mutation`: The probability of performing point mutation on a tournament winner
- `max_samples`: The fraction of samples to draw from *x* to evaluate each program on
- `verbose`: The verbosity of the evolution-building process
- `parsimony_coefficient`: The constant that penalizes large programs by adjusting their fitness to be less favorable for selection
- `random_state`: The seed used by the random number generator

5. Now, we can train the model:

```
sr_model.fit(input_train, output_train)
```

Since we have requested to view the evolution of the procedure, we will see the following data printed on the screen:

```
     |    Population Average     |              Best Individual              |
---- ----------------------------- ------------------------------------------- ----------
Gen    Length      Fitness     Length      Fitness      OOB Fitness   Time Left
 0      36.62       1.157         7         0.237825      0.184028       9.00s
 1       9.83       0.63691      11         0.0597335     0.0556074      5.76s
 2       7.14       0.468389      7         0                0           4.80s
```

Figure 8.6 – The evolution of the symbolic regression procedure

Finally, we just have to print the results on the screen:

```
print(sr_model._program)
print('R2:',sr_model.score(input_test,output_test))
```

The following results are returned:

```
add(mul(X1, X1), mul(X0, X0))
R2: 1.0
```

The rhyme line gives us a mathematical model. Note that the formula has been provided in symbolic format; we can, in fact, see the operators (add and mul) and the variables (X1 and X0). If we apply operators to variables, we can easily obtain the following formula:

$$x1^2 + x0^2$$

R2 refers to the coefficient of determination (R-squared). R-squared measures how well a model can predict the data and falls between zero and one. The higher the value of the coefficient of determination, the better the model is at predicting the data. We got the maximum value, indicating that the mathematical model fits the data perfectly.

After analyzing in detail how to obtain a mathematical model from data using symbolic regression, let's now examine the oldest computation models inspired by the physical world – CA.

Exploring the CA model

The first studies on CA were by John von Neumann in the 1950s and were motivated by his interests in the field of biology, with the aim of defining artificial systems capable of simulating the behavior of natural systems. If the self-reproductive process can be simulated with a machine, then there must be an algorithm that can describe the operation of the machine. The research was aimed at replicating models that were computationally universal. We wanted to study computational models that had the property of not distinguishing the computation component from the memory component. The interactions in a cellular automaton are local, deterministic, and synchronous. The model is parallel, homogeneous, and discrete, both in space and in time. The connection of CA with the physical world and their simplicity are the basis of their success in the field of simulating natural phenomena.

CA are mathematical models that are used in the study of self-organizing systems in statistical mechanics. In fact, given a random starting configuration, thanks to the irreversible character of the automaton, it is possible to obtain a great variety of self-organizing phenomena. CA are treated using a topological approach; this mathematical method considers automata as continuous functions in a compact metric space. Being complex systems, they refer to the second law of thermodynamics, according to which, given a microscopic and isolated physical system, it tends to reach a state of maximum entropy and, therefore, disorder over time. Microscopic, irreversible, and dissipative systems, or open systems that are allowed to exchange energy and matter with the environment, are allowed to evolve from a state of disorder to a more orderly one through self-organizing phenomena.

CA, therefore, mathematically reproduce physical systems through discrete space-time variables, whose measurements are limited to a limited set of values, which are also discrete. The cellular automaton evolves over time, producing a certain number of generations. A new generation replaces the previous one, which affects the values of the new variables. The mere fact that the values do not have a linear dependency relationship generates non-trivial automata. The formation of patterns in the development of natural organisms is governed by very simple local rules and could be well described by the adoption of a cellular automaton model. Each spatial unit of the lattice, forming a regular matrix, is accompanied by discrete values describing the types of living cells. Short-range interactions can lead to the emergence of varied genetic characteristics.

A cellular automaton is defined on an infinite regular grid of any size in which the shape of the cells is always the same. Each point of the grid is called a cell. At every instant, each cell of the grid is in a state among the finite possible states. Each cell is characterized by at least two states – in the simplest cases, 0 and 1, or true or false. At discrete time intervals, all cells simultaneously update their state, based on the state of a finite number of other cells called neighbors. Usually, the neighborhood of each cell contains all those that have a distance of less than a certain value. The rule applied by each cell to determine its next state is the same. The global evolution of a cellular automaton is the evolution of the overall configuration of the cell network. Each configuration is sometimes called a generation, if related to the evolution of the overall state of the cell network or a pattern. We can define CA that differ from each other by varying the number of dimensions of the grid, the set of states, and so on.

The overall behavior of the automaton, at the t + 1 instance, is the reflection of the state at the previous t instant and of the transition operation, which usually affects the state transition of all cells, in synchrony. The rules that influence the behavior of the single cell are called "local" and the comprehensive behavior is called "global". The changes affect the generations and the so-called neighborhood (the cells present in the surroundings that mutually influence the behavior of the others). The starting configuration is called a "seed" and can be made up of a cell or a pattern of cells. The choice of the initial situation can be casual or linked to design choices.

In summary, a cellular automaton is defined by the following elements:

- **Cell**: Represents the minimum unit to define the 2D and 3D space. An unlimited network of cells can describe the universe.
- **Cellular state**: Describes the state, at t instant, of the considered space unit, which can be full or empty. The cellular state can be defined according to a project's needs.

Game-of-life

An example of a cellular automaton is the one conceived by Conway, which takes the name of *game-of-life*. The grid is two-dimensional, and the cells are square. A cell can be alive (represented with a black square) or dead (represented with a white square) (see *Figure 8.7*). Each cell changes its state according to the state of the eight cells around it (the neighbors) and the following rules:

- A living cell dies if the number of living neighbors is less than two (death by isolation)
- A living cell dies if the number of living neighbors is greater than three (death from overcrowding)
- A dead cell becomes alive if the number of living neighbors is exactly three (birth) 1
- In any other case, a cell retains its state

We can see that the rules are quite simple, but the behavior of the configurations, surprisingly, is anything but trivial. Indeed, Conway has shown that it is undecidable whether a given pattern will eventually lead to a configuration in which all cells are dead.

A classification has been made to refer to the various classes of patterns that can be observed in a configuration of this cellular automaton. Some of them are proposed here:

- **Still life**: These are patterns that remain unchanged over time and are fixed points of an updating rule
- **Oscillator**: Temporally periodic structures that do not remain unchanged, but after a finite number of updates, they reappear identically and at the same point of the grid
- **Spaceship**: Patterns that, after a finite number of generations, are repeated identically but are translated into space
- **Cannon**: A pattern that, like an oscillator, returns to its initial state after a finite number of generations but that also emits spaceships

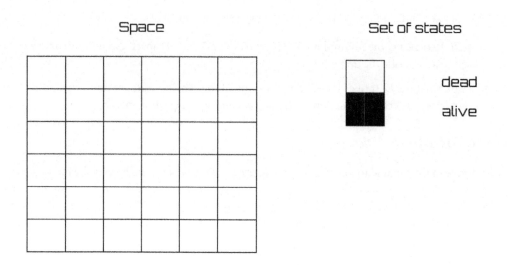

Figure 8.7 – A two-dimensional grid and states of the game-of-life CA

Note that many of the observed patterns were not designed ad hoc but emerged from initial random configurations. In fact, during the evolution of a pattern, the objects collide with each other, sometimes destroying themselves and sometimes creating new patterns.

Now, let's see how to implement a simple example of a CA model in a Python environment.

Wolfram code for CA

Elementary CA models are, in a certain sense, the simplest CA that we can imagine (or almost). They have one dimension and each cell can be in one of two states; moreover, the neighborhood of each cell contains itself and the two adjacent cells. The only thing that distinguishes an elementary cellular automaton from the other is the local update rule. In general, it is possible to calculate the number of possible rules with s ^ (s ^ n), where s = number of states and n = number of neighbors (or, rather, of the cells taken into consideration, the neighbors plus the central node). In the case of elementary CA, s = 2 and n = 3, and then 2 ^ (2 ^ 3) = 2 ^ 8 = 256 elementary CA. Many of these are equivalent if you rename the states or swap left and right in cellular space. A common way of referring to an elementary cellular automaton is to use its Wolfram number. Wolfram proposed a classification of one-dimensional CAs with a few states and neighborhoods of a few cells, based on their qualitative behavior, and he identified four complexity classes (Wolfram classification):

- **Class 1**: CA that quickly converge to a uniform state after a defined number of steps.
- **Class 2**: CAs that quickly converge to stable or repetitive states. The structures are short-term.

- **Class 3**: CA evolving in states that appear to be completely random. Starting from initially different schemes, they create statistically interesting generations, comparable to fractal curves. In this class, the greatest disorder is perceived both globally and locally.
- **Class 4**: CA that form not only repetitive or stable states but also structures that interact with each other in a complicated way. At the local level, a certain order is observed.

Among the complexity classes, class 4 was particularly interesting due to the presence of structures (gliders) capable of propagating in space and time, so much so that it led Wolfram to put forward the hypothesis that the CA of this class may be capable of universal computation.

According to Wolfram, a rule is uniquely specified by the value it takes on each possible combination of input values – that is, by the following eight-bit sequence:

$f(1, 1, 1), f(1, 1, 0), f(1, 0, 1), f(1, 0, 0),$

$f(0, 1, 1), f(0, 1, 0), f(0, 0, 1), f(0, 0, 0)$

We can think of each bit as the state of the three cells adjacent to the next cell.

For example, consider the update rule of the next cell, 126. The 8-bit binary expansion of the decimal number 126 is 0111110, so the local update rule of the elementary cellular automaton with Wolfram number 126 is f, such that we get the following:

$f(1, 1, 1) = 0, f(1, 1, 0) = 1, f(1, 0, 1) = 1, f(1, 0, 0) = 1,$

$f(0, 1, 1) = 1, f(0, 1, 0) = 1, f(0, 0, 1) = 1, f(0, 0, 0) = 0$

So, if we consider the three adjacent cells, the next cell update rule will be like the one shown in *Figure 8.8*:

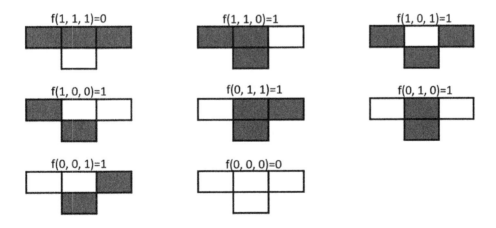

Figure 8.8 – Wolfram rules (126) of the updating cell state

The way to update the cells will then be indicated by these rules:

1. The first row at the top of the square grid will be initialized randomly.
2. Starting from these values, and following the rules indicated in *Figure 8.8*, the cells of the next row will be updated.
3. This procedure will be repeated until the last row of the grid is reached.

So, let's move on to a practical example (`cellular_automata.py`). As always, we analyze the code line by line:

1. Let's start by importing the libraries:

    ```
    import numpy as np
    import matplotlib.pyplot as plt
    ```

    The NumPy library is a Python library that contains numerous functions to help us manage multidimensional matrices. Furthermore, it contains a large collection of high-level mathematical functions we can use on these matrices. The matplotlib library is a Python library for printing high-quality graphics. With matplotlib, it is possible to generate graphs, histograms, bar graphs, power spectra, error graphs, scatter graphs, and more using just a few commands. This includes a collection of command-line functions, such as those provided by MATLAB software.

2. Let's now set the grid size and Wolfram update rules to follow:

    ```
    cols_num=100
    rows_num=100
    wolfram_rule=126
    bin_rule = np.array([int(_) for _ in np.binary_
    repr(wolfram_rule, 8)])
    print('Binary rule is:',bin_rule)
    ```

    We first set the number of rows and columns of the grid (100 x 100). We then set up the Wolfram rule (126) that enforces the updates shown in *Figure 8.8*. Subsequently, we represented the Wolfram number in binary form; to do this, we used the NumPy `binary_repr()` function, which returns the binary representation of the input number as a string. Since we are interested in having a result in the form of an array of integers, we have also used the `int(_)` function, which converts the specified value into an integer number. Finally, we have displayed the result on the screen:

    ```
    Binary rule is: [0 1 1 1 1 1 1 0]
    ```

    The result is exactly what we expected and represents the 8-bit binary expansion of the decimal number 126.

3. Now, let's initialize the grid:

```
cell_state = np.zeros((rows_num, cols_num),dtype=np.int8)
cell_state[0, :] = np.random.randint(0,2,cols_num)
```

We first created a grid of size (`rows_num`, `cols_num`) and then (100 x 100) and populated it all with zeros. We then initialized the first row, which, as previously mentioned, will contain random values. We used the NumPy `random.randint()` function, which returns an integer number's selected element from the specified range. The randomly generated values will be only 0 and 1 and in a number equal to `cols_num` (100).

4. Now, let's move on to update the status of the grid cells:

```
update_window= np.array([[4], [2], [1]])
for j in range(rows_num - 1):
    update = np.vstack((np.roll(cell_state[j, :], 1), cell_state[j, :],
                       np.roll(cell_state[j, :], -1))).astype(np.int8)
    rule_up = np.sum(update * update_window, axis=0).astype(np.int8)
    cell_state[j + 1, :] = bin_rule[7 - rule_up]
```

We first defined a window (`update_window`) that will allow us to obtain the cell update rule, starting from the state of the neighbors' cells. We then updated the status of the cells row by row. We first retrieved the status of the three neighboring cells (left, center, and right) of the previous row immediately (update). So, we calculated the position of the update rule by calculating the sum of the states of the neighboring cells multiplied by the update window. The obtained number (0–7) will give us the position of the Wolfram sequence, as indicated here:

$7 = f(1, 1, 1), 6 = f(1, 1, 0), 5 = f(1, 0, 1), 4 = f(1, 0, 0),$

$3 = f(0, 1, 1), 2 = f(0, 1, 0), 1 = f(0, 0, 1), 0 = f(0, 0, 0)$

Finally, we retrieved the cell status in binary format from Wolfram's update rules, simply subtracting from 7 the number obtained in the previous instruction (`rule_up`).

5. We just have to print the grid with the updated cell status:

```
ca_img= plt.imshow(cell_state,cmap=plt.cm.binary)
plt.show()
```

The following diagram is shown:

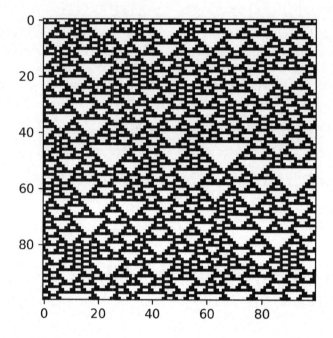

Figure 8.9 – The first 100 lines of the space-time diagrams of an elementary cellular automaton for Wolfram's 126 class; the initial configuration is random

Rule 126 generates a regular pattern with an average growth faster than any polynomial model. It represents a special case by showing how chaotic behavior can be broken down into a wide range of figures: gliders, gliding guns, and still-life structures.

Summary

In this chapter, we learned the basic concepts of SC to exploit the tolerance of imprecision, uncertainty, and rough reasoning to achieve behavior like human decision-making. We analyzed the basic techniques of SC: fuzzy logic, neural networks, evolutionary computation, and GAs. We understood how these technologies can be exploited to support us in our choices. We then deepened the concepts behind GAs and saw how genetic programming based on human evolution can be valuable in the optimization of processes.

Subsequently, we learned how to use GA techniques to implement SR. Symbolic equations represent an important resource for scientific research. The identification of a mathematical model that can represent a complete system is far from an easy task. The use of data can come to our aid; in fact, through SR, we can discover the equation underlying a set of input-output pairs. After analyzing the basics of the SR procedure through genetic programming, we saw a practical case of SR for the identification of a mathematical model, starting from the data.

Finally, we explored the CA model, which is a mathematical model used in the study of self-organizing systems in statistical mechanics. Given a random starting configuration, thanks to the irreversible character of the automaton, it is possible to obtain a great variety of self-organizing phenomena. CA are treated using a topological approach. After analyzing the basics of CA, we learned how to develop an elementary cellular automaton based on Wolfram's rules.

In the next chapter, we will learn how to use simulation models to handle financial problems. We will explore how the geometric Brownian motion model works, and we will discover how to use Monte Carlo methods for stock price prediction. Finally, we will understand how to model credit risks using Markov chains.

Part 3: Simulation Applications to Solve Real-World Problems

In this section, we will use the techniques that we introduced in the previous chapters to deal with practical cases. By the end of this section, you will be well-versed in the real-world applications of simulation models.

This part covers the following chapters:

- *Chapter 9, Using Simulation Models for Financial Engineering*
- *Chapter 10, Simulating Physical Phenomena Using Neural Networks*
- *Chapter 11, Modeling and Simulation for Project Management*
- *Chapter 12, Simulation Models for Fault Diagnosis in Dynamic Systems*
- *Chapter 13, What's Next?*

9
Using Simulation Models for Financial Engineering

The massive use of systems based on artificial intelligence and machine learning has opened up new scenarios for the financial sector. These methods can increase benefits, not only, for example, by protecting user rights but also in terms of macroeconomics.

Monte Carlo methods find a natural application in finance for the numerical resolution of pricing and option coverage problems. Essentially, these methods consist of simulating a given process or phenomenon using a given mathematical law and a sufficiently large set of data, created randomly from distributions that adequately represent real variables. The idea is that, if an analytical study is not possible, or adequate experimental sampling is not possible or convenient, the numerical simulation of the phenomenon is used. In this chapter, we will look at practical cases of using simulation methods in a financial context. You will learn how to use Monte Carlo methods to predict stock prices and how to assess the risk associated with a portfolio of shares.

In this chapter, we're going to cover the following topics:

- Understanding the geometric Brownian motion model
- Using Monte Carlo methods for stock price prediction
- Studying risk models for portfolio management

Technical requirements

In this chapter, we will learn how to use simulation models for financial engineering. In order to understand these topics, a basic knowledge of algebra and mathematical modeling is needed.

To work with the Python code in this chapter, you need the following files (available on GitHub at https://github.com/PacktPublishing/Hands-On-Simulation-Modeling-with-Python-Second-Edition):

- standard_brownian_motion.py

- `amazon_stock_montecarlo_simulation.py`
- `value_at_risk.py`

Understanding the geometric Brownian motion model

The name **Brownian** comes from the Scottish botanist Robert Brown who, in 1827, observed, under the microscope, how pollen particles suspended in water moved continuously in a random and unpredictable way. In 1905, it was Einstein who gave a molecular interpretation of the phenomenon of movement observed by Brown. He suggested that the motion of the particles was mathematically describable, assuming that the various jumps were due to the random collisions of pollen particles with water molecules.

Today, Brownian motion is, above all, a mathematical tool in the context of probability theory. This mathematical theory has been used to describe an ever-widening set of phenomena, studied by disciplines that are very different from physics. For instance, the prices of financial securities, the spread of heat, animal populations, bacteria, illness, sound, and light are modeled using the same theory.

> **Important note**
> **Brownian motion** is a phenomenon that consists of the uninterrupted and irregular movement made by small particles or grains of colloidal size – that is, particles that are far too small to be observed with the naked eye but are significantly larger than atoms when immersed in a fluid.

Defining a standard Brownian motion

There are various ways of constructing a Brownian motion model and various equivalent definitions of Brownian motion. Let's start with the definition of a standard Brownian motion (**the Wiener process**). The essential properties of the standard Brownian motion include the following:

1. The standard Brownian motion starts from zero.
2. The standard Brownian motion takes a continuous path.
3. The change in motion from the previous step (the increments suffer) by the Brownian process is independent.
4. The increases suffered by the Brownian process in the time interval, dt, indicate a Gaussian distribution, with an average that is equal to zero and a variance that is equal to the time interval, dt.

Based on these properties, we can consider the process as the sum of an unlimited large number of extremely small increments. After choosing two instants, t and s, the random variable, $Y(s) - Y(t)$, follows a normal distribution, with a mean of μ (s-t) and variance of σ^2 (s-t), which we can represent using the following equation:

$$Y(s) - Y(t) \sim \mathcal{N}(\mu(s-t), \sigma^2(s-t))$$

The hypothesis of normality is very important in the context of linear transformations. In fact, the standard Brownian motion takes its name from the standard normal distribution type, with parameters of $\mu = 0$ and $\sigma^2 = 1$.

Therefore, it can be said that the Brownian motion, $Y(t)$, with a unit mean and variance can be represented as a linear transformation of a standard Brownian motion, according to the following equation:

$$Y(t) = Y(0) + \mu * t + \sigma * Z(t)$$

In the previous equation, we can see that $Z(t)$ is the standard Brownian motion.

The weak point of this equation lies in the fact that the probability that $Y(t)$ assumes a negative value is positive; in fact, since $Z(t)$ is characterized by independent increments, which can assume a negative sign, the risk of the negativity of $Y(t)$ is not zero.

Now, let's consider the standard Brownian motion (the Wiener process) for sufficiently small time intervals. An infinitesimal increment of this process is obtained in the following form:

$$Z_{(t+dt)} - Z_{(t)} = \delta Z_t = N * \sqrt{dt}$$

Here, N is the number of samples. The previous equation can be rewritten as follows:

$$\frac{Z_{(t+dt)} - Z_{(t)}}{dt} = \frac{N}{\sqrt{dt}}$$

This process is not limited in variation and, therefore, cannot be differentiated in the context of classical analysis. In fact, the previous one tends to infinity or tends to zero of the interval, *dt*.

Addressing the Wiener process as random walk

A Wiener process can be considered a borderline case of random walk. We dealt with random walk in *Chapter 5, Simulation-Based Markov Decision Processes*. We have seen that the position of a particle at instant *n* will be represented by the following equation:

$$Y_n = Y_{n-1} + Z_n \; ; \quad n = 1, 2, \ldots$$

In the previous formula, we can observe the following:

- Y_n is the next value in the walk
- Y_{n-1} is the observation in the previous time phase
- Z_n is the random fluctuation in that step

If the n random numbers, Z_n, have a mean equal to zero and a variance equal to 1, then, for each value of n, we can define a stochastic process using the following equation:

$$Y_n(t) = \frac{1}{\sqrt{n}} * \sum_k Z_k$$

The preceding formula can be used in an iterative process. For very large values of n, we can write the following:

$$Y_n(s) - Y_n(t) \sim \mathcal{N}(0, (s-t))$$

The previous formula is due to the central limit theorem that we covered in *Chapter 4, Exploring Monte Carlo Simulations*.

Implementing a standard Brownian motion

So, let's demonstrate how to generate a simple Brownian motion in the Python environment. Let's start with the simplest case, in which we define the time interval, the number of steps to be performed, and the standard deviation:

1. We start by importing the following libraries:

   ```
   import numpy as np
   import matplotlib.pyplot as plt
   ```

   The numpy library is a Python library containing numerous functions that can help us in the management of multidimensional matrices. Additionally, it contains a large collection of high-level mathematical functions that we can use to operate on these matrices.

   The matplotlib library is a Python library used for printing high-quality graphics. With matplotlib, it is possible to generate graphs, histograms, bar graphs, power spectra, error graphs, scatter graphs, and more using just a few commands. This includes a collection of command-line functions similar to those provided by MATLAB software.

2. Now, let's proceed with some initial settings:

   ```
   np.random.seed(4)
   n = 1000
   sqn = 1/np.math.sqrt(n)
   z_values = np.random.randn(n)
   Yk = 0
   sb_motion=list()
   ```

In the first line of code, we used the random.seed() function to initialize the seed of the random number generator. This way, the simulation that uses a random number generator will be reproducible. The reproducibility of the experiment is possible because the randomly generated numbers are always the same. We set the number of iterations (*n*), and we calculated the first term of the following equation:

$$Y_n(t) = \frac{1}{\sqrt{n}} * \sum_k Z_k$$

Then, we generated the *n* random numbers using the random.randn() function. This function returns a standard normal distribution of *n* samples with a mean of 0 and a variance of 1. Finally, we set the first value of the Brownian motion as required from the properties, (Y(0)=0), and we initialized the list that will contain the Brownian motion positions.

3. At this point, we will use a for loop to calculate all of the *n* positions:

```
for k in range(n):
    Yk = Yk + sqn*z_values[k]
    sb_motion.append(Yk)
```

We simply added the current random number, multiplied by sqn, to the variable that contains the cumulative sum. The current value is then appended to the SBMotion list.

4. Finally, we draw a graph of the Brownian motion created:

```
plt.plot(sb_motion)
plt.show()
```

The following plot is printed:

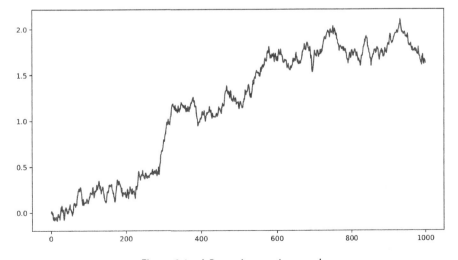

Figure 9.1 – A Brownian motion graph

So, we have created our first simulation of Brownian motion. Its use is particularly suitable for financial simulations. In the next section, we will demonstrate how this is done using a Monte Carlo simulation.

Using Monte Carlo methods for stock price prediction

As we explored in *Chapter 4*, *Exploring Monte Carlo Simulations*, Monte Carlo methods simulate different evolutions of the process under examination, using different probabilities that an event may occur under certain conditions. These simulations explore the entire parameter space of the phenomenon and return a representative sample. For each sample obtained, measures of the quantities of interest are carried out to evaluate their performance. A correct simulation means that the average value of the result of the process converges to the expected value.

Exploring the Amazon stock price trend

The stock market provides an opportunity to quickly earn large amounts of money – that is, in the eyes of an inexperienced user at least. Exchanges on the stock market make really large amounts, attracting in turn the attention of speculators from all over the world. In order to obtain revenues from investments in the stock market, it is necessary to have solid knowledge obtained from years of in-depth study of the phenomenon. In this context, the possibility of having a tool to predict stock market securities represents a popular need.

Let's demonstrate how to develop a simulation model of the stock of one of the most famous companies in the world. Amazon was founded by Jeff Bezos in the 1990s, and it was one of the first companies in the world to sell products via the internet. Amazon stock has been listed on the stock exchange since 1997 under the symbol AMZN. The historical values of AMZN stock can be obtained from various internet sites that have been dealing with the stock market over the past 10 years. We will refer to the performance of AMZN stock on the NASDAQ GS stock quote from 2012-10-03 to 2022-09-30. To select this date range, simply click on the **Time Period** item and select it.

Data can be downloaded in the `.csv` format from the Yahoo Finance website at `https://finance.yahoo.com/quote/AMZN/history/`.

In the following screenshot, you can see the Yahoo Finance section for AMZN stock with a highlighted button to download the data:

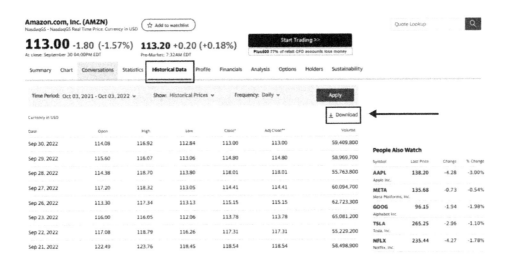

Figure 9.2 – Amazon data on Yahoo Finance

The downloaded `AMZN.csv` file contains a lot of features, but we will only use two of them, as follows:

- **Date**: The date of the quote
- **Close**: The close price

We will analyze the code, line by line, to fully understand the whole process that will lead us to simulate a series of predictions of Amazon stock price performance:

1. As always, we start by importing the libraries:

    ```
    import numpy as np
    import pandas as pd
    import matplotlib.pyplot as plt
    from scipy.stats import norm
    from pandas.plotting import
             register_matplotlib_converters
    ```

    The following libraries were imported:

    - The `numpy` library is a Python library that contains numerous functions to help us in the management of multidimensional matrices. Furthermore, it contains a large collection of high-level mathematical functions to operate on these matrices.
    - The `pandas` library is an open source BSD-licensed library that contains data structures and operations to manipulate high-performance numeric values for the Python programming language.

- The `matplotlib` library is a Python library used for printing high-quality graphics. With `matplotlib`, it is possible to generate graphs, histograms, bar graphs, power spectra, error graphs, scatter graphs, and more with just a few commands. This includes a collection of command-line functions such as those provided by MATLAB software.
- SciPy is a collection of mathematical algorithms and functions based on NumPy. It has a series of commands and high-level classes to manipulate and display data. With SciPy, functionality is added to Python, making it a data processing and system prototyping environment similar to commercial systems such as MATLAB.

The `pandas.plotting.register_matplotlib_converters()` function makes pandas formatters and converters compatible in `matplotlib`. Let's use it now:

```
register_matplotlib_converters()
```

2. Now, let's import the data contained in the `AMZN.csv` file:

```
AmznData = pd.read_csv('AMZN.csv',header=0,
          usecols = ['Date',Close'],parse_dates=True,
          index_col='Date')
```

We used the `read_csv` module of the `pandas` library, which loads the data in a `pandas` object called `DataFrame`. The following arguments are passed:

- `'AMZN.csv'`: The name of the file.
- `header=0`: The row number containing the column names and the start of the data. By default, if a non-header row is passed (header=0), the column names are inferred from the first line of the file.
- `usecols=['Date','Close']`: This argument extracts a subset of the dataset by specifying the column names.
- `parse_dates=True`: A Boolean value; if `True`, try parsing the index.
- `index_col='Date'`: This allows us to specify the name of the column that will be used as the index of the DataFrame.

3. Now, we will explore the imported dataset to extract preliminary information. To do this, we will use the `info()` function, as follows:

```
print(AmznData.info())
```

The following information is printed:

```
<class 'pandas.core.frame.DataFrame'>
DatetimeIndex: 2515 entries, 2012-10-03 to 2022-09-30
Data columns (total 1 columns):
```

```
 #   Column  Non-Null Count  Dtype
---  ------  --------------  -----
 0   Close   2515 non-null   float64
dtypes: float64(1)
memory usage: 39.3 KB
None
```

Here, a lot of useful information is returned: the object class, the number of records present (2,515), the start and end values of the index (2012-10-03 to 2022-09-30), the number of columns and the type of data they contain, and other information.

We can also print the first five lines of the dataset, as follows:

```
print(AmznData.head())
```

The following data is printed:

```
                Close
                Close
Date
2012-10-03    12.7960
2012-10-04    13.0235
2012-10-05    12.9255
2012-10-08    12.9530
2012-10-09    12.5480
```

If we wanted to print a different number of records, it would be enough to specify it by indicating the number of lines to be printed. Similarly, we can print the last 10 records of the dataset:

```
print(AmznData.tail())
```

The following records are printed:

```
                 Close
Date
2022-09-26    115.150002
2022-09-27    114.410004
2022-09-28    118.010002
2022-09-29    114.800003
2022-09-30    113.000000
```

An initial quick comparison between the head and the tail allows us to verify that Amazon stock in the last 10 years has gone from a value of about $12.79 to about $113.00. This is an excellent deal for Amazon shareholders.

Using the `describe()` function, we will extract a preview of the data using basic statistics:

```
print(AmznData.describe())
```

The following results are returned:

```
              Close
count   2515.000000
mean      71.683993
std       53.944668
min       11.030000
25%       19.921750
50%       50.187000
75%      104.449249
max      186.570496
```

We can confirm the significant increase in value in the last 10 years, but we can also see how the stock has undergone significant fluctuations, given the very high value of the standard deviation. This tells us that the shareholders who were loyal to the share and maintained it over time benefited the most from its increase.

4. After analyzing the preliminary data statistics, we can take a look at the performance of Amazon's share in the last 10 years by drawing a simple graph:

```
plt.figure(figsize=(10,5))
plt.plot(AmznData)
plt.show()
```

The following `matplotlib` functions were used:

- `figure()`: This function creates a new figure, which is empty for now. We set the size of the frame using the `figsize` parameter, which sets the width and height in inches.
- `plot()`: This function plots the `AmznData` dataset.
- `show()`: This function, when running in iPython in PyLab mode, displays all the figures and returns to the iPython prompt.

The following diagram is printed:

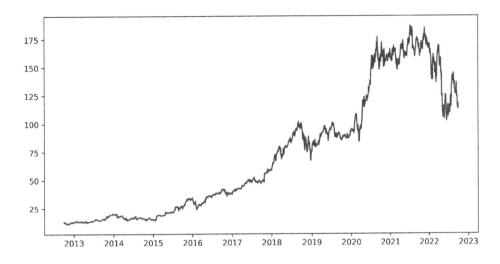

Figure 9.3 – An Amazon share graph

Now it is much clearer, and the significant increase undergone by Amazon stock over the past 10 years is evident. Furthermore, it should be noted that the greatest increase has been recorded since 2015, but we will try to extract more information from the data. We can also see that in the period of the COVID-19 pandemic, there was a further and considerable increase in prices, essentially due to the fact that Amazon was a store accessible to everyone, even during the lockdown period. A significant decrease can also be noted at the beginning of 2022, due to the problems of the war in Ukraine and the increase in energy costs.

Handling the stock price trend as a time series

The trend over time of the Amazon stock price, represented in the previous diagram, is configured as a sequence of ordered data. This type of data can be conveniently handled as a time series. Let's consider a simple definition: a time series contains a chronological sequence of experimental observations of a variable. This variable can relate to data of different origins. Very often, it concerns financial data such as unemployment rates, spreads, stock market indices, and stock price trends.

The usefulness of dealing with the problem as a time series will allow us to extract useful information from the data in order to develop predictive models for the management of future scenarios. It may be useful to compare the trend of stock prices in the same periods for different years or, more simply, between contiguous periods.

Let's use $Y_1, ..., Y_t, ...,$ and Y_n as the elements of a time series. Let's start by comparing the data for two different times, indicated with t and t + 1. It is, therefore, two contiguous periods. We are interested in evaluating the variation undergone by the phenomenon under observation, which can be defined by the following ratio:

$$\frac{Y_{t+1} - Y_t}{Y_t} * 100$$

This percentage ratio is called a percentage change. It can be defined as the percentage change rate of Y of $t + 1$ time compared to the previous time, t. This descriptor returns information about how the data underwent a change over a period. The percentage change allows you to monitor both the stock prices and the market indices, not just compare currencies from different countries:

1. To evaluate this useful descriptor, we will use the pct_change() function contained in the pandas library:

```
AmznDataPctChange = AmznData.pct_change()
```

This function returns the percentage change between the current element and a previous element. By default, the function calculates the percentage change from the immediately preceding row.

The concept of the percentage variation of a time series is linked to the concept of the return of a stock price. The returns-based approach allows for the normalization of data, which is an operation of fundamental importance when evaluating the relationships between variables characterized by different metrics.

We will deal with the return on a logarithmic scale, as this choice will give us several advantages: normally distributed results, values returned (the logarithm of the return) very close to the initial return value (at least for very small values), and additive results over time.

2. To pass the return on a logarithmic scale, we will use the log() function of the numpy library, as follows:

```
AmznLogReturns = np.log(1 + AmznDataPctChange)
print(AmznLogReturns.tail(10))
```

The following results are printed:

```
                Close
Date
2022-09-19   0.009106
2022-09-20  -0.020013
2022-09-21  -0.030327
2022-09-22  -0.010430
2022-09-23  -0.030553
2022-09-26   0.011969
2022-09-27  -0.006447
2022-09-28   0.030981
2022-09-29  -0.027578
2022-09-30  -0.015804
```

3. To better understand how the return is distributed over time, let's draw a diagram:

```
plt.figure(figsize=(10,5))
plt.plot(AmznLogReturns)
plt.show()
```

The following diagram is printed:

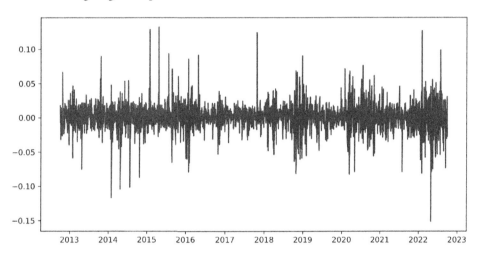

Figure 9.4 – The logarithmic values of the returns

The previous diagram shows us that the logarithmic return is normally distributed over the entire period and the mean is stable.

Introducing the Black-Scholes model

The **Black-Scholes** (**BS**) model certainly represents the most important and revolutionary work in the history of quantitative finance. In traditional financial literature, it is assumed that almost all financial asset prices (stocks, currencies, and interest rates) are driven by a Brownian drift motion.

This model assumes that the expected return of an asset is equal to the non-risky interest rate, r. This approach is capable of simulating returns on a logarithmic scale of an asset. Suppose we observe an asset in the instants: t (0), t (1) ,. . ., and t (n). We denote, using $s(i) = S(t_i)$, the value of an asset at $t(i)$. Based on these hypotheses, we can calculate the return using the following equation:

$$y(i) = \frac{[s(i) - s(i-1)]}{s(i-1)}, i = 1, 2, \ldots, n$$

Then, we will transform the return on the logarithmic scale, as follows:

$$x(i) = \ln s(i) - \ln s(i-1), i = 1, 2, \ldots, n$$

By applying the BS approach to Brownian geometric motion, the stock price will satisfy the following stochastic differential equation:

$$dS(t) = \mu * S(t) * dt + \sigma * S(t) * dB(t)$$

In the previous equation, dB(t) is a standard Brownian motion and μ and σ are real constants. The previous equation is valid in the hypothesis that s (i) - s (i - 1) is small, and this happens when the stock prices undergo slight variations. This is because ln (1 + z) is roughly equal to z if z is small. The analytical solution of the previous equation is the following equation:

$$S(t) = S(0) * e^{(\alpha(t) + \sigma * B(t))}$$

By passing the previous equation on a logarithmic scale, we obtain the following equation:

$$\ln \frac{S(t)}{S(0)} = \alpha(t) + \sigma * B(t)$$

In the previous equation, we can observe the following:

- α is the drift
- B(t) is a standard Brownian motion
- σ is the standard deviation

We introduced the concept of drift, which represents the trend of a long-term asset in the stock market. To understand drift, we will use an analogy of river currents. If we pour liquid color into a river, it will spread by following the direction imposed by the river current. Similarly, drift represents the tendency of a stock to follow the trend of a long-term asset.

Applying the Monte Carlo simulation

Using the BS model discussed in the previous section, we can evaluate the daily price of an asset starting from that of the previous day, multiplied by an exponential contribution based on a coefficient, r. This coefficient is a periodic rate of return. It translates into the following equation:

$$StockPrice(t) = StockPrice(t-1) * e^r$$

The second term in the previous equation, e^r, is called the **daily return**, and according to the BS model, it is given by the following formula:

$$e^{(\alpha(t) + \sigma * B(t))}$$

There is no way to predict the rate of return of an asset. The only way to represent it is to consider it as a random number. So, to predict the price trend of an asset, we can use a model based on random movement such as that represented by BS equations.

The BS model assumes that changes in the stock price depend on the expected return over time. The daily return has two terms: the fixed drift rate and the random stochastic variable. The two terms provide for the certainty of movement and uncertainty caused by volatility.

To calculate the drift, we will use the expected rate of return, which is the most likely rate to occur, using the historical average of the log returns and variance, as follows:

$$drift = mean(\log(returns)) - 0.5 * variance(\log(returns))$$

According to the previous equation, the daily change rate of the asset is the mean of the returns, which are less than half of the variance over time. Let's continue our work, calculating the drift for the return of the Amazon security calculated in the *Handling the stock price trend as time series* section:

1. To evaluate the drift, we need the mean and variance of the returns. Since we are also calculating the standard deviation, we will need the calculation of the daily return:

    ```
    MeanLogReturns = np.array(AmznLogReturns.mean())
    VarLogReturns = np.array(AmznLogReturns.var())
    StdevLogReturns = np.array(AmznLogReturns.std())
    ```

    Three numpy functions were used:

    - `mean()`: This computes the arithmetic mean along the specified axis and returns the average of the array elements.
    - `var()`: This computes the variance along the specified axis. It returns the variance of the array elements, which is a measure of the spread of a distribution.
    - `std()`: This computes the standard deviation along the specified axis.

    Now, we can calculate the drift as follows:

    ```
    Drift = MeanLogReturns - (0.5 * VarLogReturns)
    print("Drift = ",Drift)
    ```

    The following result is returned:

    ```
    Drift =    [0.0006643]
    ```

    This is the fixed part of the Brownian motion. The drift returns the annualized change in the expected value and compensates for the asymmetry in the results compared to the straight Brownian motion.

2. To evaluate the second component of the Brownian motion, we will use the random stochastic variable. This corresponds to the distance between the mean and the events, expressed as the

number of standard deviations. Before doing this, we need to set the number of intervals and iterations. The number of intervals will be equal to the number of observations, which is 2,515, while the number of iterations that represents the number of simulation models that we intend to develop is 20:

```
NumIntervals = 2515
Iterations = 20
```

3. Before generating random values, it is recommended that you set the seed to make the experiment reproducible:

```
np.random.seed(7)
```

Now, we can generate the random distribution:

```
SBMotion = norm.ppf(np.random.rand(NumIntervals, Iterations))
```

A 2515 x 20 matrix is returned, containing the random contribution for the 20 simulations that we want to perform and for the 2,515 time intervals that we want to consider. Recall that these intervals correspond to the daily prices of the last 10 years.

Two functions were used:

- `norm.ppf()`: This SciPy function gives the value of the variate for which the cumulative probability has the given value.
- `np.random.rand()`: This NumPy function computes random values in a given shape. It creates an array of the given shape and populates it with random samples from a uniform distribution over [0, 1].

We will calculate the daily return as follows:

```
DailyReturns = np.exp(Drift + StdevLogReturns * SBMotion)
```

The daily return is a measure of the change that occurred in a stock's price. It is expressed as a percentage of the previous day's closing price. A positive return means the stock has grown in value, while a negative return means it has lost value. The `np.exp()` function was used to calculate the exponential value of all elements in the input array.

4. After a long preparation, we have arrived at a crucial moment. We will be able to carry out predictions based on the Monte Carlo method. The first thing to do is to recover the starting point of our simulation. Since we want to predict the trend of Amazon stock prices, we recover the first value present in the AMZN.csv file:

```
StartStockPrices = AmznData.iloc[0]
```

The pandas `iloc()` function is used to return a pure integer using location-based indexing for selection. Then, we will initialize the array that will contain the predictions:

```
StockPrice = np.zeros_like(DailyReturns)
```

The NumPy `zeros_like()` function is used to return an array of zeros with the same shape and type as a given array. Now, we will set the starting value of the `StockPrice` array, as follows:

```
StockPrice[0] = StartStockPrices
```

5. To update the predictions of Amazon stock prices, we will use a `for` loop that iterates for a number that is equal to the time intervals we are considering:

```
for t in range(1, NumIntervals):
    StockPrice[t] = StockPrice[t - 1] * DailyReturns[t]
```

For the update, we will use the BS model according to the following equation:

$$StockPrice(t) = StockPrice(t-1) * e^r$$
$$= StockPrice(t-1) * e^{(\alpha(t) + \sigma * B(t))}$$

Finally, we can view the results:

```
plt.figure(figsize=(10,5))
plt.plot(StockPrice)
AMZNTrend = np.array(AmznData.iloc[:, 0:1])
plt.plot(AMZNTrend,'k*')
plt.show()
```

The following diagram is printed:

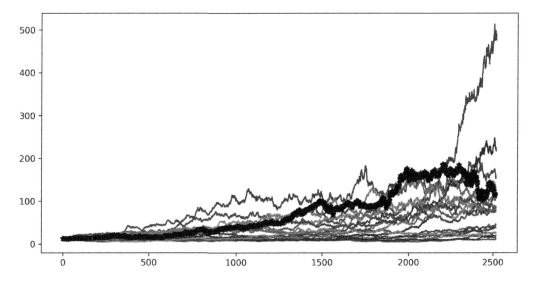

Figure 9.5 – An Amazon trend graph

In the previous diagram, the curve highlighted in black represents the trend of Amazon stock prices in the last 10 years. The other curves are our simulations. We can see that some curves move away from the expected curve, while others appear much closer to the actual trend.

Monte Carlo methods concern the use of random sampling techniques and computer simulation to obtain an approximate solution to a mathematical or financial problem. In other words, we can consider them a family of simulation methods through which it is possible to reproduce and study empirical systems in a controlled form. In practice, these are processes in which data is generated based on a model specified a priori. In this case, we generated data based on the performance of Amazon shares, starting from the prices in the previous period, and then we generated data starting from other data. These methods have now become an integral part of the standard techniques used in statistical methods.

After having seen how it is possible to predict the performance of stock, let's move on to study how to develop a risk model to manage financial portfolios.

Studying risk models for portfolio management

Having a good risk measure is of fundamental importance in finance, as it is one of the main tools for evaluating financial assets. This is because it allows you to monitor securities and provides a criterion for the construction of portfolios. One measure that has been widely used over the years more than any other is **variance**.

Using variance as a risk measure

The advantage of a diversified portfolio in terms of risk and expected value is that it allows us to find the right allocation for securities. Our aim is to obtain the highest expected value at the same risk or to minimize the risk of obtaining the same expected value. To achieve this, it is necessary to trace the concept of risk back to a measurable quantity, which is generally referred to as the variance. Therefore, by maximizing the expected value of the portfolio returns for each level of variance, it is possible to reconstruct a curve called the efficient frontier, which determines the maximum expected value that can be obtained with the securities available for the construction of the portfolio for each level of risk.

The minimum variance portfolio represents the portfolio with the lowest possible variance value, regardless of the expected value. This parameter has the purpose of optimizing the risk represented by the variance of the portfolio. Tracing the risk exclusively to the measure of variance is optimal only if the distribution of returns is normal. In fact, the normal distribution enjoys some properties that make the variance a measure that is enough to represent the risk. It is completely determinable through only two parameters (mean and variance). It is, therefore, enough to know the mean and the variance to determine any other point of the distribution.

Introducing the Value-at-Risk metric

Consider the variance as the only risk measure in the case of non-normal and limiting values. A risk measure that has been widely used for over two decades is **Value at Risk** (**VaR**). The birth of VaR is linked to the growing need for financial institutions to manage risk and, therefore, to be able to measure it. This is due to the increasingly complex structure of financial markets.

Actually, this measure was not introduced to stem the limits of variance as a risk measure, since an approach to calculate the VaR value starts precisely from the assumptions of normality. However, to make it easier to understand, let's enclose the overall risk of security into a single number or a portfolio of financial assets by adopting a single metric for different types of risk.

In a financial context, VaR is an estimate, given a confidence interval, of how high the losses of a security or portfolio may be in each time horizon. VaR, therefore, focuses on the left tail of the distribution of returns, where events with a low probability of realization are located. Indicating the losses and not the dispersion of the returns around their expected value makes it a measure closer to the common idea of risk than of variance.

> **Important note**
> J.P. Morgan is credited as the bank that made VaR a widespread measure. In 1990, the president of J.P. Morgan, Dennis Weatherstone, was dissatisfied with the lengthy risk analysis reports he received every day. He wanted a simple report that summarized the bank's total exposure across its entire trading portfolio.

After calculating VaR, we can say that, with a probability given by the confidence interval, we will not lose more than the VaR of the portfolio in the next N days. VaR is the level of loss that will not be exceeded with a probability given by the confidence interval.

For example, a VaR of €1 million over a year with a 95% confidence level means that the maximum loss for the portfolio for the next year will be €1 million in 95% of cases. Nothing tells us what happens to the remaining 5% of cases.

The following diagram shows the probability distribution of portfolio returns with the indication of the value of VaR:

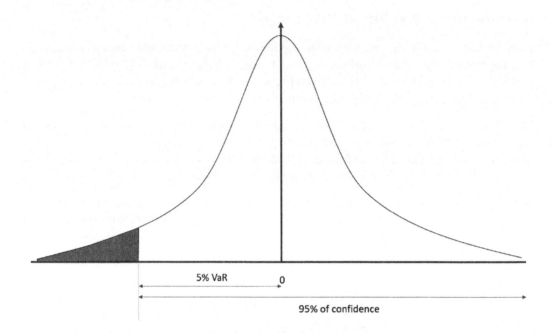

Figure 9.6 – The probability distribution of the portfolio returns

VaR is a function of the following two parameters:

- Time horizon
- Level of confidence

Some characteristics of the VaR must be specified:

- VaR does not describe the worst loss
- VaR says nothing about the distribution of losses in the left tail
- VaR is subject to sampling errors

> **Important note**
> The sampling error tells us how much the sampled value deviates from the real population value. This deviation is because the sample is not representative of the population or has distortions.

VaR is a widely used risk measure that summarizes, in a single number, important aspects of the risk of a portfolio of financial instruments. It has the same unit of measurement as the returns of the portfolio on which it is calculated, and it is easy to understand, answering the simple question, "How bad can financial investments go?"

Let's now examine a practical case of calculating VaR.

Estimating VaR for some NASDAQ assets

NASDAQ is one of the most famous stock market indices in the world. Its name is an acronym for the **National Association of Securities Dealers Quotation**. This is the index that represents the stocks of the technology sector in the US. Thinking of NASDAQ in the investor's mind, the brands of the main technological and social houses of the US can easily emerge. Just think of companies such as Google, Amazon, and Facebook; they are all covered by the NASDAQ listing.

Here, we will learn how to recover the data of the quotes of six companies listed by NASDAQ, and then we will demonstrate how to estimate the risk associated with the purchase of a portfolio of shares of these securities:

1. As always, we start by importing the libraries:

   ```
   import datetime as dt
   import numpy as np
   import pandas_datareader.data as wb
   import matplotlib.pyplot as plt
   from scipy.stats import norm
   ```

   The following libraries were imported:

   - The `datetime` library contains classes for manipulating dates and times. The functions it contains allow us to easily extract attributes to format and manipulate dates.

   - `numpy` is a Python library that contains numerous functions that help us in the management of multidimensional matrices. Additionally, it contains a large collection of high-level mathematical functions that we can use to operate on these matrices.

   - The `pandas_datareader.data` module of the `pandas` library contains functions that allow us to extract financial information, not just from a series of websites that provide these types of data. The collected data is returned in the pandas DataFrame format. The `pandas` library is an open source BSD-licensed library that contains data structures and operations to manipulate high-performance numeric values for the Python programming language.

   - `matplotlib` is a Python library used for printing high-quality graphics. With `matplotlib`, it is possible to generate graphs, histograms, bar graphs, power spectra, error graphs, scatter graphs, and more with just a few commands. This is a collection of command-line functions similar to those provided by MATLAB software.

   - SciPy is a collection of mathematical algorithms and functions based on NumPy. It has a series of commands and high-level classes to manipulate and display data. With SciPy, functionality is added to Python, making it a data processing and system prototyping environment like commercial systems such as MATLAB.

2. We will set the stocks we want to analyze by defining them with tickers. We also decide on the time horizon:

```
StockList = ['ADBE','CSCO','IBM','NVDA','MSFT','HPQ']
StartDay = dt.datetime(2021, 1, 1)
EndDay = dt.datetime(2021, 12, 31)
```

Six tickers have been included in a DataFrame. A ticker is an abbreviation used to uniquely identify the shares listed on the stock exchange of a particular security on a specific stock market. It is made up of letters, numbers, or a combination of both. The tickers are used to refer to six leading companies in the global technology sector:

- ADBE: **Adobe Systems Inc** – One of the largest and most differentiated software companies in the world.
- CSCO: **Cisco Systems Inc** – The production of **Internet Protocol** (IP)-based networking and other products for communications and information technology.
- IBM: **International Business Machines** – The production and consultancy of information technology-related products.
- NVDA: **NVIDIA Corp** – Visual computing technologies. This is the company that invented the GPU.
- MSFT: **Microsoft Corp** – This is one of the most important companies in the sector, as well as one of the largest software producers in the world by turnover.
- HPQ: **HP Inc** – The leading global provider of products, technologies, software, solutions, and services to individual consumers and large enterprises.

After deciding on the tickers, we set the time horizon of our analysis. We simply set the start date and end date of our analysis by defining the whole year of 2019.

3. Now, we can recover the data:

```
StockData =  wb.DataReader(StockList, 'yahoo',
                        StartDay,EndDay)
StockClose = StockData["Adj Close"]
print(StockClose.describe())
```

To retrieve the data, we used the `DataReader()` function of the `pandas_datareader.data` module. This function extracts data from various internet sources into a pandas DataFrame. The following topics have been passed:

- `StockList`: The list of stocks to be recovered
- `'yahoo'`: The website from which to collect data
- `StartDay`: The start date of monitoring

- EndDay: The end date of monitoring

The recovered data is entered in a pandas DataFrame that will contain 36 columns, corresponding to 6 pieces of information for each of the 6 stocks. Each record will contain the following information for each day: the high value, the low value, the open value, the close value, the volume, and the adjusted close.

For the risk assessment of a portfolio, only one value will suffice: the adjusted close. This column was extracted from the starting DataFrame and stored in the `StockData` variable. We then developed basic statistics for each stock using the `describe()` function. The following statistics have been returned:

Symbols	ADBE	CSCO	IBM	NVDA	MSFT	HPQ
count	252.000000	252.000000	252.000000	252.000000	252.000000	252.000000
mean	560.613652	51.630536	121.504238	195.025938	273.160699	29.193442
std	76.180505	4.661304	8.332261	58.695392	37.209664	3.222804
min	421.200012	42.131500	104.060684	115.752731	209.119324	22.972631
25%	488.294998	49.579469	114.431425	142.970486	240.849697	27.233361
50%	569.324982	52.063898	124.711437	191.879829	274.337830	28.697907
75%	632.419998	54.721060	128.620552	222.241852	299.335548	30.853754
max	688.369995	62.589500	136.033936	333.492676	340.882812	37.263916

Figure 9.7 – The statistics of the portfolios

Analyzing the previous table, note that there are 252 records. These are the days when the stock exchange was opened in 2021. Let's take note of it, as this data will be useful later on. We also note that the values in the columns have very different ranges due to the different values of the stocks. To better understand the trend of stocks, it is better to draw graphs. Let's do this next:

```
fig, axs = plt.subplots(3, 2, figsize=(20,10))
axs[0, 0].plot(StockClose['ADBE'])
axs[0, 0].set_title('ADBE')
axs[0, 1].plot(StockClose['CSCO'])
axs[0, 1].set_title('CSCO')
axs[1, 0].plot(StockClose['IBM'])
axs[1, 0].set_title('IBM')
axs[1, 1].plot(StockClose['NVDA'])
axs[1, 1].set_title('NVDA')
axs[2, 0].plot(StockClose['MSFT'])
axs[2, 0].set_title('MSFT')
axs[2, 1].plot(StockClose['HPQ'])
axs[2, 1].set_title('HPQ')
```

In order to make an easy comparison between the trends of the six stocks, we have traced six subplots that are ordered in three rows and two columns. We used the `subplots()` function of the `matplotlib` library. This function returns a tuple containing a figure object and axes. So, when you use `fig` and `axs = plt.subplots()`, you unpack this tuple into the variables of `fig` and `axs`. Having `fig` is useful if you want to change the attributes at the figure level or save the figure as an image file later. The variable, `axs`, allows us to set the attributes of the axes of each subplot. In fact, we called this variable to define what to draw in each subplot by calling it with the row-column indices of the chart matrix. In addition, for each chart, we also printed the title that allows us to understand which ticker it refers to.

The following plots are shown:

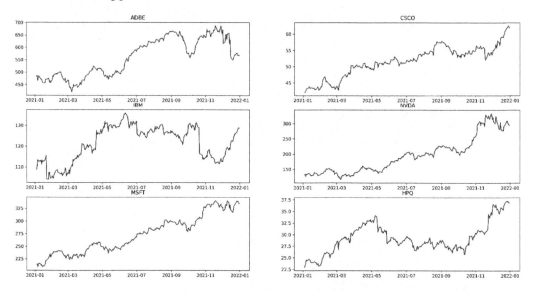

Figure 9.8 – The graphs of the statistics

4. After taking a quick look at the trend of stocks, the time has come to evaluate the returns:

```
StockReturns = StockClose.pct_change()
print(StockReturns.tail())
```

The `pct.change()` function returns the percentage change between the current close price and the previous value. By default, the function calculates the percentage change from the immediately preceding row.

The concept of the percentage variation of a time series is linked to the concept of the return of a stock price. The returns-based approach provides for the normalization of data, which is an operation of fundamental importance to evaluate the relationships between variables characterized by different metrics. These concepts have been explored in this chapter's *Using Monte Carlo methods for stock price prediction* section. Note that we have only referred to some of them.

We then printed the queue of the returned DataFrame to analyze its contents. The following results are returned:

```
Symbols        ADBE      CSCO       IBM      NVDA      MSFT       HPQ
Date
2021-12-10  0.034589  0.029540  0.004208 -0.009577  0.028340  0.007745
2021-12-13  0.005883 -0.010802 -0.012169 -0.067455 -0.009167 -0.031567
2021-12-14 -0.065988 -0.014332  0.009626  0.006250 -0.032587  0.009070
2021-12-15  0.025160  0.037390 -0.005252  0.074884  0.019218  0.018539
2021-12-16 -0.101915  0.006341  0.022906 -0.068026 -0.029135  0.012135
2021-12-17 -0.016693  0.002487  0.011673 -0.020643 -0.003386 -0.003542
2021-12-20 -0.012342 -0.001654 -0.002669 -0.002950 -0.012013 -0.012032
2021-12-21  0.014097  0.011100  0.015032  0.048919  0.023069  0.024633
2021-12-22  0.011587  0.008193  0.006048  0.011178  0.018057  0.007023
2021-12-23  0.010000  0.012189  0.006782  0.008163  0.004472  0.009925
2021-12-27  0.014150  0.018304  0.007579  0.044028  0.023186  0.011952
2021-12-28 -0.014402  0.001734  0.007674 -0.020133 -0.003504 -0.003937
2021-12-29 -0.000123  0.006768  0.005429 -0.010586  0.002051  0.000790
2021-12-30  0.002178 -0.005316  0.004199 -0.013833 -0.007691 -0.006056
2021-12-31 -0.006082 -0.003930 -0.001867 -0.005915 -0.008841 -0.002119
```

Figure 9.9 – The stock returns DataFrame

In the previous table, the minus sign indicates a negative return with a loss, at least daily.

5. Now, we are ready to assess the investment risk of a substantial portfolio of stocks of these prestigious companies. To do this, we need to set some variables and calculate others:

```
PortfolioValue = 1000000000.00
ConfidenceValue = 0.95
Mu = np.mean(StockReturns)
Sigma = np.std(StockReturns)
```

To start, we set the value of our portfolio; it is a billion dollars. These figures should not frighten you. For a bank that manages numerous investors, achieving this investment value is not difficult. So, we set the confidence interval. Previously, we said that VaR is based on this value. Subsequently, we started to calculate some fundamental quantities for the VaR calculation. I am referring to the mean and standard deviation of returns. To do this, we used the related numpy functions: `np.mean()` and `np.std`.

We continue to set the parameters necessary for calculating VaR:

```
WorkingDays2021 = 252.
AnnualizedMeanStockRet = MeanStockRet/WorkingDays2021
AnnualizedStdStockRet =
          StdStockRet/np.sqrt(WorkingDays2021)
```

Previously, we saw that the data extracted from the finance section of the Yahoo website contained 252 records. This is the number of working days of the stock exchange in 2021, so we set this value. So, let's move on to annualizing the mean and the standard deviation just calculated. This is because we want to calculate the annual risk index of the stocks. For the annualization of the average, it is enough to divide by the number of working days, while for the standard deviation, we must divide by the square root of the number of working days.

6. Now, we have all the data we need to calculate VaR:

```
INPD = norm.ppf(1-ConfidenceValue,AnnualizedMeanStockRet,
                AnnualizedStdStockRet)
VaR = PortfolioValue*INPD
```

To start, we calculate the inverse normal probability distribution with a risk level of 1 confidence, mean, and standard deviation. This technique involves the construction of a probability distribution, starting from the three parameters we have mentioned. In this case, we work backward, starting from some distribution statistics, and try to reconstruct the starting distribution. To do this, we used the `norm.ppf()` function of the SciPy library.

The `norm()` function returns a normal continuous random variable. The acronym, **ppf**, stands for **percentage point function**, which is another name for the quantile function. The quantile function, associated with a probability distribution of a random variable, specifies the value of the random variable so that the probability that the variable is less than or equal to that value is equal to the given probability.

At this point, VaR is calculated by multiplying the inverse normal probability distribution obtained by the value of the portfolio. To make the value obtained more readable, it was rounded to the first two decimal places:

```
RoundVaR=np.round_(VaR,2)
```

Finally, the results obtained were printed, one for each row, to make the comparison simple:

```
for i in range(len(StockList)):
    print("Value-at-Risk for", StockList[i],
          "is equal to ",RoundVaR[i])
```

The following results are returned:

```
Value-at-Risk for ADBE is equal to  -1901394.1
Value-at-Risk for CSCO is equal to  -1244064.53
Value-at-Risk for IBM is equal to  -1485931.31
Value-at-Risk for NVDA is equal to  -2922034.18
Value-at-Risk for MSFT is equal to  -1358218.7
Value-at-Risk for HPQ is equal to  -2020466.29
```

The stocks that returned the highest risk were NVDA and HPQ, while the one that returned the lowest risk was the CSCO stock.

Summary

In this chapter, we applied the concepts of simulation based on Monte Carlo methods and, more generally, on the generation of random numbers to real cases related to the world of financial engineering. We started by defining the model based on Brownian motion, which describes the uninterrupted and irregular movement of small particles when immersed in a fluid. We learned how to describe the mathematical model, and then we derived a practical application that simulates a random walk as a Wiener process.

Afterward, we dealt with another practical case of considerable interest – that is, how to use Monte Carlo methods to predict the stock prices of the famous company Amazon. We started to explore the trend of Amazon shares in the last 10 years, and we performed simple statistics to extract preliminary information on any trends that we confirmed through visual analysis. Subsequently, we learned to treat the trend of stock prices as a time series, calculating the daily return. We then addressed the problem with the BS model, defining the concepts of drift and standard Brownian motion. Finally, we applied the Monte Carlo method to predict possible scenarios relating to the trend of stock prices.

As a final practical application, we assessed the risk associated with a portfolio of shares of some of the most famous technology companies listed on the NASDAQ market. We first defined the concept of referrals connected to a financial asset, and then we introduced the concept of VaR. Subsequently, we implemented an algorithm that, given a confidence interval and a time horizon, calculates VaR based on the daily returns returned by the historical data of the stock prices.

In the next chapter, we will learn about the basic concepts of artificial neural networks, how to apply feedforward neural network methods to our data, and how the neural network algorithm works. Then, we will take a look at the basic concepts of a deep neural network and how to use neural networks to simulate a physical phenomenon.

10
Simulating Physical Phenomena Using Neural Networks

Neural networks are exceptionally effective at getting good characteristics for highly structured data. Physical phenomena are conditioned by numerous variables that can be easily measured through modern sensors. In this way, big data is produced that is difficult to deal with using classic techniques. Neural networks lend themselves to simulating complex environments.

In this chapter, we will learn how to develop models based on **artificial neural networks** (**ANNs**) to simulate physical phenomena. We will start by exploring the basic concepts of neural networks, and then we will examine their architecture and main elements. We will demonstrate how to train a network to update its weights. Then, we will apply these concepts to a practical use case to solve a regression problem. In the last part of the chapter, we will analyze deep neural networks.

In this chapter, we're going to cover the following topics:

- Introducing the basics of neural networks
- Understanding feedforward neural networks
- Simulating airfoil self-noise using ANNs
- Approaching deep neural networks
- Exploring **graph neural networks** (**GNNs**)
- Simulation modeling using neural network techniques

Technical requirements

In this chapter, we will learn how to use ANNs to simulate complex environments. To understand the topics, a basic knowledge of algebra and mathematical modeling is needed.

To work with the Python code in this chapter, you need the following files (available on GitHub at https://github.com/PacktPublishing/Hands-On-Simulation-Modeling-with-Python-Second-Edition):

- `airfoil_self_noise.py`
- `concrete_quality.py`

Introducing the basics of neural networks

ANNs are numerical models developed with the aim of reproducing simple neural activities of the human brain, such as object identification and voice recognition. The structure of an ANN is composed of nodes that, similar to the neurons present in a human brain, are interconnected with each other through weighted connections, which reproduce the synapses between neurons.

The system output is updated until it iteratively converges via the connection weights. The information deriving from experimental activities is used as input data and the result processed by the network is returned as an output. The input nodes represent the predictive variables, and the output neurons are represented by the dependent variables. We use the predictive variables to process the dependent variables.

ANNs are very versatile in simulating regression and classification problems. They can learn the process of working out the solution to a problem by analyzing a series of examples. In this way, the researcher is released from the difficult task of building a mathematical model of the physical system, which, in some cases, is impossible to represent.

Understanding biological neural networks

ANNs are based on a model that draws inspiration from the functioning principles of the human brain and how the human brain processes the information that comes to it from the peripheral organs. In fact, ANNs consist of a series of neurons that can be thought of as individual processors, since their basic task is to process the information that is provided to them at the input. This processing is similar to the functioning of a biological neuron, which receives electrical signals, processes them, and then transmits the results to the next neuron. The essential elements of a biological neuron include the following:

- Dendrites
- Synapses
- Body cells
- Axon

The information is processed by the biological neuron according to the following steps:

1. Dendrites get information from other neurons in the form of electrical signals.
2. The flow of information occurs through the synapses.
3. The dendrites transmit this information to the cell body.
4. In the cell body, the information is added together.
5. If the result exceeds a threshold limit, the cell reacts by passing the signal to another cell. The passage of information takes place through the axon.

The following diagram shows the essential elements of the structure of a biological neuron:

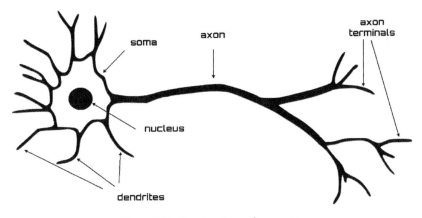

Figure 10.1: The structure of a neuron

Synapses assume the role of neurotransmitters; in fact, they can exert an excitatory or inhibitory action against the neuron that is immediately after it. This effect is regulated by the synapses through the weight that is associated with them. In this way, each neuron can perform a weighted sum of the inputs, and if this sum exceeds a certain threshold, it activates the next neuron.

> **Important note**
> The processing performed by the neuron lasts for a few milliseconds. From a computational point of view, it represents a relatively long time. So, we could say that this processing system, taken individually, is relatively slow. However, as we know, it is a model based on quantity; it is made up of a very high number of neurons and synapses that work simultaneously and in parallel.

In this way, the processing operations that are performed are very effective, and they allow us to obtain results in a relatively short period of time. We can say that the strength of neural networks lies in the teamwork of neurons. Taken individually, they do not represent a particularly effective processing system; however, taken together, they represent an extremely high-performing simulation model.

The functioning of a brain is regulated by neurons and represents an optimized machine that can solve even complex problems. It is a simple structure, improved over time through the evolution of the species. It has no central control; the areas of the brain are all active in carrying out a task, which is aimed at solving a problem. The workings of all parts of the brain take place in a contributory way, and each part contributes to the result. In addition to this, the human brain is equipped with a very effective error regulation system. In fact, if a part of the brain stops working, the operations of the entire system continue to be performed, even if with a lower performance.

Exploring ANNs

As we have anticipated, a model based on ANNs draws inspiration from the functioning of the human brain. In fact, an artificial neuron is similar to a biological neuron in that it receives information as input derived from another neuron. A neuron's input represents the output of the neuron that is found immediately before in the architecture of a model based on neural networks.

Each input signal to the neuron is then multiplied by the corresponding weight. It is then added to the results obtained by the other neurons to process the activation level of the next neuron. The essential elements of the architecture of a model based on ANNs include the neurons that are distinguished from the input neurons and the output neurons by the number of layers of synapses and by the connections between these neurons. The following diagram shows the typical architecture of an ANN:

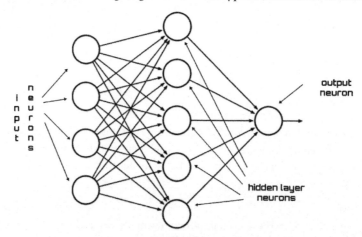

Figure 10.2: Architecture of an ANN

The input signals, which represent the information detected by the environment, are sent to the input layer of the ANN. In this way, they travel, in parallel, along with the connections through the internal nodes of the system and up to the output. The architecture of the network, therefore, returns a response from the system. Put simply, in a neural network, each node is able to process only local information with no knowledge of the final goal of the processing and not keeping any memory of the latter. The result obtained depends on the architecture of the network and the values assumed by the artificial synapses.

There are cases in which a single synapse layer is unable to return an adequate network response to the signal supplied at the input. In these cases, multiple layers of synapses are required because a single layer is not enough. These networks are called deep neural networks. The network response is obtained by treating the activation of one layer of neurons at a time and then proceeding from the input to the output, passing through the intermediate layers.

> **Important note**
> The ANN target is the result of the calculation of the outputs performed for all the neurons, so an ANN is presented as a set of mathematical function approximations.

The following elements are essential in an ANN architecture:

- Weights
- Bias
- Layers
- Activation functions

In the following sections, we will deepen our understanding of these concepts.

Describing the structure of the layers

In the architecture of an ANN, it is possible to identify the nodes representing the neurons distributed in a form that provides a succession of layers. In a simple structure of an ANN, it is possible to identify an input layer, an intermediate layer (hidden layer), and an output layer, as shown in the following diagram:

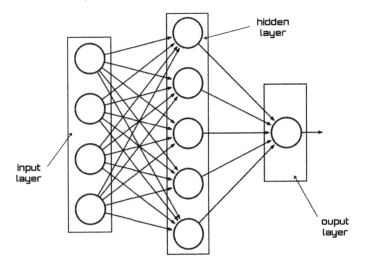

Figure 10.3: The various layers of an ANN

Each layer has its own task, which it performs through the actions of the neurons it contains. The input layer is intended to introduce the initial data into the system for further processing by the subsequent layers. From the input level, the workflow of the ANN begins.

> **Important note**
>
> In the input layer, artificial neurons have a different role to play in some passive way because they do not receive information from the previous levels. In general, they receive a series of inputs and introduce information into the system for the first time. This level then sends the data to the next levels where the neurons receive weighted inputs.

The hidden layer in an ANN is interposed between input levels and output levels. The neurons of the hidden layer receive a set of weighted inputs and produce an output according to the indications received from an activation function. It represents the essential part of the entire network since it is here that the magic of transforming the input data into output responses takes place.

Hidden levels can operate in many ways. In some cases, the inputs are weighted randomly, while in others they are calibrated through an iterative process. In general, the neuron of the hidden layer functions as a biological neuron in the brain. That is, it takes its probabilistic input signals, processes them, and converts them into an output corresponding to the axon of the biological neuron.

Finally, the output layer produces certain outputs for the model. Although they are very similar to other artificial neurons in the neural network, the type and number of neurons in the output layer depend on the type of response the system must provide. For example, if we are designing a neural network for the classification of an object, the output layer will consist of a single node that will provide us with this value. In fact, the output of this node must simply provide a positive or negative indication of the presence or absence of the target in the input data. For example, if our network must perform an object classification task, then this layer will contain only one neuron destined to return this value. This is because this neuron must return a binary signal, that is, a positive or negative response that signals the presence or absence of the object among the data provided as input.

Analyzing weights and biases

In a neural network, weights represent a crucial factor in converting an input signal into the system response. They represent a factor such as the slope of a linear regression line. In fact, the weight is multiplied by the inputs and the result is added to the other contributions. These are numerical parameters that determine the contribution of a single neuron in the formation of the output.

If the inputs are *x1, x2, … xn*, and the synaptic weights to be applied to them are denoted as *w1, w2, … w_n*, the output returned by the neuron is expressed through the following formula:

$$y = f(x) = \sum_{i=1}^{n} x_i * w_i$$

The previous formula is a matrix multiplication to reach the weighted sum. The neuron elaboration can be denoted as follows:

$$Output = \sum_{i=1}^{n} x_i * w_i + bias$$

In the previous formula, the bias assumes the role of the intercept in a linear equation. The bias represents an additional parameter that is used to regulate the output along with the weighted sum of the inputs to the neuron.

The input of the next layer is the output of the neurons in the previous layer, as shown in the following diagram:

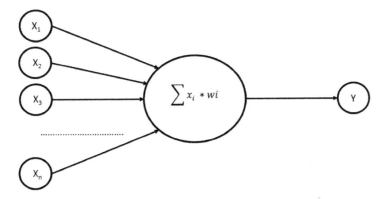

Figure 10.4: Output of the neurons

The schema presented in the previous diagram explains the role played by the weight in the formation of a neuron. Note that the input provided to the neuron is weighed with a real number that reproduces the activity of the synapse of a biological neuron. When the weight value is positive, the signal has an excitatory activity. If, on the other hand, the value is negative, the signal is inhibitory. The absolute value of the weight indicates the strength of the contribution to the formation of the system response.

Explaining the role of activation functions

The abstraction of neural network processing is primarily obtained via the activation functions. This is a mathematical function that transforms the input into output and controls a neural network process. Without the contribution of activation functions, a neural network can be assimilated into a linear function. A linear function occurs when the output is directly proportional to the input. For example, let's analyze the following equation:

$$y = 5 * x + 3$$

In the previous equation, the exponent of x is equal to 1. This is the condition for the function to be linear: it must be a first-degree polynomial. It is a straight line with no curves. Unfortunately, most real-life problems are nonlinear and complex in nature. To treat nonlinearity, activation functions are introduced in neural networks. Recall that, for a function to be nonlinear, it is sufficient that it is a polynomial function of a degree higher than the first. For example, the following equation defines a nonlinear function of the third degree:

$$y = 5 * x^3 + 3$$

The graph of a nonlinear function is curved and adds to the complexity factor. Activation functions give the nonlinearity property to neural networks and make them true universal function approximators.

There are many activation functions available for a neural network to use. The most used activation functions are listed here:

- **Sigmoid**: The sigmoid function is a mathematical function that produces a sigmoidal curve, which is a characteristic curve for its S shape. This is one of the earliest and most used activation functions. This squashes the input to any value between *0* and *1* and makes the model logistic in nature.

- **Unit step**: A unit step activation function is a much-used feature in neural networks. The output assumes a value of *0* for a negative argument and a value of *1* for a positive argument. The range is between *0* and *1* and the output is binary in nature. These types of activation functions are useful for binary schemes.

- **Hyperbolic tangent**: This is a nonlinear function, defined in the range of values *-1* to *1*, so you do not need to worry about activations blowing up. One thing to clarify is that the gradient is stronger for tanh than sigmoid. Deciding between sigmoid and tanh will depend on your gradient strength requirement. Like sigmoid, tanh also has the missing slope problem.

- **Rectified Linear Unit (ReLU)**: This is a function with linear characteristics for parts of the existing domain that will output the input directly if it is positive; otherwise, it will output *0*. The range of output is between *0* and *infinity*. ReLU finds applications in computer vision and speech recognition using deep neural networks.

The previously listed activation functions are shown in the following diagram:

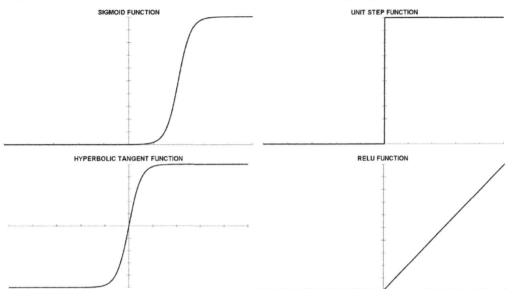

Figure 10.5: Representation of activation functions

Now we will look at the simple architecture of a neural network, which shows us how the flow of information proceeds.

Let's see how a feedforward neural network is structured.

Understanding feedforward neural networks

The processing from the input layer to the hidden layer(s) and then to the output layer is called feedforward propagation. The transfer function is applied at each hidden layer, and then the activation function value is propagated to the next layer. The next layer can be another hidden layer or the output layer.

> **Important note**
> The term "feedforward" is used to indicate the networks in which each node receives connections only from the lower layers. These networks emit a response for each input pattern but fail to capture the possible temporal structure of the input information or exhibit endogenous temporal dynamics.

Exploring neural network training

The learning ability of an ANN is manifested in the training procedure. This represents the crucial phase of the whole algorithm as it is through the characteristics extracted during training that the network acquires the ability to generalize. The training takes place through a comparison between a series of inputs corresponding to known outputs and those supplied by the model. At least, this is what happens in the case of supervised algorithms in which the labeled data is compared with that provided by the model.

The results achieved by the model are influenced by the data used in the training phase—to obtain good performance, the data used must be sufficiently representative of the phenomenon. From this, we can understand the importance of the dataset used in this phase, which is called the training set.

To understand the procedure through which a neural network learns from data, let's consider a simple example. We will analyze the training of a neural network with a single hidden level. Let's say that the input level has a neuron, and the network will have to face a binary classification problem—only two output values of either *0* or *1*.

The training of the network will take place according to the following steps:

1. Enter the input in the form of a data matrix.
2. Initialize the weights and biases with random values. This step will be performed once, at the beginning of the procedure only. Later, the weights and biases will be updated through the error propagation process.
3. Repeat *steps 4 to 9* until the error is minimized.
4. Apply inputs to the network.
5. Calculate the output for each neuron from the input level to the hidden levels to the output level.
6. Calculate the error on the outputs: less real than expected.
7. Use the output error to calculate the error signals for the previous layers. The partial derivative of the activation function is used to calculate the error signals.
8. Use the error signals to calculate the weight adjustments.
9. Apply the weight adjustments.

Steps 4 and 5 represent the direct propagation phase, while *steps 6 to 9* represent the backpropagation phase.

The most used method to train a network through the adjustment of neuron weights is the delta rule, which compares the network output with the desired values. Subtract the two values and the difference is used to update all of the input weights, which have different values of zero. The process is repeated until convergence is achieved.

The following diagram shows the weight adjustment procedure:

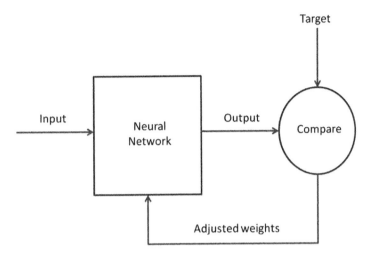

Figure 10.6: The weight adjustment procedure

The training procedure is extremely simple: it is a simple comparison between the calculated values and the labeled values. The difference between the weighted input values and the expected output values is calculated from the comparison—the difference, which represents the evaluation error between the calculated and expected values, is used to recalculate all of the input weights. It is an iterative procedure that is repeated until the error between the expected and calculated value approaches zero.

The training phase can be performed with the backpropagation technique in combination with an optimization method such as stochastic gradient descent. Backpropagation is one of the most widely used algorithms to train neural networks, both single-layer and multi-layer. This algorithm aims to modify the weights of the neurons in the network to minimize the error given by the difference between the output vector obtained and the desired one. Numerical optimization is iterative and requires the calculation of the gradient of the cost function with respect to weights and biases. For each iteration of the algorithm, the gradient calculation requires several operations equal to the number of coefficients. The backpropagation algorithm requires the calculation of the gradient only in the last layer, which is then propagated backward to obtain the gradient in the previous layers. The stochastic descent of the gradient, as seen in detail in *Chapter 7, Using Simulation to Improve and Optimize Systems*, aims to reach the lowest point of the cost function. In more technical terms, the gradient represents a derivative indicating the slope or slope of the objective function.

After having analyzed in detail the architecture of the ANNs, we will now see a practical case of elaboration of an ANN-based model to solve a physical problem.

Simulating airfoil self-noise using ANNs

The noise generated by an airfoil is due to the interaction between a turbulent airflow and the aircraft's airfoil blades. Predicting the acoustic field in these situations requires an aeroacoustics methodology that can operate in complex environments. Additionally, the method that is used must avoid the formulation of coarse hypotheses regarding geometry, compactness, and content of the frequency of sound sources. The prediction of the sound generated by a turbulent flow must, therefore, correctly model both the physical phenomena of sound propagation and the turbulence of the flow. Since these two phenomena manifest energy and scales of very different lengths, the correct prediction of the sound generated by a turbulent flow is not easy to model.

Aircraft noise is a crucial topic for the aerospace industry. The NASA Langley Research Center has funded several strands of research to effectively study the various mechanisms of a self-noise airfoil. Interest was motivated by its importance for broadband helicopter rotors, wind turbines, and cell noises. The goal of these studies, then, focused on the challenge of reducing external noises generated by the entire cell of an aircraft by 10 decibels.

In this example, we will elaborate on a model based on ANNs to predict self-noise airfoils from a series of airfoil data measured in a wind tunnel.

> **Important note**
> The dataset we will use was developed by NASA in 1989 and is available on the UCI Machine Learning Repository site. The UCI Machine Learning Repository is available at https://archive.ics.uci.edu/ml/datasets/Airfoil+Self-Noise.

The dataset was built using the results of a series of aerodynamic and acoustic tests on sections of aerodynamic blades performed in an anechoic wind tunnel.

The following list shows the features of the dataset:

- Number of instances: 1,503
- Number of attributes: 6
- Dataset characteristics: multivariate
- Attribute characteristics: real
- Dataset date: 2014-03-04

The following list presents a brief description of the attributes:

- `Frequency`: **Frequency in Hertz (Hz)**
- `AngleAttack`: angle of attack in degrees
- `ChordLength`: chord length in meters

- `FSVelox`: free-stream velocity in meters per second
- `SSDT`: **suction-side displacement thickness (SSDT)** in meters
- `SSP`: scaled sound pressure level in decibels

In the six attributes we have listed, the first five represent the predictors, and the last one represents the response of the system that we want to simulate. It is, therefore, a regression problem because the answer has continuous values. In fact, it represents the self-noise airfoil, in decibels, measured in the wind tunnel.

Importing data using pandas

The first operation we will perform is the importing of data, which, as we have already mentioned, is available on the UCI website. As always, we will analyze the code line by line:

1. We start by importing the libraries. In this case, we will operate differently from what has been done so far. We will not import all of the necessary libraries at the beginning of the code, but we will introduce them in correspondence with their use and we will illustrate their purposes in detail:

    ```
    import pandas as pd
    ```

    The `pandas` library is an open source, BSD-licensed library that provides high-performance, easy-to-use data structures, and data analysis tools for the Python programming language. It offers data structures and operations for manipulating numerical data in a simple way. We will use this library to import the data contained in the dataset retrieved from the UCI website.

    The UCI dataset does not contain a header, so it is necessary to insert the names of the variables in another variable. Now, let's put these variable names in the following list:

    ```
    ASNNames= ['Frequency','AngleAttack','ChordLength',
    'FSVelox','SSDT','SSP']
    ```

2. Now we can import the dataset. This is available in `.dat` format, and, to make your job easier, it has already been downloaded and is available in this book's GitHub repository:

    ```
    ASNData = pd.read_csv('airfoil_self_noise.dat', delim_
    whitespace=True, names=ASNNames)
    ```

    To import the `.dat` dataset, we used the `read_csv` module of the `pandas` library. In this function, we passed the filename and two other attributes, namely `delim_whitespace` and `names`. The first specifies whether or not whitespace will be used as `sep`, and the second specifies a list of column names to use.

> **Important note**
> Remember to set the path so that Python can find the `.dat` file to open.

Before beginning with data analysis through ANN regression, we perform an exploratory analysis to identify how data is distributed and extract preliminary knowledge. To display the first 20 rows of the imported DataFrame, we can use the `head()` function, as follows:

```
print(ASNData.head(20))
```

The pandas `head()` function gets the first *n* rows of a pandas DataFrame. In this case, it returns the first 20 rows for the `ASNData` object based on position. It is used for quickly testing whether our dataset has the right type of data in it. This function, with no arguments, gets the first five rows of data from the DataFrame.

The following data is printed:

	Frequency	AngleAttack	ChordLength	FSVelox	SSDT	SSP
0	800	0.0	0.3048	71.3	0.002663	126.201
1	1000	0.0	0.3048	71.3	0.002663	125.201
2	1250	0.0	0.3048	71.3	0.002663	125.951
3	1600	0.0	0.3048	71.3	0.002663	127.591
4	2000	0.0	0.3048	71.3	0.002663	127.461
5	2500	0.0	0.3048	71.3	0.002663	125.571
6	3150	0.0	0.3048	71.3	0.002663	125.201
7	4000	0.0	0.3048	71.3	0.002663	123.061
8	5000	0.0	0.3048	71.3	0.002663	121.301
9	6300	0.0	0.3048	71.3	0.002663	119.541
10	8000	0.0	0.3048	71.3	0.002663	117.151
11	10000	0.0	0.3048	71.3	0.002663	115.391
12	12500	0.0	0.3048	71.3	0.002663	112.241
13	16000	0.0	0.3048	71.3	0.002663	108.721
14	500	0.0	0.3048	55.5	0.002831	126.416
15	630	0.0	0.3048	55.5	0.002831	127.696
16	800	0.0	0.3048	55.5	0.002831	128.086
17	1000	0.0	0.3048	55.5	0.002831	126.966
18	1250	0.0	0.3048	55.5	0.002831	126.086
19	1600	0.0	0.3048	55.5	0.002831	126.986
20	2000	0.0	0.3048	55.5	0.002831	126.616

Figure 10.7: DataFrame output

To extract further information, we can use the `info()` function, as follows:

```
print(ASNData.info())
```

The `info()` method returns a concise summary of the `ASNData` DataFrame, including the `dtypes` index and `dtypes` column, non-null values, and memory usage.

The following results are returned:

```
<class 'pandas.core.frame.DataFrame'>
RangeIndex: 1503 entries, 0 to 1502
Data columns (total 6 columns):
Frequency      1503 non-null int64
```

```
AngleAttack     1503 non-null float64
ChordLength     1503 non-null float64
FSVelox         1503 non-null float64
SSDT            1503 non-null float64
SSP             1503 non-null float64
dtypes: float64(5), int64(1)
memory usage: 70.5 KB
None
```

By reading the information returned by the `info` method, we can confirm that it is 1,503 instances and six variables. In addition to this, the types of variables returned to us are five `float64` variables and one `int64` variable.

3. To obtain a first screening of the data contained in the `ASNData` DataFrame, we can compute a series of basic statistics. We can use the `describe()` function in the following way:

```
BasicStats = ASNData.describe()
BasicStats = BasicStats.transpose()
print(BasicStats)
```

The `describe()` function produces descriptive statistics that return the central tendency, dispersion, and shape of a dataset's distribution, excluding **Not-a-Number** (**NaN**) values. It is used for both numeric and object series, as well as the DataFrame column sets of mixed data types. The output will vary depending on what is provided. In addition to this, we have transposed the statistics to appear better on the screen and to make it easier to read the data.

The following statistics are printed:

```
              count          mean          std         min          25%  \
Frequency    1503.0   2886.380572  3152.573137  200.000000   800.000000
AngleAttack  1503.0      6.782302     5.918128    0.000000     2.000000
ChordLength  1503.0      0.136548     0.093541    0.025400     0.050800
FSVelox      1503.0     50.860745    15.572784   31.700000    39.600000
SSDT         1503.0      0.011140     0.013150    0.000401     0.002535
SSP          1503.0    124.835943     6.898657  103.380000   120.191000

                    50%          75%          max
Frequency   1600.000000  4000.000000  20000.000000
AngleAttack    5.400000     9.900000     22.200000
ChordLength    0.101600     0.228600      0.304800
FSVelox       39.600000    71.300000     71.300000
SSDT           0.004957     0.015576      0.058411
SSP          125.721000   129.995500    140.987000
```

Figure 10.8: Basic statistics from the DataFrame

From the analysis of the previous table, we can extract useful information. First of all, we can note that the data shows great variability. The average of the values ranges from approximately 0.14 to 2,886. Not only that, but some variables have a very large standard deviation. For each variable, we can easily recover the minimum and maximum. In this way, we can note that the interval of the analyzed frequencies goes from 200 to 20,000 Hz. These are just some considerations; we can recover many others.

Scaling the data using sklearn

In the statistics extracted using the `describe()` function, we have seen that the predictor variables (frequency, angle of attack, chord length, free-stream velocity, and SSDT) have an important variability. In the case of predictors with different and varied ranges, the influence on the system response by variables with a larger numerical range could be greater than those with a lower numeric range. This different impact could affect the accuracy of the prediction. Actually, we want to do exactly the opposite, that is, improve predictive accuracy and reduce the sensitivity of the model from features that can affect prediction due to a wide range of numerical values.

To avoid this phenomenon, we can reduce the values so that they fall within a common range, guaranteeing the same characteristics of variability possessed by the initial dataset. In this way, we will be able to compare variables belonging to different distributions and variables expressed in different units of measurement.

> **Important note**
> Recall how we rescale the data before training a regression algorithm. This is a good practice. Using a rescaling technique, data units are removed, allowing us to compare data from different locations easily.

We then proceed with data rescaling. In this example, we will use the min-max method (usually called **feature scaling**) to get all of the scaled data in the range 0–1. The formula to achieve this is as follows:

$$x_{scaled} = \frac{x - x_{min}}{x_{max} - x_{min}}$$

To do feature scaling, we can apply a preprocessing package offered by the `sklearn` library. This library is a free software machine learning library for the Python programming language. The `sklearn` library offers support for various machine learning techniques, such as classification, regression, and clustering algorithms, including **support vector machines** (**SVMs**), random forests, gradient boosting, k-means, and DBSCAN. `sklearn` is created to work with various Python numerical and scientific libraries, such as NumPy and SciPy:

> **Important note**
> To import a library that is not part of the initial distribution of Python, you can use the `pip install` command, followed by the name of the library. This command should be used only once and not every time you run the code.

1. The `sklearn.preprocessing` package contains numerous common utility functions and transformer classes to transform the features in a way that works with our requirements. Let's start by importing the package:

   ```
   from sklearn.preprocessing import MinMaxScaler
   ```

   To scale features between the minimum and maximum values, the `MinMaxScaler` function can be used. In this example, we want to rescale the data between zero and one so that the maximum absolute value of each feature is scaled to unit size.

2. Let's start by setting the scaler object:

   ```
   ScalerObject = MinMaxScaler()
   ```

3. To get validation of what we are doing, we will print the object just created in order to check the set parameters:

   ```
   print(ScalerObject.fit(ASNData))
   ```

   The following result is returned:

   ```
   MinMaxScaler(copy=True, feature_range=(0, 1))
   ```

4. Now, we can apply the `MinMaxScaler()` function, as follows:

   ```
   ASNDataScaled = ScalerObject.fit_transform(ASNData)
   ```

   The `fit_transform()` method fits to the data and then transforms it. Before applying the method, the minimum and maximum values that are to be used for later scaling are calculated. This method returns a NumPy array object.

5. Recall that the initial data had been exported in the pandas `DataFrame` format. Scaled data should also be transformed into the same format in mdoo to be able to apply the functions available for pandas DataFrames. The transformation procedure is easy to perform; just apply the pandas `DataFrame()` function as follows:

   ```
   ASNDataScaled = pd.DataFrame(ASNDataScaled,
   columns=ASNNames)
   ```

6. At this point, we can verify the results obtained with data scaling. Let's compute the statistics using the `describe()` function once again:

   ```
   summary = ASNDataScaled.describe()
   summary = summary.transpose()
   print(summary)
   ```

The following statistics are printed:

```
              count      mean       std  min       25%       50%       75%  \
Frequency    1503.0  0.135676  0.159221  0.0  0.030303  0.070707  0.191919
AngleAttack  1503.0  0.305509  0.266582  0.0  0.090090  0.243243  0.445946
ChordLength  1503.0  0.397810  0.334791  0.0  0.090909  0.272727  0.727273
FSVelox      1503.0  0.483857  0.393252  0.0  0.199495  0.199495  1.000000
SSDT         1503.0  0.185125  0.226687  0.0  0.036794  0.078550  0.261594
SSP          1503.0  0.570531  0.183441  0.0  0.447018  0.594065  0.707727

             max
Frequency    1.0
AngleAttack  1.0
ChordLength  1.0
FSVelox      1.0
SSDT         1.0
SSP          1.0
```

Figure 10.9: Output with scaled data

From the analysis of the previous table, the result of the data scaling appears evident. Now all six variables have values between *0* and *1*.

Viewing the data using Matplotlib

Now, we will try to have a confirmation of the distribution of the data through a visual approach:

1. To start, we will draw a boxplot, as follows:

   ```
   import matplotlib.pyplot as plt
   boxplot = ASNDataScaled.boxplot(column=ASNNames)
   plt.show()
   ```

A boxplot, also referred to as a whisker chart, is a graphical description used to illustrate the distribution of data using dispersion and position indices. The rectangle (box) is delimited by the first quartile (25th percentile) and the third quartile (75th percentile) and divided by the median (50th percentile). In addition, there are two whiskers, one upper and one lower, indicating the maximum and minimum distribution values, excluding any anomalous values.

Matplotlib is a Python library for printing high-quality graphics. With `matplotlib`, it is possible to generate graphs, histograms, bar graphs, power spectra, error graphs, scatter graphs, and more with just a few commands. This is a collection of command-line functions such as those provided by the MATLAB software.

As we mentioned earlier, the scaled data is in pandas DataFrame format. So, it is advisable that you use the `pandas.DataFrame.boxplot` function. This function makes a boxplot from the DataFrame columns, which are optionally grouped by some other columns.

The following diagram is printed:

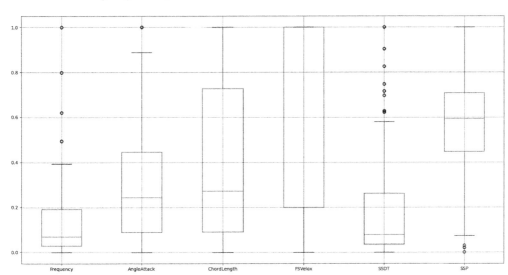

Figure 10.10: Boxplot of the DataFrame

In the previous diagram, you can see that there are some anomalous values indicated by small circles at the bottom and side of the extreme mustache of each box. Three variables have these values, called outliers, and their presence can create problems in the construction of the model. Furthermore, we can verify that all the variables are contained in the extreme values that are equal to *0* and *1*; this is the result of data scaling. Finally, some variables such as `FSVelox` show a great variability of values compared to others, for example, SSDT.

We will now measure the correlation between the predictors and the response variable. A technique for measuring the relationship between two variables is offered by correlation, which can be obtained using covariance. To calculate the correlation coefficients in Python, we can use the `pandas.DataFrame.corr()` function. This function computes the pairwise correlation of columns, excluding NA/null values. Three procedures are offered, as follows:

- `pearson` (standard correlation coefficient)
- `kendall` (Kendall Tau correlation coefficient)
- `spearman` (Spearman rank correlation)

> **Important note**
> Remember that the correlation coefficient of two random variables is a measure of their linear dependence.

2. Let's calculate the correlation coefficients for the data scaled:

```
CorASNData = ASNDataScaled.corr(method='pearson')
with pd.option_context('display.max_rows', None,
    'display.max_columns', CorASNData.shape[1]):
print(CorASNData)
```

To show all data columns on a screenshot, we used the `option_context` function. The following results are returned:

```
             Frequency  AngleAttack  ChordLength   FSVelox       SSDT        SSP
Frequency     1.000000    -0.272765    -0.003661  0.133664  -0.230107  -0.390711
AngleAttack  -0.272765     1.000000    -0.504868  0.058760   0.753394  -0.156108
ChordLength  -0.003661    -0.504868     1.000000  0.003787  -0.220842  -0.236162
FSVelox       0.133664     0.058760     0.003787  1.000000  -0.003974   0.125103
SSDT         -0.230107     0.753394    -0.220842 -0.003974   1.000000  -0.312670
SSP          -0.390711    -0.156108    -0.236162  0.125103  -0.312670   1.000000
```

Figure 10.11: Data columns in the DataFrame

We are interested in studying the possible correlation between the predictors and the system response. So, to do this, it will be sufficient to analyze the last row of the previous table. Recall that the values of the various correlation indices vary between -1 and +1; both extreme values represent perfect relationships between the variables, while 0 represents the absence of a relationship. This is if we consider linear relationships. Based on this, we can say that the predictors that show a greater correlation with the response **SSP** are **Frequency** and **SSDT**. Both show a negative correlation.

To see visual evidence of the correlation between the variables, we can plot a correlogram. A correlogram graphically presents a correlation matrix. It is used to focus on the most correlated variables in a data table. In a correlogram, correlation coefficients are shown in a nuance that depends on our values. Next to the graph, a colored bar will be proposed in which the corresponding nuance values of the correlation coefficient can be read. To plot a correlogram, we can use the `matplotlib.pyplot.matshow()` function, which shows a DataFrame as a matrix in a new figure window.

3. Let's plot a correlogram, as follows:

```
plt.matshow(CorASNData)
plt.xticks(range(len(CorASNData.columns)), CorASNData.columns)
plt.yticks(range(len(CorASNData.columns)), CorASNData.columns)
plt.colorbar()
plt.show()
```

The following diagram is returned:

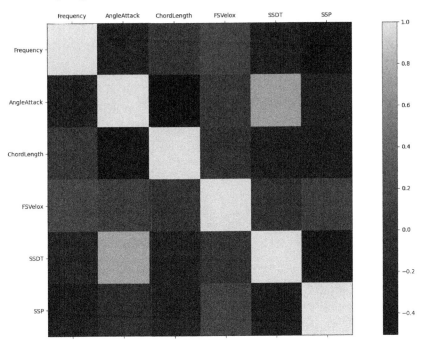

Figure 10.12: Correlogram of the DataFrame

As we already did in the case of the correlation matrix, in this case too, to analyze the correlation between predictors and the system response, it will be sufficient to consider the bottom row of the graph. The trends already obtained from the correlation matrix are confirmed.

Splitting the data

The training of an algorithm, based on machine learning, represents the crucial phase of the whole process of elaboration of the model. Performing the training of an ANN on the same dataset, which will subsequently be used to test the network, represents a methodological error. This is because the model will be able to perfectly predict the data used for testing, having already seen them in the training phase. However, when it will then be used to predict new cases that have never been seen before, it will inexorably commit evaluation errors. This problem is called data overfitting. To avoid this error, it is a good practice to train a neural network to use a different set of data from the one used in the test phase. Therefore, before proceeding with the training phase, it is recommended that you perform a correct division of the data.

Data splitting is used to split the original data into two sets: one is used to train a model and the other to test the model's performance. The training and testing procedures represent the starting point for the model setting in predictive analytics. In a dataset that has 100 observations for predictor and response variables, a data splitting example occurs that divides this data into 70 rows for training and 30 rows for testing. To perform good data splitting, the observations must be selected randomly. When the training data is extracted, the data will be used to upload the weights and biases until an appropriate convergence is achieved.

The next step is to test the model. To do this, the remaining observations contained in the test set will be used to check whether the actual output matches the predicted output. To perform this check, several metrics can be adopted to validate the model:

1. We will use the `sklearn` library to split the `ASNDataScaled` DataFrame. To start, we will import the `train_test_split` function:

   ```
   from sklearn.model_selection import train_test_split
   ```

2. Now, we can divide the starting data into two sets: the X set containing the predictors and the Y set containing the target. We will use the `pandas.DataFrame.drop()` function, as follows:

   ```
   X = ASNDataScaled.drop('SSP', axis = 1)
   print('X shape = ',X.shape)
   Y = ASNDataScaled['SSP']
   print('Y shape = ',Y.shape)
   ```

3. Using the `pandas.DataFrame.drop()` function, we can remove rows or columns indicating label names and the corresponding axis or the index or column names. In this example, we have removed the target column (SSP) from the starting DataFrame.

   The following shapes are printed:

   ```
   X shape = (1503, 5)
   Y shape = (1503,)
   ```

   As is our intention, the two datasets, X and Y, now contain the five predictors and the system response, respectively.

4. Now we can divide the two datasets, X and Y, into two further datasets that will be used for the training phase and the test phase, respectively:

   ```
   X_train, X_test, Y_train, Y_test = train_test_split(X, Y,
   test_size = 0.30, random_state = 5)
   print('X train shape = ',X_train.shape)
   print('X test shape = ', X_test.shape)
   print('Y train shape = ', Y_train.shape)
   print('Y test shape = ',Y_test.shape)
   ```

The `train_test_split()` function was used by passing the following four parameters:

- X: The predictors.
- Y: The target.
- `test_size`: This parameter represents the proportion of the dataset to include in the test split. The following types are available: float, integer, or none, and optional (default = 0.25).
- `random_state`: This parameter sets the seed used by the random number generator. In this way, the repetitive splitting of the operation is guaranteed.

The following results are returned:

```
X train shape  =   (1052, 5)
X test shape   =   (451, 5)
Y train shape  =   (1052,)
Y test shape   =   (451,)
```

As expected, we finally divided the initial dataset into four subsets. The first two, `X_train` and `Y_train`, will be used in the training phase. The remaining two, `X_test` and `Y_test`, will be used in the testing phase.

Explaining multiple linear regression

In this section, we will deal with a regression problem using ANNs. To evaluate the results effectively, we will compare them with a model based on a different technology. Here, we will make a comparison between the model based on multiple linear regression and a model based on ANNs.

In multiple linear regression, the dependent variable (response) is related to two or more independent variables (predictors). The following equation is the general form of this model:

$$y = \beta_0 + \beta_1 * x_1 + \beta_2 * x_2 + \cdots + \beta_n * x_n$$

In the previous equation, $x_1, x_2, \ldots x_n$ are the predictors, and y is the response variable. The β_i coefficients define the change in the response of the model related to the changes that occurred in x_i, when the other variables remain constant. In the simple linear regression model, we are looking for a straight line that best fits the data. In the multiple linear regression model, we are looking for the plane that best fits the data. So, in the last one, our aim is to minimize the overall squared distance between this plane and the response variable.

To estimate the coefficients β, we want to minimize the following term:

$$\sum_i [y_i - (\beta_0 + \beta_1 * x_1 + \beta_2 * x_2 + \cdots + \beta_n * x_n)]^2$$

To execute a multiple linear regression study, we can easily use the `sklearn` library. The `sklearn.linear_model` module is a module that contains several functions to resolve linear problems as a `LinearRegression` class that achieves an ordinary least squares linear regression:

1. To start, we will import the function as follows:

   ```
   from sklearn.linear_model import LinearRegression
   ```

   Then, we set the model using the `LinearRegression ()` function with the following command:

   ```
   Lmodel = LinearRegression()
   ```

2. Now, we can use the `fit()` function to fit the model:

   ```
   Lmodel.fit(X_train, Y_train)
   ```

   The following parameters are passed:

   - `X_train`: The training data
   - `Y_train`: The target data
   - Eventually, a third parameter can be passed; this is the `sample_weight` parameter, which contains the individual weights for each sample

   This function fits a linear model using a series of coefficients to minimize the residual sum of squares between the expected targets and the predicted targets predicted.

3. Finally, we can use the linear model to predict the new values using the predictors contained in the test dataset:

   ```
   Y_predLM = Lmodel.predict(X_test)
   ```

   At this point, we have the predictions.

   Now, we must carry out a first evaluation of the model to verify how much the prediction approached the expected value.

   There are several descriptors for evaluating a prediction model. In this example, we will use the **mean squared error** (**MSE**).

   > **Important note**
   > MSE returns the average of the squares of the errors. This is the average squared difference between the expected values and the value that is predicted. MSE returns a measure of the quality of an estimator; this is a non-negative value and, the closer the values are to zero, the better the prediction.

4. To calculate the MSE, we will use the `mean_squared_error()` function contained in the `sklearn.metrics` module. This module contains score functions, performance metrics and pairwise metrics, and distance computations. We start by importing the function, as follows:

   ```
   from sklearn.metrics import mean_squared_error
   ```

5. Then, we can apply the function to the data:

   ```
   MseLM = mean_squared_error(Y_test, Y_predLM)
   print('MSE of Linear Regression Model')
   print(MseLM)
   ```

6. Two parameters were passed: the expected values (`Y_test`) and the values that were predicted (`Y_predLM`). Then, we print the results, as follows:

   ```
   MSE of Linear Regression Model
   0.015826467113949756
   ```

The value obtained is low and very close to zero. However, for now, we cannot add anything. We will use this value, later on, to compare it with the value that we calculate for the model based on neural networks.

Understanding a multilayer perceptron regressor model

A multilayer perceptron contains at least three layers of nodes: input nodes, hidden nodes, and output nodes. Apart from the input nodes, each node is a neuron that uses a nonlinear activation function. A multilayer perceptron works with a supervised learning technique and a backpropagation method for training the network. The presence of multiple layers and nonlinearity distinguish a multilayer perceptron from a simple perceptron. A multilayer perceptron is applied when data cannot be separated linearly:

1. To build a multilayer-perceptron-based model, we will use the `sklearn MLPRegressor` function. This regressor proceeds iteratively in the data training. At each step, it calculates the partial derivatives of the loss function with respect to the model parameters and uses the results obtained to update the parameters. There is a regularization term added to the loss function to reduce the model parameters to avoid data overfitting.

   First, we will import the function:

   ```
   from sklearn.neural_network import MLPRegressor
   ```

   The `MLPRegressor()` function implements a multilayer perceptron regressor. This model optimizes the squared loss by using a limited-memory version of the **Broyden–Fletcher–Goldfarb–Shanno** (**LBFGS**) algorithm or stochastic gradient descent algorithm.

2. Now, we can set the model using the `MLPRegressor` function, as follows:

```
MLPRegModel = MLPRegressor(hidden_layer_sizes=(50),
              activation='relu', solver='lbfgs',
              tol=1e-4, max_iter=10000, random_state=0)
```

The following parameters are passed:

- `hidden_layer_sizes=(50)`: This parameter sets the number of neurons in the hidden layer; the default value is `100`.
- `activation='relu'`: This parameter sets the activation function. The following activation functions are available: identity, logistic, tanh, and ReLU. The last one is set by default.
- `solver='lbfgs'`: This parameter sets the solver algorithm for weight optimization. The following solver algorithms are available: LBFGS, **stochastic gradient descent (SGD)**, and the SGD optimizer (`adam`).
- `tol=1e-4`: This parameter sets the tolerance for optimization. By default, `tol` is equal to `1e-4`.
- `max_iter=10000`: This parameter sets the maximum number of iterations. The solver algorithm iterates until convergence is imposed by tolerance or by this number of iterations.
- `random_state=1`: This parameter sets the seed used by the random number generator. In this way, it will be possible to reproduce the same model and obtain the same results.

3. After setting the parameters, we can use the data to train our model:

```
MLPRegModel.fit(X_train, Y_train)
```

The `fit()` function fits the model using the training data for predictors (`X_train`) and the response (`Y_train`). Finally, we can use the model trained to predict new values:

```
Y_predMLPReg = MLPRegModel.predict(X_test)
```

In this case, the test dataset (`X_test`) was used.

4. Now, we will evaluate the performance of the MLP model using the `MSE` metric, as follows:

```
MseMLP = mean_squared_error(Y_test, Y_predMLPReg)
print(' MSE of the SKLearn Neural Network Model')
print(MseMLP)
```

The following result is returned:

```
MSE of the SKLearn Neural Network Model
0.003315706807624097
```

At this point, we can make an initial comparison between the two models that we have set up: the multiple-linear-regression-based model and the ANN-based model. We will do this by comparing the results obtained by evaluating the MSE for the two models.

5. We obtained an MSE of 0.0158 for the multiple-linear-regression-based model, and an MSE of 0.0033 for the ANN-based model. The last one returns a smaller MSE than the first by an order of magnitude, confirming the prediction that we made where neural networks return values much closer to the expected values.

Finally, we make the same comparison between the two models; however, this time, we adopt a visual approach. We will draw two scatter plots in which we will report on the two axes the actual values (expected) and the predicted values, respectively:

```
# SKLearn Neural Network diagram
plt.figure(1)
plt.subplot(121)
plt.scatter(Y_test, Y_predMLPReg)
plt.plot((0, 1), "r--")
plt.xlabel("Actual values")
plt.ylabel("Predicted values")
plt.title("SKLearn Neural Network Model")

# SKLearn Linear Regression diagram
plt.subplot(122)
plt.scatter(Y_test, Y_predLM)
plt.plot((0, 1), "r--")
plt.xlabel("Actual values")
plt.ylabel("Predicted values")
plt.title("SKLearn Linear Regression Model")
plt.show()
```

By reporting the actual and expected values on the two axes, it is possible to check how this data is arranged. To help with the analysis, it is possible to trace the bisector of the quadrant, that is, the line of the equation, $Y = X$.

The following diagrams are printed:

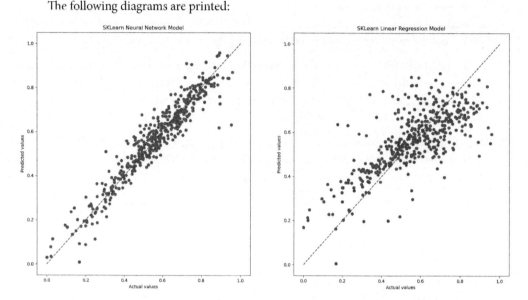

Figure 10.13: Scatterplots of the neural network models

Hypothetically, all observations should be positioned exactly on the bisector line (the dotted line in the diagram), but we can be satisfied when the data is close to this line. About half of the data points must be below the line and the other half must be above the line. Points that move significantly away from this line represent possible outliers.

Analyzing the results reported in the previous diagram, we can see that, in the graph related to the ANN-based model, the points are much closer to the dotted line. This confirms the idea that this model returns better predictions than the multiple-linear-regression-based model.

The recent history of ANNs has been enriched with models in which hidden layers extract information in different ways, in the next section we will explore deep learning.

Approaching deep neural networks

Deep learning is defined as a class of machine learning algorithms with certain characteristics. These models use multiple, hidden, nonlinear cascade layers to perform feature extraction and transformation jobs. Each level takes in the outputs from the previous level. These algorithms can be supervised, to deal with classification problems, or unsupervised, to deal with pattern analysis. The latter is based on multiple hierarchical layers of data characteristics and representations. In this way, the features of the higher layers are obtained from those of the lower layers, thus forming a hierarchy. Moreover, they learn multiple levels of representation corresponding to various levels of abstraction until they form a hierarchy of concepts.

The composition of each layer depends on the problem that needs to be solved. Deep learning techniques mainly adopt multiple hidden levels of an ANN but also sets of propositional formulas. The ANNs adopted have at least two hidden layers, but the applications of deep learning contain many more layers, for example, 10 or 20 hidden levels.

The development of deep learning in this period certainly depended on the exponential increase in data, with the consequent introduction of big data. In fact, with this exponential increase in data, there has been an increase in performance due to the increase in the level of learning, especially with respect to algorithms that already exist. In addition to this, the increase in computer performance also contributed to the improvement of obtainable results and to the considerable reduction in calculation times. There are several models based on deep learning. In the following sections, we will analyze the most popular ones.

Getting familiar with convolutional neural networks

A consequence of the application of deep learning algorithms to ANNs is the development of a new model that is much more complex but with amazing results, which is the **convolutional neural network** (**CNN**).

A CNN is a particular type of artificial feedforward neural network in which the connectivity pattern between neurons is inspired by the organization of the visual cortex of the human eye. Here, individual neurons are arranged in such a way as to devote themselves to the various regions that make up the visual field as a whole.

The hidden layers of this network are classified into various types: convolutional, pooling, ReLU, fully connected, and loss layers, depending on the role played. In the following sections, we will analyze them in detail.

Convolutional layers

This is the main layer of this model. It consists of a set of learning filters with a limited field of vision but extended along the entire surface of the input. Here, there are convolutions of each filter along the surface dimensions, making the scalar products between the filter inputs and the input image. Therefore, this generates a two-dimensional activation function that is activated if the network recognizes a certain pattern.

Pooling layers

In this layer, there is a nonlinear decimation that partitions the input image into a set of non-overlapping rectangles that are determined according to the nonlinear function. For example, with max pooling, the maximum of a certain value is identified for each region. The idea behind this layer is that the exact position of a feature is less important than its position compared to the others; therefore, superfluous information is omitted, also avoiding overfitting.

ReLU layers

The ReLU layer allows you to linearize the two-dimensional activation function and to set all negative values to zero.

Fully connected layers

This layer is generally placed at the end of the structure. It allows you to carry out the high-level reasoning of the neural network. It is called "fully connected" because the neurons in this layer have all been completely connected to the previous level.

Loss layers

This layer specifies how much the training penalizes the deviation between the predictions and the true values in output; therefore, it is always found at the end of the structure.

Examining recurrent neural networks

One of the tasks considered standard for a human, but of great difficulty for a machine, is the understanding of a piece of text. Given an ordered set of words, how can a machine be taught to understand its meaning? It is evident that, in this task, there is a more subtle relationship between the input data than in other cases. In the case of the process of classifying the content of an image, the whole image is processed by the machine simultaneously. This does not make sense in the elaboration of a piece of text, since the meaning of the words does not depend only on the words themselves, but also on their context.

Therefore, to understand a piece of text, it is not enough to know the set of words that are needed in it, it is necessary to relate them to respect the order in which they are read. It is necessary to consider, and subsequently remember, the temporal context.

Recurrent neural networks essentially allow us to remember data that could be significant during the process we want to study.

This depends on the propagation rule that is used. To understand the functioning of this type of propagation, and of memory, we can consider a case of a recurrent neural network: the Elman network.

An Elman network is very similar to a hidden single-layer feedforward neural network, but a set of neurons called context units is added to the hidden layer. For each neuron present in the hidden layer, a context unit is added, which receives, as input, the output of the corresponding hidden neuron and returns its output to the same hidden neuron.

A type of network very similar to that of Elman is that of Jordan, in which the context units save the states of the output neurons instead of those of the hidden neurons. The idea behind it, however, is the same as Elman, and it is the same as many other recurrent neural networks that are based on the following principle: receiving a sequence of data as input and processing a new sequence of output data, which is obtained by subsequently recalculating the data of the same neurons.

The recurrent neural networks that are based on this principle are manifold, and the individual topologies are chosen to face different problems. For example, in case it is not enough to remember the previous state of the network, but information processed many steps before may be necessary, **long short-term memory** (**LSTM**) neural networks can be used.

Analyzing long short-term memory networks

An LSTM is a special architecture of recurrent neural networks. These models are particularly suited to the context of deep learning because they offer excellent results and performance.

LSTM-based models are perfect for prediction and classification in the time series field, and they are replacing several traditional machine learning approaches. This is because LSTM networks can treat long-term dependencies between data. For example, this allows us to keep track of the context within a sentence to improve speech recognition.

An LSTM-based model contains cells, named LSTM blocks, which are linked together. Each cell provides three types of ports: the input gate, the output gate, and the forget gate. These ports execute the write, read, and reset functions, respectively, on the cell memory. These ports are analogical and are controlled by a sigmoid activation function in the range of 0–1. Here, 0 means total inhibition, and 1 means total activation. The ports allow the LSTM cells to remember information for an unspecified amount of time.

Therefore, if the input port reads a value below the activation threshold, the cell will maintain the previous state; whereas, if the value is above the activation threshold, the current state will be combined with the input value. The forget gate restores the current state of the cell when its value is zero, while the exit gate decides whether the value inside the cell should be removed or not.

Now let's see a new technology developed using the ANNs, but this time, graph theory comes to the rescue.

Exploring GNNs

Using structured data as graphs in an ML model is problematic due to dimensionality and non-Euclidean properties. Researchers have tried to train ML models on graph-structured data by summarizing or representing information in a simplified way. But, this feels more like preprocessing than a real training process. GNNs help us create an end-to-end ML model trained to learn a representation of structured data in graphs and to fit a predictive model into it. In order to understand the operating principle of these algorithms, it is necessary to start from the basic concepts of graph theory.

Introducing graph theory

Graphs are rational mathematical structures that are used in various fields of study including mathematics, physics, and computer science up to topology, chemistry, and engineering. A graph is represented graphically by a structure of vertices and edges. The vertices can be seen as events from which different alternatives (the edge) start. Typically, graphics are used to represent a network in an unambiguous way: the vertices can represent PCs in a LAN, road junctions, or bus stops, while the edges can represent electrical connections or roads. Edges can connect vertices in any way possible. Graph theory is a branch of mathematics that allows us to describe sets of objects together with their relations; it was birthed in 1700 by Leonhard Euler.

A graph can be represented by the following equation:

$$G = (V, E)$$

In the previous formula, we can observe the following:

- V: set of vertices
- E: set of edges

A multigraph is a graph in which it is possible that the same pair of vertices is connected by several edges, while a hypergraph is a graph in which an edge can connect one or more vertices (even more than two).

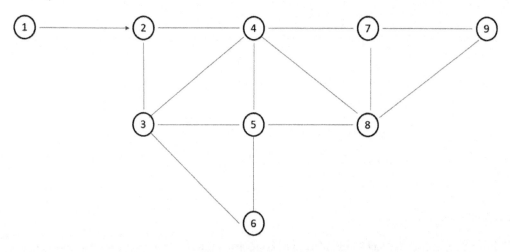

Figure 10.14: Visual representation of a graph

The number of vertices of the graph is the fundamental quantity to define its dimensions, while the number and distribution of edges describe its connectivity. Two vertices are said to be adjacent if they are connected by an arc. Adjacent edges are those edges that have a common vertex. There are several types of edges: direct edges and undirected edges without a direction. A direct edge is called an arc and its graph is called a digraph. For example, undirected edges are used to represent computer networks with synchronous links for data transmission while direct edges can represent road networks, allowing for the representation of two-way and one-way streets.

Connected graphs are defined as those for which we can reach all the other vertices of the graph from any vertex. Weighted graphs are defined as those for which a weight is associated with each arc, normally defined by a **weight function** (**w**). Weight can be seen as a cost or as the distance between the two nodes. The cost may depend on the flow that crosses the border through a law.

A vertex is characterized by its degree, which is equal to the number of edges that end on the vertex itself. According to the degree, the vertices are named as follows: isolated vertex if of order 0, leaf vertex if of order 1.

In an oriented graph, we can distinguish the outdegree (number of outgoing arcs), from the indegree (number of incoming arcs). Based on this hypothesis, the following vertices are named: source if it has zero degrees, sink if it has a degree greater than zero.

To represent a graph, we can adopt the visual representation that is obtained by drawing a point or a circle for each vertex and drawing a line between two vertices if they are connected to each other (*Figure 10.14*). If the edge is oriented, the direction is indicated by drawing an arrow.

Adjacency matrix

The graphical representation of a graph proposed in *Figure 10.14* is very intuitive indeed, but it was a simple graph with only 9 vertices. What happens if, on the other hand, the leaders become centinaia? In this case, the graphical representation becomes confused, and it is convenient to adopt an alternative method: the adjacency matrix represents a suitable solution. The representation through the adjacency matrix of an undirected graph of n nodes and arcs occurs through a matrix A of size $n \times n$ where each row and each column correspond to the nodes and the elements of the a_{ij} *matrix* are equal to 0, if not there is an arc joining the two vertices (i and j), or they are equal to the sum of the number of edges joining the vertices (*i and j*). Obviously, if only one edge is allowed between two vertices, then the matrix will contain binary values (0 and 1).

The adjacency matrix for undirected graphs is symmetric and has n^2 elements. Consequently, this representation is efficient only if the graph is significantly dense, while for sparse graphs there is a considerable waste of memory.

GNNs

GNNs learn from graph representation by extracting available information from graph elements and making it available in a low-dimensional space. Vertices collect information from their neighbors as they regularly exchange messages with each other. In this way, the information is transmitted and included in the properties of the respective vertex. The structural properties of the graph are iteratively englobed by aggregating information from the neighbors of each vertex. This procedure has an onerous computational cost and resampling can play a decisive role. The incorporation can be exploited by ML algorithms for classification or regression activities both at the top level and at the level of the entire graph. In the first case, the incorporation involves the representation of each vertex. In the second case, the grouping techniques are used for the incorporation.

In GNNs, the graph is processed by a set of units, one for each vertex, connected to each other according to the topology of the graph. Each unit calculates a status for the associated vertex, based on local information about the vertex and information about its neighbors. The state function is recursive and if the graph is cyclical, cyclical dependencies are also created in the connections of the neural network. One solution is to calculate the state through an iterative process, which will stop when an equilibrium point is reached. If the state function is a contraction mapping (that is, if it shortens the distances between points, but in a weaker way than a contraction), the system of equations admits a unique solution. Therefore, in order for the model to be applicable to a graph that is not direct or in any case containing cycles, it is necessary that constraints are imposed on the state function (that is, on the weights of the neural network) to ensure the contraction of the state function.

Let's now analyze a practical application of using ANN-based models to simulate the behavior of materials.

Simulation modeling using neural network techniques

In this example, we will try to predict the quality of the concrete starting from its ingredients. Reinforced concrete, after about a century of life, has shown its vulnerability to the action of time, atmospheric agents, and earthquakes. The proportions of the phenomenon are such as to prevent any attempt at a solution based on the planned replacement of existing buildings; therefore, in recent years, the interest in the degradation and recovery of reinforced concrete structures has increased both to safeguard the existing building heritage and, when necessary, to increase the structural safety coefficients. The problem, already relevant in itself, immediately causes a second one, that of the correct design of new buildings, aimed at minimizing deterioration, avoiding, for the future, huge and unexpected recovery costs. All this is convincing the technicians that it is no longer enough to traditionally design concrete structures, based mainly on the mechanical verification of the sections. It is essential to spread a new concept of reinforced concrete, more complete and articulated, which leads to the choice of materials on the basis of durability and no longer only on resistance. In this perspective, a simulation model of the properties of the concrete starting from its basic constituents can represent an important resource to support the designer.

Concrete quality prediction model

Concrete is a mixture of binder (cement), water, and aggregates. In its fresh state, concrete has a plastic consistency, after laying and with time it hardens and, depending on the percentage quantity of the individual components, reaches lithoid characteristics, similar to those of a natural conglomerate. The mixture may also contain additives and mineral additions. Concrete can be classified either according to its characteristics or according to its composition. The mechanical characteristics of the concrete are evaluated through tests in the laboratory or in situ. Compressive strength is one of the most important characteristics of concrete. Based on the compressive strength, concrete can be classified into one of the classes provided. The evaluation is carried out on the basis of tests after 28 days on cylindrical specimens of 30 cm in length and 15 cm in diameter or cubic specimens with a side of 15 cm. Such tests are destructive, so having a tool that allows us to predict these characteristics is crucial.

In this section, we will elaborate on a model of prediction of the compressive strength of a concrete starting from the composition of the mixture of its constituents.

As always, we will analyze the code line by line:

1. Let's start by importing the libraries and functions:

    ```
    import pandas as pd
    import seaborn as sns
    from sklearn.preprocessing import MinMaxScaler
    from sklearn.model_selection import train_test_split
    from keras.models import Sequential
    from keras.layers import Dense
    from sklearn.metrics import r2_score
    ```

    The `pandas` library is an open source, BSD-licensed library that provides high-performance, easy-to-use data structures and data analysis tools for the Python programming language. It offers data structures and operations for manipulating numerical data in a simple way. In the `seaborn` module, there are several features we can use to graphically represent our data. There are methods that facilitate the construction of statistical graphs with `matplotlib`. The `sklearn.preprocessing` package provides several common utility functions and transformer classes to modify the features available in a representation that best suits our needs. The `sklearn.model_selection.train_test_split()` function computes a random split into training and test sets. We then imported the Keras models that we will use to build the ANN architecture. Finally, we imported the metrics for evaluating the network's performance.

2. Now we need to import the data:

    ```
    features_names=
    ['Cement','BFS','FLA','Water','SP','CA','FA','Age','CCS']
    concrete_data = pd.read_excel('concrete_data.xlsx',
    names=features_names)
    ```

    As we have already anticipated, our simulation will try to predict the compressive strength of the concrete starting from its basic components. To do this, we will use a dataset available on the UCI Machine Learning Repository at https://archive.ics.uci.edu/ml/datasets/concrete+compressive+strength. The dataset contains nine features: cement, **blast furnace slag (BFS)**, **fly ash (FLA)**, water, **superplasticizer (SP)**, **coarse aggregate (CA)**, **fine aggregate (FA)**, age, and **concrete compressive strength (CCS)**. The last feature is precisely the variable that we want to estimate. We, therefore, first defined the feature labels and then we used the pandas `read_excel()` function to import the file available in the `.xlsx` format.

3. So, let's look at the statistics of the dataset:

```
summary = concrete_data.describe()
print(summary)
```

The following data is returned:

```
              Cement          BFS    ...          Age          CCS
count    1029.000000  1029.000000    ...  1029.000000  1029.000000
mean      280.914091    73.967298    ...    45.679300    35.774912
std       104.245542    86.290255    ...    63.198226    16.656880
min       102.000000     0.000000    ...     1.000000     2.331808
25%       192.000000     0.000000    ...     7.000000    23.696601
50%       272.800000    22.000000    ...    28.000000    34.397958
75%       350.000000   143.000000    ...    56.000000    45.939786
max       540.000000   359.400000    ...   365.000000    82.599225
```

Figure 10.15: Summary of data

We can see that there are 1,029 records; moreover, the numerical values of the features have different ranges.

4. In order to better appreciate these differences, let's draw a boxplot:

```
sns.set(style="ticks")
sns.boxplot(data = concrete_data)
```

The following chart is plotted:

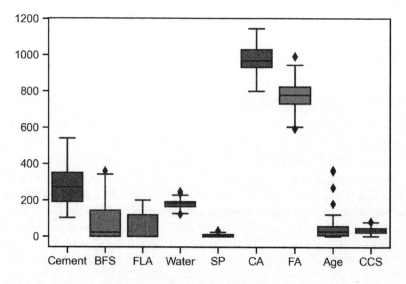

Figure 10.16: A boxplot of the features

As already mentioned, we can verify that the features have different ranges of values. This represents a factor that can be influential in the elaboration of a model based on ANNs: the variables that assume higher numerical values can assume a relevant weight in the prediction of the data, even if this is not true from the point of view of the phenomenon. In these cases, it is advisable to scale the data.

5. So, let's try to scale the data:

```
scaler = MinMaxScaler()
print(scaler.fit(concrete_data))
scaled_data = scaler.fit_transform(concrete_data)
scaled_data = pd.DataFrame(scaled_data,
                columns=features_names)
```

To do this, we used the `MinMaxScaler()` function of the `sklearn.preprocessing` library. The scaling is carried out using the following formula:

$$x_{scaled} = \frac{x - x_{min}}{x_{max} - x_{min}}$$

In this way, all features will have the same range of variability (0-1).

To confirm this, we proceed to carry out a new statistic on the scaled data:

```
summary = scaled_data.describe()
print(summary)
```

The following results are returned:

```
              Cement          BFS   ...          Age          CCS
count    1029.000000  1029.000000   ...  1029.000000  1029.000000
mean        0.408480     0.205808   ...     0.122745     0.416646
std         0.238004     0.240095   ...     0.173621     0.207517
min         0.000000     0.000000   ...     0.000000     0.000000
25%         0.205479     0.000000   ...     0.016484     0.266170
50%         0.389954     0.061213   ...     0.074176     0.399491
75%         0.566210     0.397885   ...     0.151099     0.543284
max         1.000000     1.000000   ...     1.000000     1.000000
```

Figure 10.17: Statistics of the scaled features

6. Now each feature is scaled between 0–1. Let's see the effect on the variables through the boxplot:

```
sns.boxplot(data = scaled_data)
```

The following chart is plotted:

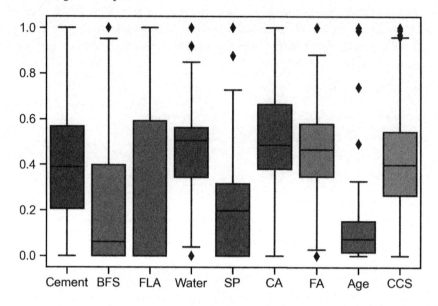

Figure 10.18: A boxplot of the scaled features

Now everything is clearer, we can easily make a visual check of the variability of the features in a much easier way. But above all, now we will not run the risk of introducing addictions to specific variables that are not true in reality.

7. Let's reconstruct the pandas DataFrames for the input and output data:

```
input_data = pd.DataFrame(scaled_data.iloc[:,:8])
output_data = pd.DataFrame(scaled_data.iloc[:,8])
```

8. Now, we need to split the data we will use to train the model and test it:

```
inp_train, inp_test, out_train, out_test = train_test_split(input_data,output_data, test_size = 0.30, random_state = 1)
print(inp_train.shape)
print(inp_test.shape)
print(out_train.shape)
print(out_test.shape)
```

Here, we used the `train_test_split()` function of the `sklearn.model_selection` library. This function quickly computes a random split into training and test sets. We decided to split the dataset into 70% for training and the remaining 30% to test the model. We also set the `random_state` to make the simulation reproducible. Finally, we printed the dimensions of the new datasets on the screen. The following values were printed:

```
(720, 8)
(309, 8)
(720, 1)
(309, 1)
```

9. Now, we can build the ANN-based model:

    ```
    model = Sequential()
    model.add(Dense(20, input_dim=8, activation='relu'))
    model.add(Dense(10, activation='relu'))
    model.add(Dense(10, activation='relu'))
    model.add(Dense(1, activation='linear'))
    model.compile(optimizer='adam',loss='mean_squared_error',metrics=['accuracy'])
    model.fit(inp_train, out_train, epochs=1000, verbose=1)
    ```

    The sequential class is used to define a linear stack of network layers that make up a model. The Dense class is used to instantiate a Dense layer, which is the basic feedforward fully connected layer. Fully connected levels are defined using the Dense class. The ANN architecture is a completely connected network structure with four layers (two hidden layers):

    - The first layer added defines the input, three parameters are passed: 20, input_dim = 8, and activation = ' relu '. 20 (units) is a positive integer representing the dimensionality of the output space, denoting the number of neurons in the level. input_dim = 8 is the number of the input features. Finally, activation = ' relu ' is used to set the activation function (**Rectified Linear Unit (ReLU)** activation function.)

    - The second layer (first hidden layer) has 10 neurons and the relu activation function

    - The third layer (second hidden layer) has 10 neurons and the relu activation function, again

    - Finally, the output layer has a single neuron (output) and a linear activation function.

10. Before training a model, we need to configure the learning process, which is done via the compile() method. Three arguments are passed: the adam optimizer, the mean_squared_error loss function, and the accuracy metric. The first parameter set the optimization as first-order gradient-based optimization of stochastic objective functions, based on adaptive estimates of lower-order moments. The second parameter measured the average of the squares of the errors. Finally, the third parameter set the evaluation metric as accuracy.

Finally, to train the model, the `fit()` method is used passing four arguments: the array of predictors training data, the array of response data, the number of epochs to train the model, and the verbosity mode.

Therefore, we summarize the characteristics of the model:

```
model.summary()
```

The following data is returned:

```
Model: "sequential_1"
_____
Layer (type)                 Output Shape              Param #
=================================================================
dense (Dense)                (None, 20)                180

dense_1 (Dense)              (None, 10)                210

dense_2 (Dense)              (None, 10)                110

dense_3 (Dense)              (None, 1)                 11
=================================================================
Total params: 511
Trainable params: 511
Non-trainable params: 0
_____
```

Figure 10.19: Architecture of ANN-based model

11. Now, we can finally use the model to make predictions of the characteristics of the material:

```
output_pred = model.predict(inp_test)
```

We, therefore, predicted the compressive strength of the concrete starting from the mixture of ingredients.

12. Now, we just have to evaluate the performance of the model:

```
print('Coefficient of determination = ')
print(r2_score(out_test, output_pred))
```

To do this, we used the coefficient of determination. This metric indicates the proportion of total variance of the response values around its mean that is explained by the model. Precisely because it is a proportion, its value will always be between 0 and 1. The following result is returned:

```
Coefficient of determination =
0.8591542120174626
```

We obtained a very high value to indicate the good predictive ability of the model.

Summary

In this chapter, we learned how to develop models based on ANNs to simulate physical phenomena. We started by analyzing the basic concepts of neural networks and the principles they are based on that are derived from biological neurons. We examined, in detail, the architecture of an ANN, understanding the concepts of weights, bias, layers, and the `activation` function.

Subsequently, we analyzed the architecture of a feedforward neural network. We saw how the training of the network with data takes place, and we understood the weight adjustment procedure that leads the network to correctly recognize new observations.

After that, we applied the concepts learned by tackling a practical case. We developed a model based on neural networks to solve a regression problem. We learned to scale data and then to subset the data for training and testing. We learned how to develop a model based on linear and MLP regression. We learned how to evaluate the performance of these models to make a comparison. Then, we explored deep neural networks. We defined them by analyzing their basic concepts. We analyzed the basics of CNNs, recurrent neural networks, and LSTM networks.

Finally, after introducing the theory of graphs and GNNs, we have faced a practical case of prediction of the physical characteristics of a material using the deep learning model.

In the next chapter, we will explore practical cases of project management using the tools we learned about in the previous chapters. We will learn how to evaluate the results of the actions we take when managing a forest using Markov processes, and then move on and learn how to evaluate a project using Monte Carlo simulation.

11
Modeling and Simulation for Project Management

Sometimes, monitoring resources, budgets, and milestones for various projects and divisions can present a challenge. Simulation tools help us to improve planning and coordination in the various phases of a project so that we always keep control of it. In addition, the preventive simulation of a project can highlight the critical issues related to a specific task. This helps us evaluate the cost of any actions to be taken. Through the preventive evaluation of the development of a project, errors that increase the costs of a project can be avoided.

In this chapter, we will deal with practical cases of project management using the tools we learned about in the previous chapters. We will learn how to evaluate the results of the actions we take when managing a forest using Markov processes, and then we will learn how to evaluate a project using the Monte Carlo simulation.

In this chapter, we're going to cover the following main topics:

- Introducing project management
- Managing a tiny forest problem
- Scheduling project time using the Monte Carlo simulation

Technical requirements

In this chapter, we will address modeling examples of project management. To deal with these topics, it is necessary that you have a basic knowledge of algebra and mathematical modeling.

To work with the Python code in this chapter, you'll need the following files (available on GitHub at the following URL: https://github.com/PacktPublishing/Hands-On-Simulation-Modeling-with-Python-Second-Edition):

- tiny_forest_management.py

- `tiny_forest_management_modified.py`
- `monteCarlo_tasks_scheduling.py`

Introducing project management

To assess the consequences of a strategic or tactical move in advance, companies need reliable predictive systems. Predictive analysis systems are based on data collection and the projection of reliable scenarios in the medium and long term. In this way, we can provide indications and guidelines for complex strategies, especially those that must consider numerous factors from different entities.

This allows us to examine the results of the evaluation in a more complete and coordinated way since we can simultaneously consider a range of values and, consequently, a range of possible scenarios. Finally, when managing complex projects, the use of artificial intelligence to interpret data has increased, thus giving these projects meaning. This is because we can perform sophisticated analyses of the information to improve the strategic decision-making process we will undertake. This methodology allows us to search and analyze data from different sources so that we can identify patterns and relationships that may be relevant.

Understanding what-if analysis

What-if analysis is a type of analysis that can contribute significantly to making managerial decisions more effective, safe, and informed. It is also the most basic level of predictive analysis that is based on data. What-if analysis is a tool capable of elaborating different scenarios to offer different possible outcomes. Unlike advanced predictive analysis, what-if analysis has the advantage of only requiring basic data to be processed.

This type of activity falls into the category of predictive analytics, that is, those that produce forecasts for the future, starting from a historical basis or trends. By varying some parameters, it is possible to simulate different scenarios and, therefore, understand what impact a given choice would have on costs, revenues, profits, and so on.

What-if analysis is therefore a structured method to determine which predictions related to strategy changes can go wrong, thereby judging the probability and consequences of the studies carried out before they happen. Through the analysis of historical data, it is possible to create such predictive systems capable of estimating future results following the assumptions that were made about a group of variables of independent inputs, thus allowing us to formulate some forecasting scenarios with the aim of evaluating the behavior of a real system.

Analyzing the scenario at hand allows us to determine the expected values related to a management project. These analysis scenarios can be applied in different ways, the most typical of which is to perform multi-factor analysis, that is, analyze models containing multiple variables:

- The realization of a fixed number of scenarios by determining the maximum and minimum difference and creating intermediate scenarios through risk analysis. Risk analysis aims to determine the probability that a future result will be different from the average expected result. To show this possible variation, an estimate of the less likely positive and negative results is performed.
- A random factorial analysis using Monte Carlo methods, thus solving a problem by generating appropriate random numbers and observing the fraction of the numbers that obey one or more properties. These methods are useful for obtaining numerical solutions for problems that are too complicated to solve analytically.

After having adequately introduced the basic concepts of project management, let's now see how to implement a simple project by treating it as a Markovian process.

Managing a tiny forest problem

As we mentioned in *Chapter 5, Simulation-Based Markov Decision Processes*, a stochastic process is called **Markovian** if it starts from an instant *t* in which an observation of the system is made. The evolution of this process will depend only on *t*, so it will not be influenced by the previous instants. So, a process is called Markovian when the future evolution of the process depends only on that of instant of observing the system and does not depend in any way on the past. **Markov Decision Processes** (MDP) is characterized by five elements: decision epochs, states, actions, transition probability, and reward.

Summarizing the Markov decision process

The crucial elements of a Markovian process are the states in which the system finds itself and the available actions that the decision-maker can carry out in that state. These elements identify two sets: the set of states in which the system can be found, and the set of actions available for each specific state. The action chosen by the decision maker determines a random response from the system, which ultimately brings it into a new state. This transition returns a reward that the decision-maker can use to evaluate the efficacy of their choice.

> **Important note**
> In a Markovian process, the decision maker has the option of choosing which action to perform in each state of the system. The action chosen takes the system to the next state and the reward for that choice is returned. The transition from one state to another enjoys the property of Markov; the current state depends only on the previous one.

A Markov process is defined by four elements, as follows:

- S: System states
- A: Actions available for each state

- P: Transition matrix. This contains the probabilities that an action a takes the system from s state to s' state
- R: Rewards obtained in the transition from s state to s' state with an action a

In an MDP problem, it becomes crucial to take action to obtain the maximum reward from the system. Therefore, this is an optimization problem in which the sequence of choices that the decision-maker will have to make is called an optimal policy.

A policy maps both the states of the environment and the actions to be chosen for those states, representing a set of rules or associations that respond to a stimulus. The policy's goal is to maximize the total reward received through the entire sequence of actions performed by the system. The total reward that's obtained by adopting a policy is calculated as follows:

$$R_T = \sum_{i=0}^{T} r_{t+1} = r_t + r_{t+1} + \cdots + r_T$$

In the previous equation, r_T is the reward of the action that brings the environment into the terminal state s_T. To get the maximum total reward, we can select the action that provides the highest reward to each individual state. This leads to us choosing the optimal policy that maximizes the total reward.

Exploring the optimization process

As we mentioned in *Chapter 5, Simulation-Based Markov Decision Processes*, an MDP problem can be addressed using **dynamic programming** (**DP**). DP is a programming technique that aims to calculate an optimal policy based on a knowing model of the environment. The core of DP is to utilize the state value and action value to identify good policies.

In DP methods, two processes called policy evaluation and policy improvement are used. These processes interact with each other, as follows:

- Policy evaluation is done through an iterative process that seeks to solve Bellman's equation. The convergence of the process for $k \to \infty$ imposes approximation rules, thus introducing a stop condition.
- Policy improvement improves the current policy based on the current values.

In the DP technique, the previous phases alternate and end before the other begins via an iteration procedure. This procedure requires a policy evaluation at each step, which is done through an iterative method, whose convergence is not known a priori and depends on the starting policy; that is, we can stop evaluating the policy at some point, while still ensuring convergence to an optimal value.

> **Important note**
> The iterative procedure we have described uses two vectors that preserve the results obtained from the policy evaluation and policy improvement processes. We indicate the vector that will contain the value function with *V*, that is, the discounted sum of the rewards obtained. We indicate the carrier that will contain the actions chosen to obtain those rewards with *Policy*.

The algorithm then, through a recursive procedure, updates these two vectors. In the policy evaluation, the value function is updated as follows:

$$V(s) = \sum_{s'} P_{Policy(s)}(s, s') \left(R_{Policy(s)}(s, s') + \gamma * V(s') \right)$$

In the previous equation, we have the following:

- $V(s)$ is the function value at the state *s*
- $R(s, s')$ is the reward returned in the transition from state *s* to state *s'*
- γ is the discount factor
- $V(s')$ is the function value at the next state

In the policy improvement process, the policy is updated as follows:

$$Policy(s) = \arg\max_a \left\{ \sum_{s'} P_a(s, s')(R_a(s, s') + \gamma * V(s')) \right\}$$

In the previous equation, we have the following:

- $V(s')$ is the function value at state *s'*
- $R_a(s, s')$ is the reward returned in the transition from state *s* to state *s'* with action *a*
- γ is the discount factor
- $P_a(s, s')$ is the probability that an action *a* in the *s* state is carried out in the *s'* state

Now, let's see what tools we have available to deal with MDP problems in Python.

Introducing MDPtoolbox

The `MDPtoolbox` package contains several functions connected to the resolution of discrete-time Markov decision processes, such as value iteration, finite horizon, policy iteration, linear programming algorithms with some variants, and several functions we can use to perform reinforcement learning analysis.

This toolbox was created by researchers from the Applied Mathematics and Computer Science Unit of INRA Toulouse (France), in the MATLAB environment. The toolbox was presented by the authors in the following article: Chadès, I., Chapron, G., Cros, M. J., Garcia, F., & Sabbadin, R. (2014). *MDPtoolbox: a multi-platform toolbox to solve stochastic dynamic programming problems. Ecography*, 37 (9), 916-920.

> **Important note**
>
> The `MDPtoolbox` package was subsequently made available in other programming platforms, including GNU Octave, Scilab, and R. It was later made available for Python programmers by S. Cordwell.
>
> You can find out more at the following URL: https://github.com/sawcordwell/pymdptoolbox.

To use the `MDPtoolbox` package, we need to install it. The different installation procedures are indicated on the project's GitHub website. As recommended by the author, you can use the default Python pip package manager. **Python Package Index** (**pip**) is the largest and most official Python package repository. Anyone who develops a Python package, in 99% of cases, makes it available on this repository.

To install the `MDPtoolbox` package using `pip`, just write the following command:

```
pip install pymdptoolbox
```

Once installed, just load the library to be able to use it immediately.

Defining the tiny forest management example

To analyze in detail how to deal with a management problem using Markovian processes, we will use an example already available in the `MDPtoolbox` package. It deals with managing a small forest in which there are two types of resources: wild fauna and trees. The trees of the forest can be cut, and the wood that's obtained can be sold. The decision maker has two actions: wait and cut. The first action is to wait for the tree to grow fully before cutting it to obtain more wood. The second action involves cutting the tree to get money immediately. The decision maker has the task of making their decision every 20 years.

The tiny forest environment can be in one of the following three states:

- **State 1**: Forest age 0-20 years
- **State 2**: Forest age 21-40 years
- **State 3**: Forest age over 40 years

We might think that the best action is to wait until we have the maximum amount of wood to come and thus obtain the greatest gain. Waiting can lead to the loss of all the wood available. This is because as the trees grow, there is also the danger that a fire could develop, which could cause the wood to be lost completely. In this case, the tiny forest would be returned to its initial state (state 1), so we would lose what we would have gained by waiting until the forest reached full maturity (state 3).

In the case a fire does not occur, at the end of each period t (20 years), if the state is s and the wait action is chosen, the forest will move to the next state, which will be the minimum of the following pair $(s + 1, 3)$. If there are no fires, the age of the forest will never assume a state higher than 3, since state 3 matches the oldest age class. Conversely, if a fire occurs after the action is applied, the forest returns the system to its initial state (state 1), as shown in the following figure:

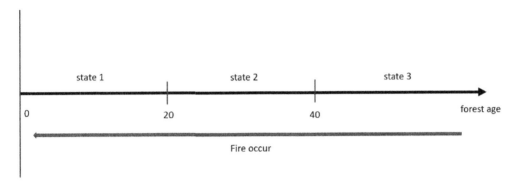

Figure 11.1 – States of the age of a forest

Set $P = 0.1$ as the probability that a fire occurs during a period t: the problem is how to manage this in the long term to maximize the reward. This problem can be treated as a Markov decision process.

Now, let's move on and define the problem as an MDP. We have said that the elements of an MDP are state, action, transition matrix (P), and reward (R). We must then define these elements. We have defined the states already – there are three. We also defined the actions, which are wait or cut. We pass these to define the transition matrix $P(s, s', a)$. It contains the chances of the system going from one state to another. We have two actions available (wait or cut), so we will define two transition matrices. If we indicate with p the probability that a fire occurs, then in the case of the wait action, we will have the following transition matrix:

$$P(,,1) = \begin{bmatrix} p & 1-p & 0 \\ p & 0 & 1-p \\ p & 0 & 1-p \end{bmatrix}$$

Now, let's analyze the content of the transition matrix. Each row is relative to a state, in the sense that row 1 returns the probabilities that, starting from state 1, the tiny forest will remain in state 1 or pass to state 2 or 3. If we are in state 1, we will have a probability p that the tiny forest will remain in that state, which happens if a fire occurs. Beginning in state 1, if no fire occurs, we have the remaining $1-p$ probability of moving to the next state, which is state 2. When starting from state 1, the probability of passing to state 3 is equal to 0 – it's impossible to do so.

Row 2 of the transition matrix contains the transition probabilities when starting from state 2. If a fire occurs when starting from state 2, there will be an equal probability p to pass into state 1. If no fire occurs when starting from state 2, we have the remaining $1-p$ probability of moving to the next state, which is state 3. In this case, once again, the probability of remaining in state 2 is equal to 0.

To better understand the transition of states we can analyze the transition diagram first introduced in *Chapter 5, Simulation-Based Markov Decision Processes* (*Figure 11.2*).

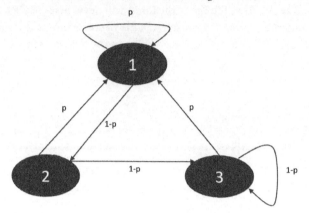

Figure 11.2 – Transition diagram

Finally, if we are in state 3 and a fire occurs, we will have a probability equal to p that we will go back to state 1, and the remaining *1-p* probability of remaining in state 3, which happens if no fire occurs. The probability of going to state 2 is equal to *0*.

Now, let's define the transition matrix in the case of choosing the cut action:

$$P(,,2) = \begin{bmatrix} 1 & 0 & 0 \\ 1 & 0 & 0 \\ 1 & 0 & 0 \end{bmatrix}$$

In this case, the analysis of the previous transition matrix is much more immediate. In fact, the cut action brings the state of the system to 1 in each instance. Therefore, the probability is always 1. Then, that *1* goes to state 1 and *0* for all the other transitions as they are not possible.

Now, let's define the vectors that contain the rewards, that is, the vector R (s, s', as we have defined it), starting from the rewards returned by the wait action:

$$R(,1) = \begin{bmatrix} 0 \\ 0 \\ 4 \end{bmatrix}$$

The action of waiting for the growth of the forest will bring a reward of *0* for the first two states, while the reward will be the maximum for state 3. The value of the reward in state 3 is equal to 4, which represents the value provided by the system by default. Let's see how the vector of rewards is modified if you choose the cut action:

$$R(,2) = \begin{bmatrix} 0 \\ 1 \\ 2 \end{bmatrix}$$

In this case, cutting in state 1 does not bring any reward since the trees are not able to supply wood yet. The decision to cut the trees in state 2 brings a reward, but this is lower than the maximum one, which is obtainable if we wait for the end of the three periods *t* before cutting. A similar situation arises if the cut is made at the beginning of the third period. In this case, the reward is greater than that of the previous state but still less than the maximum.

Addressing management problems using MDPtoolbox

Our goal is to develop a policy that allows us to manage the tiny forest to obtain the maximum prize. We will do this using the MDPtoolbox package, which we introduced in the previous section, and analyzing the code line by line:

1. Let's start as always by importing the necessary library:

   ```
   import mdptoolbox.example
   ```

   By doing this, we imported the MDPtoolbox module, which contains the data for this example.

2. To begin, we will extract the transition matrix and the reward vectors:

   ```
   P, R = mdptoolbox.example.forest()
   ```

   This command retrieves the data stored in the example. To confirm the data is correct, we need to print the content of these variables, starting from the transition matrix:

   ```
   print(P[0])
   ```

   The following matrix is printed:

   ```
   [[0.1 0.9 0. ]
    [0.1 0.  0.9]
    [0.1 0.  0.9]]
   ```

   This is the transition matrix for the wait action. Consistent with what is indicated in the *Defining the tiny forest management example* section, after having set $p = 0.1$ we can confirm the transition matrix.

3. Now, we will print the transition matrix related to the cut action:

   ```
   print(P[1])
   ```

   The following matrix is printed:

   ```
   [[1. 0. 0.]
    [1. 0. 0.]
    [1. 0. 0.]]
   ```

4. This matrix is also consistent with what was previously defined. Now, let's check the shape of the reward vectors, starting from the wait action:

   ```
   print(R[:,0])
   ```

   Let's see the content of this vector:

   ```
   [0. 0. 4.]
   ```

   If the cut action is chosen, we will have the following rewards:

   ```
   print(R[:,1])
   ```

   The following vector is printed:

   ```
   [0. 1. 2.]
   ```

5. Finally, let's fix the discount factor:

   ```
   gamma=0.9
   ```

   All the problem data has now been defined. We can now move on and look at the model in greater detail.

6. The time has come to apply the policy iteration algorithm to the problem we have just defined:

   ```
   PolIterModel = mdptoolbox.mdp.PolicyIteration(P, R, gamma)
   PolIterModel.run()
   ```

   The `mdptoolbox.mdp.PolicyIteration()` function performs a discounted MDP that's solved using the policy iteration algorithm. Policy iteration is a dynamic programming method that adopts a value function in order to model the expected return for each pair of action-state. These techniques update the value functions by using the immediate reward and the (discounted) value of the next state in a process called **bootstrapping**. The results are stored in tables or with approximate function techniques.

   The function, starting from an initial *P0* policy, updated the function value and the policy through an iterative procedure, alternating the following two phases:

   - **Policy evaluation**: Given the current policy *P*, estimate the action-value function.
   - **Policy Improvement**: If we calculate a better policy based on the action-value function, then this policy is made the new policy and we return to the previous step.

   When the value function can be calculated exactly for each action-state pair, the policy iteration we performed with the greedy policy improvement leads to convergence by returning the optimal policy. Essentially, repeatedly executing these two processes converges the general process toward the optimal solution.

In the `mdptoolbox.mdp.PolicyIteration()` function, we have passed the following arguments:

- **P**: Transition probability
- **R**: Reward
- **gamma**: Discount factor

The following results are returned:

- **V**: The optimal value function. *V* is an *S* length vector.
- **Policy**: The optimal policy. The policy is an *S* length vector. Each element is an integer corresponding to an action that maximizes the value function. In this example, only two actions are foreseen: *0* = wait, and *1* = cut.
- **iter**: The number of iterations.
- **time**: The CPU time used to run the program.

7. Now that the model is ready, we must evaluate the results by checking the obtained policy.

    To begin, we will check the updates of the value function:

    ```
    print(PolIterModel.V)
    ```

    The following results are returned:

    ```
    (26.244000000000014, 29.484000000000016,
    33.484000000000016)
    ```

    A value function specifies how good a state is for the system. This value represents the total reward expected for a system from the status *s*. The value function depends on the policy that the agent selects for the actions to be performed.

8. Let's move on and extract the policy:

    ```
    print(PolIterModel.policy)
    ```

    A policy suggests the behavior of the system at a given time. It maps the detected states of the environment and the actions to take when they are in those states. This corresponds to what, in psychology, would be called a set of rules or associations of the stimulus-response. The policy is the crucial element of an MDP model since it defines the behavior. The following results are returned:

    ```
    (0, 0, 0)
    ```

    Here, the optimal policy is to not cut the forest in all three states. This is due to the low probability of a fire occurring, which causes the wait action to be the best action to perform. In this way, the forest has time to grow and we can achieve both goals: maintain an old forest for wildlife and earn money by selling the cut wood.

9. Let's see how many iterations have been made:

   ```
   print(PolIterModel.iter)
   ```

   The following result is returned:

   ```
   2
   ```

10. Finally, let's print the CPU time:

    ```
    print(PolIterModel.time)
    ```

    The following result is returned:

    ```
    0.0009965896606445312
    ```

Only *0.0009* seconds are required to perform the value iteration procedure, but obviously, this value depends on the machine in use.

Changing the probability of a fire starting

The analysis of the previous example has clarified how to derive an optimal policy from a well-posed problem. We can now define a new problem by changing the initial conditions of the system. Under the default conditions provided in our example, the probability of a fire occurring is low. In this case, we have seen that the optimal policy advises us to wait and not cut the forest. But what if we increase the probability of a fire occurring? This is a real-life situation; just think of warm places particularly subject to strong winds. To model this new condition, simply change the problem settings by changing the probability value *p*. The `mdptoolbox.example.forest()` module allows us to modify the basic characteristics of the problem. Let's get started:

1. Let's start by importing the example module:

   ```
   import mdptoolbox.example
   P, R = mdptoolbox.example.forest(3,4,2,0.8)
   ```

   Compared to the example discussed in the previous section, *Addressing management problems using MDPtoolbox*, we have passed some parameters. Let's analyze them in detail.

   In the `mdptoolbox.example.forest ()` function, we passed the following parameters (3, 4, 2, 0.8).

   Let's analyze their meaning:

   - 3: The number of states. This must be an integer greater than *0*.
   - 4: The reward when the forest is in the oldest state and the wait action is performed. This must be an integer greater than *0*.

- 2: The reward when the forest is in the oldest state and the cut action is performed. This must be an integer greater than 0.
- 0.8: The probability of a fire occurring. This must be in [0, 1].

By analyzing the past data, we can see that we confirmed the first three parameters, while we only changed the probability of a fire occurring, thus increasing this possibility from *0.1* to *0.8*.

2. Let's see this change to the initial data as it changed the transition matrix:

```
print(P[0])
```

The following matrix is printed:

```
[[0.8 0.2 0. ]
 [0.8 0.  0.2]
 [0.8 0.  0.2]]
```

As we can see, the transition matrix linked to the wait action has changed. Now, the probability that the transition is in a state other than 1 has significantly decreased. This is due to the high probability that a fire will occur. Let's see what happens when the transition matrix is linked to the cut action:

```
print(P[1])
```

The following matrix is printed:

```
[[1. 0. 0.]
 [1. 0. 0.]
 [1. 0. 0.]]
```

This matrix remains unchanged. This is due to the cut action, which returns the system to its initial state. Likewise, the reward vectors may be unchanged since the rewards that were passed are the same as the ones provided by the default problem. Let's print these values:

```
print(R[:,0])
```

This is the reward vector connected to the wait action. The following vector is printed:

```
[0. 0. 4.]
```

In the case of the cut action, the following vector is returned:

```
print(R[:,1])
```

The following vector is printed:

```
[0. 1. 2.]
```

As anticipated, nothing has changed. Finally, let's fix the discount factor:

```
gamma=0.9
```

All the problem data has now been defined. We can now move on and look at the model in greater detail.

3. We will now apply the value iteration algorithm:

    ```
    PolIterModel = mdptoolbox.mdp.PolicyIteration(P, R,
    gamma)
    PolIterModel.run()
    ```

4. Now, we can extract the results, starting from the value function:

    ```
    print(PolIterModel.V)
    ```

    The following results are printed:

    ```
    (1.5254237288135597, 2.3728813559322037,
    6.217445225299711)
    ```

5. Let's now analyze the crucial part of the problem: let's see what policy the simulation model suggests:

    ```
    print(PolIterModel.policy)
    ```

    The following results are printed:

    ```
    (0, 1, 0)
    ```

    In this case, the changes we've made here, compared to the default example, are substantial. It's suggested that we adopt the wait action if we are in state 1 or 3, while if we are in state 2, it is advisable to try to cut the forest. Since the probability of a fire is high, it is convenient to cut the wood already available and sell it before a fire destroys it.

6. Then, we print the number of iterations of the problem:

    ```
    print(PolIterModel.iter)
    ```

    The following result is printed:

    ```
    1
    ```

    Finally, we print the CPU time, which is as follows:

    ```
    print(PolIterModel.time)
    ```

    The following result is printed:

    ```
    0.0009970664978027344
    ```

These examples have highlighted how simple the modeling procedure of a management problem is through using MDPs.

Now, let's see how to design a scheduling grid using the Monte Carlo simulation.

Scheduling project time using the Monte Carlo simulation

Each project requires a time of realization, and the beginning of some activities can be independent or dependent on previous activities ending. Scheduling a project means determining the time of realization for the project itself. A project is a temporary effort undertaken to create a unique product, service, or result. The term **project management** refers to the application of knowledge, skills, tools, and techniques for the purpose of planning, managing, and controlling a project and the activities that it is composed of.

The key figure in this area is the project manager, who has the task and responsibility of coordinating and controlling the various components and actors involved in the project, with the aim of reducing the probability of project failure. The main difficulty in this series of activities is achieving the objectives set in compliance with any constraints, such as the scope of the project, time, costs, quality, and resources. In fact, these are limited aspects that are linked to each other and need effective optimization.

The definition of these activities constitutes one of the key moments of the planning phase. After defining what the project objectives are with respect to time, costs, and resources, it is necessary to proceed with identifying and documenting the activities that must be carried out to successfully complete the project.

For complex projects, it is necessary to create an ordered structure by decomposing the project into simpler tasks. For each task, it will be necessary to define the activities and their execution times. This starts with the main objective and breaks down the project to the immediately lower level in all those deliverables or main sub-projects that make it up.

These will, in turn, be broken down. This will continue until you are satisfied with the degree of detail in the resulting final items. Each breakdown results in a reduction in the size, complexity, and cost of the interested party.

Defining the scheduling grid

A fundamental part of project management is constructing the scheduling grid. This is an oriented graph that represents the temporal succession and the logical dependencies between the activities that are involved in the realization of the project. In addition to constructing the grid, the scheduling process also determines the start and end times of the activities based on factors such as duration, resources, and so on.

In the example we are dealing with, we will evaluate the times necessary for the realization of a complex project. Let's start by defining the scheduling grid. Suppose that, by decomposing the project structure, we have defined six tasks. For each task, the activities, the personnel involved, and the time needed to finish the job were defined.

Some tasks must be performed in a series, in the sense that the activities of the previous task must be completed so that they can start the activities of the next task. Others, however, can be performed in parallel, in the sense that two teams can simultaneously work on two different tasks to reduce project delivery times. This sequence of tasks is defined in the scheduling grid, as follows:

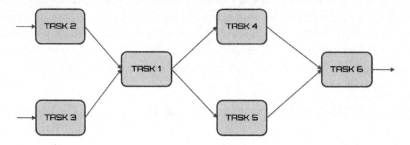

Figure 11.3 – The sequence of tasks in the grid

The preceding diagram shows us that the first two tasks developed in parallel, which means that the time required to finish these two tasks will be provided by the time-consuming task. The third task develops in a series, while the next two are, again, in parallel. The last task is still in the series. This sequence will be necessary when we evaluate the project times.

Estimating the task's time

The duration of these tasks is often difficult to estimate due to the number of factors that can influence it such as the availability and/or productivity of the resources, the technical and physical constraints between the activities, and the contractual commitments.

Expert advice, supported by historical information, can be used wherever possible. The members of the project team will also be able to provide information on the duration of the task or the maximum recommended limit for the duration of the task by deriving information from similar projects.

There are several ways we can estimate tasks. In this example, we will use a three-point estimation. In a three-point estimation, the accuracy of the duration of the activity estimate can be increased in terms of the amount of risk in the original estimate. Three-point estimates are based on determining the following three types of estimates:

- **Optimistic**: The duration of the activity is based on the best scenario of what is described in the most probable estimate. This is the minimum time it will take to complete the task.
- **Pessimistic**: The duration of the activity is based on the worst-case scenario of what is described in the most probable estimate. This is the maximum time that it will take to complete the task.
- **More likely**: The duration of the activity is based on realistic expectations in terms of availability for the planned activity.

A first estimate of the duration of the activity can be constructed using an average of the three estimated durations. This average typically provides a more accurate estimate of the duration of the activity than a more likely single-value estimate. But that's not what we want to do.

Suppose that the project team used three-point estimation for each of the six tasks. The following table shows the times proposed by the team:

Task	Optimistic time	Pessimistic time	More likely time
Task 1	3	5	8
Task 2	2	4	7
Task 3	3	5	9
Task 4	4	6	10
Task 5	3	5	9
Task 6	2	6	8

Table 11.1– A time chart for individual tasks

After defining the sequence of tasks and the time it will take to perform each individual task, we can develop an algorithm for estimating the overall time of the project.

Developing an algorithm for project scheduling

In this section, we will analyze an algorithm for scheduling a project based on Monte Carlo simulation. We will look at all the commands in detail, line by line:

1. Let's start by importing the libraries that we will be using in the algorithm:

    ```
    import numpy as np
    import random
    import pandas as pd
    ```

    numpy is a Python library that contains numerous functions that help us manage multidimensional matrices. Furthermore, it contains a large collection of high-level mathematical functions we can use on these matrices.

    The random library implements pseudo-random number generators for various distributions. The random module is based on the Mersenne Twister algorithm.

    The pandas library is an open source BSD licensed library that contains data structures and operations that can be used to manipulate high-performance numeric values for the Python programming language.

2. Let's move on and initialize the parameters and the variables:

    ```
    N = 10000
    TotalTime=[]
    T =  np.empty(shape=(N,6))
    ```

    N represents the number of points that we generate. These are the random numbers that will help us define the time for that task. The TotalTime variable is a list that will contain the

N assessments of the overall time needed to complete the project. Finally, the T variable is a matrix with N rows and six columns and will contain the N assessments of the time needed to complete each individual task.

3. Now, let's set the three-point estimation matrix, as defined in the table in the *Estimating the task's time* section:

```
TaskTimes=[[3,5,8],
           [2,4,7],
           [3,5,9],
           [4,6,10],
           [3,5,9],
           [2,6,8]]
```

This matrix contains the three times for each representative row of the six tasks: optimistic, more likely, and pessimistic.

At this point, we must establish the form of the distribution for the times that we intend to adopt.

Exploring triangular distribution

When developing a simulation model, it is necessary to introduce probabilistic events. Often, the simulation process starts before you have enough information about the behavior of the input data. This forces us to decide on a distribution. Among those that apply to incomplete data is the triangular distribution. The triangular distribution is used when assumptions can be made about the minimum, maximum, and modal values.

> **Important note**
>
> A probability distribution is a mathematical model that links the values of a variable to the probabilities that these values can be observed. Probability distributions are used for modeling the behavior of a phenomenon of interest in relation to the reference population, or to all the cases in which the investigator observes a given sample.

In *Chapter 3, Probability and Data Generating Processes*, we analyzed the most widely used probability distributions. When the random variable is defined in a certain range, but there are reasons to believe that the degrees of confidence decreases linearly from the center to the extremes, there is the so-called triangular distribution. This distribution is very useful for calculating measurement uncertainties since, in many circumstances, this type of model can be more realistic than the uniform one.

The triangular distribution is a continuous probability distribution whose probability density function describes a triangle, returning a null distribution on the two extreme values and linearly between those extreme values and an intermediate value, which represents its mode as seen in *Figure 11.4*. It is used as a model when the sample available is very small, estimating the minimum, maximum, and fashion values.

Let's consider the first task of the project we are analyzing. For this, we have defined the three times estimated for the project: optimistic (3), more likely (5), and pessimistic (8). We can draw a graph in which we report these estimates three times on the abscissa and the probability of their occurrence on the ordinate. Using the triangular probability distribution, the probability that the event occurs is between the limit values, which, in our case, are optimistic and pessimistic. We do this while assuming the maximum value in correspondence with the value more likely to occur. For the intermediate values, where we know nothing, suppose that the probability increases linearly from optimistic to more likely, and then always decreases linearly from more likely to pessimistic, as shown in the following graph:

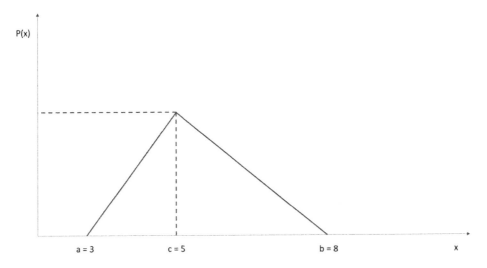

Figure 11.4 – Probability graph

Our goal is to model the times of each individual task using a random variable with uniform distribution in the interval (0, 1). If we indicate this random variable with trand, then the triangular distribution allows us to evaluate the probability that the task ends in that time is distributed. In the triangular distribution, we have identified two triangles that have the abscissa in common for $x = c$. This value acts as a separator between the two values that the distribution assumes. Let's denote it with Lh. The value is given the following formula:

$$Lh = \frac{(c - a)}{(b - a)}$$

In the previous equation, we have the following:

- a: The optimistic time
- b: The pessimistic time
- c: The more likely time

That being said, we can generate variations according to the triangular distribution with the following equations:

$$T = \begin{cases} a + \sqrt{trand * (b - a) * (c - a)} & \forall\, a < trand < Lh \\ b - \sqrt{(1 - trand) * (b - a) * (b - c)} & \forall\, Lh \leq trand < 1 \end{cases}$$

The previous equations allow us to perform the Monte Carlo simulation. Let's see how we can do this:

1. First, we generate the separation value of the triangular distribution:

   ```
   Lh= []
   for i in range(6):
       Lh.append(((TaskTimes[i][1]-TaskTimes[i][0])
                /(TaskTimes[i][2]-TaskTimes[i][0])))
   ```

   Here, we initialized a list and then populated it with a `for` loop that iterates over the six tasks, evaluating a value of Lh for each one.

2. Now, we can use two `for` loops and an `if` conditional structure to develop the Monte Carlo simulation:

   ```
   for p in range(N):
       for i in range(6):
           trand=random.random()
           if (trand < Lh[i]):
               T[p][i] = TaskTimes[i][0] +
                   np.sqrt(trand*(TaskTimes[i][1]-
                   TaskTimes[i][0])*
                   (TaskTimes[i][2]-TaskTimes[i][0]))
           else:
               T[p][i] = TaskTimes[i][2] -
                   np.sqrt((1-trand)*(TaskTimes[i][2]-
                   TaskTimes[i][1])*
                   (TaskTimes[i][2]-TaskTimes[i][0]))
   ```

   The first `for` loop continues generating the random values N times, while the second loop is used to perform the evaluation for six tasks. The conditional `if` structure, on the other hand, is used to discriminate between the two values distinct from the value of Lh so that it can use the two equations that we defined previously.

3. Finally, for each of the N iterations, we calculate an estimate of the total time for the execution of the project:

   ```
   TotalTime.append( T[p][0]+
   ```

```
                 np.maximum(T[p][1],T[p][2]) +
                 np.maximum(T[p][3],T[p][4]) + T[p][5])
```

For the calculation of the total time, we referred to the scheduling grid defined in the *Defining the scheduling grid* section. The procedure is simple: if the tasks develop in a series, then you add the times up, while if they develop in parallel, you choose the maximum value among the times of the tasks.

4. Now, let's take a look at the values we have attained:

```
Data = pd.DataFrame(T,columns=['Task1', 'Task2', 'Task3',
                               'Task4', 'Task5', 'Task6'])
pd.set_option('display.max_columns', None)
print(Data.describe())
```

For detailed statistics of the times estimated with the Monte Carlo method, we have transformed the matrix (*Nx6*) containing the times into a pandas DataFrame. The reason for this is that the `pandas` library has useful functions that allow us to extract detailed statistics from a dataset immediately. In fact, we can do this with just a line of code by using the `describe()` function.

The `describe()` function generates a series of descriptive statistics that return useful information on the dispersion and the form of the distribution of a dataset.

The pandas `set.option()` function was used to display all the statistics of the matrix and not just a part of it, as expected by default.

The following results are returned:

	Task1	Task2	Task3
count	10000.000000	10000.000000	10000.000000
mean	5.334687	4.336086	5.662239
std	1.022804	1.027895	1.250554
min	3.015679	2.027039	3.031016
25%	4.581429	3.589910	4.728432
50%	5.254768	4.274976	5.526395
75%	6.058666	5.058137	6.559695
max	7.931496	6.958586	8.962228

	Task4	Task5	Task6
count	10000.000000	10000.000000	10000.000000
mean	6.676473	5.675249	5.326908
std	1.258054	1.248865	1.254785
min	4.035221	3.053461	2.034225
25%	5.735770	4.744179	4.425259
50%	6.545980	5.553995	5.461404
75%	7.591595	6.556427	6.266919
max	9.912979	8.967686	7.973863

Figure 11.5 – The values of the DataFrame

By analyzing these statistics, we have confirmed that the estimated times are between the limit values imposed by the problem: optimistic and pessimistic. In fact, the minimum and maximum times are very close to these values. Furthermore, we can see that the standard deviation is very close to the unit. Finally, we can confirm that we have generated 10,000 values.

5. We can now trace the histograms of the distribution of the values of the times to analyze their form:

```
hist = Data.hist(bins=10)
```

The following diagram is printed:

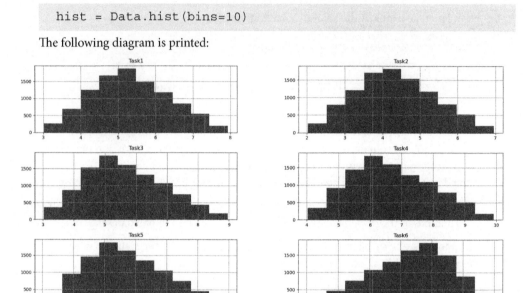

Figure 11.6 – Histograms of the values

By analyzing the previous diagram, we can confirm the triangular distribution of the time estimates, as we had imposed at the beginning of the calculations.

6. At this point, we only need to print the statistics of the total times. Let's start with the minimum value:

```
print("Minimum project completion time = ",
                    np.amin(TotalTime))
```

The following result is returned:

```
Minimum project completion time =  14.966486785163458
```

Let's analyze the average value:

```
print("Mean project completion time = ",np.
mean(TotalTime))
```

The following result is returned:

```
Mean project completion time =   23.503585938922157
```

Finally, we will print the maximum value:

```
print("Maximum project completion time = ",np.
amax(TotalTime))
```

The following result is printed:

```
Maximum project completion time =   31.90064194829465
```

In this way, we obtained an estimate of the time needed to complete the project based on Monte Carlo simulation.

Summary

In this chapter, we addressed several practical model simulation applications based on project management-related models. To start, we looked at the essential elements of project management and how these factors can be simulated to retrieve useful information.

Next, we tackled the problem of running a tiny forest for the wood trade. We treated the problem as a Markov decision process, summarizing the basic characteristics of these processes and then moved on to a practical discussion of them. We defined the elements of the problem and then we saw how to use the policy evaluation and policy improvement algorithms to obtain the optimal forest management policy. This problem was addressed using the MDPtoolbox package, which is available from Python.

Subsequently, we addressed how to evaluate the execution times of a project using Monte Carlo simulation. To start, we defined the task execution diagram by specifying which tasks must be performed in series and which can be performed in parallel. So, we introduced the times of each task through a three-point estimation. After this, we saw how to model the execution times of the project with triangular distribution using random evaluations of each phase. Finally, we performed 10,000 assessments of the overall project times.

In the next chapter, we will introduce the basic concepts of the fault diagnosis procedure. Then, we will learn how to implement a model for fault diagnosis for an induction motor. Finally, we will learn how to implement a fault diagnosis system for an unmanned aerial vehicle.

12
Simulating Models for Fault Diagnosis in Dynamic Systems

A physical system, in its life cycle, can be subject to failures or malfunctions that can compromise its normal operation. It is therefore necessary to introduce a fault diagnosis system within a plant capable of preventing critical interruptions. This is called a fault diagnosis system and it is capable of identifying the possible presence of a malfunction within the monitored system. The search for the fault is one of the most important and qualifying maintenance intervention phases and it is necessary to act in a systematic and deterministic way. To carry out a complete search for the fault, it is necessary to analyze all the possible causes that may have determined it.

In this chapter, we will learn how to approach fault diagnosis using a simulation model. We will start by exploring the basic concepts of fault diagnosis. Then, we will learn how to implement a model for fault diagnosis for a motor gearbox. In the last part of the chapter, we will analyze how to implement a fault diagnosis system for unmanned aerial vehicles.

In this chapter, we're going to cover the following topics:

- Introducing fault diagnosis
- A fault diagnosis model for a motor gearbox
- A fault diagnosis system for an unmanned aerial vehicle

Technical requirements

In this chapter, we will learn how to use ANNs to simulate complex environments. To understand the topics, a basic knowledge of algebra and mathematical modeling is needed.

To work with the Python code in this chapter, you need the following files (available on GitHub at `https://github.com/PacktPublishing/Hands-On-Simulation-Modeling-with-Python-Second-Edition`):

- `gearbox_fault_diagnosis.py`
- `UAV_detector.py`

Introducing fault diagnosis

Diagnostics is a procedure for translating information, deriven from the measurement of parameters and from the collection of data relating to a machine and turned into information on actual or incipient failures of the machine itself. Diagnostics summarizes the complexity of analysis and synthesis activities, which, using the measurements of certain physical quantities and characteristics of the monitored machine, allow us to obtain significant information on the conditions of the machine itself and on its trend over time, for evaluations and forecasts on its short- and long-term reliability.

The use of fault diagnosis techniques is becoming increasingly important to ensure high levels of safety and reliability in automated and autonomous systems. In fact, in recent years, the international scientific community has produced considerable efforts to develop systematic approaches to the diagnosis of failures in systems of various kinds. The main purpose of a fault diagnosis scheme is to monitor a system during its operation to detect the occurrence of faults (fault detection), locate faults (fault isolation), and determine their temporal evolution (fault identification).

Understanding fault diagnosis methods

Typically, the output of a fault diagnosis system returns a set of sensitive variables from the type of faults, modified by an anomaly when the system is subject to failure. Then, the information contained in the occurrences of faults is extracted and processed to detect, isolate, and identify faults. The methods used for fault diagnostics can be grouped into three basic groups: model-based, knowledge-based, and data-based, as seen in *Figure 12.1*.

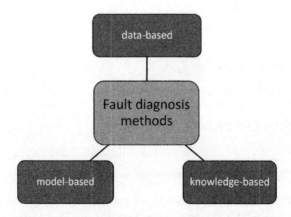

Figure 12.1: The group of methods used for fault diagnostics

Exploring the model-based approach

The model-based approach makes use of accurate mathematical models that allow both the detection and diagnosis of faults to be carried out efficiently. These models are based on the description of the actual degradation process of the components of interest. This specifically means modeling, in terms of the laws of physics, how the operating conditions affect the efficiency and longevity of assets. The most relevant variables include various thermal, mechanical, chemical, and electrical quantities. Being able to represent how they impact the health of machinery is a very complicated task. Therefore, the person who deals with creating this type of solution requires a high knowledge of the domain and modeling skills.

> **Important note**
> Once the model has been created, it is necessary to have sensors available that allows you to obtain values corresponding to the quantities considered relevant in the analysis and modeling phase to use them as inputs.

The main advantage of this type of approach is that it is descriptive; that is, it allows you to analyze the motivations of each output it provides, precisely because it is based on a physical description of the process. Consequently, it allows validation and certification. As for accuracy, it is strongly linked to the quality of analysis and modeling by domain experts. On the other hand, the negative aspects are the complexity and the high cost of implementation, together with the high specificity of the system, which entails little possibility of reuse and extension.

Describing the knowledge-based approach

The knowledge-based approach also relies on domain experts, as what you want to model with this type of approach is the skills and behavior of the experts themselves. The goal is to obtain a formalization of the knowledge they possess, to allow it to be reproduced and applied automatically.

> **Important note**
> Expert systems are in fact programs that use knowledge bases collected from people competent in each field and then apply inference and reasoning mechanisms on them to emulate thought and provide support and solutions to practical problems.

Among the most common approaches for implementing this type of model are rule-based mechanisms and fuzzy logic. The former has merits such as the simplicity of implementation and interpretability, but may not be sufficient to express complicated conditions and may suffer from a combinatorial explosion when the number of rules is very high. The use of fuzzy logic allows you to describe the state of the system through more vague and inaccurate inputs, making the process of formalization and description of the model simpler and more intuitive. Even for expert systems, as for model-based methods, the results are highly dependent on the quality and level of detail achieved by the model and are highly specific.

Discovering the data-based approach

The data-based approach applies statistical and machine learning techniques to the data collected by the machines, with the aim of then being able to recognize the status of the components.

> **Important note**
> The idea is to be able to obtain the greatest amount of information about the status of the machinery in real time, typically through sensors and from the logs of production and maintenance activities, and to correlate them with the level of degradation of the individual components or with the performance of the system.

This type of approach is currently the most used in practical cases. This is due to a series of advantages that this approach guarantees over the other two methods; data-driven approaches require large amounts of data to be effective, but with the availability of modern interconnected sensors (IIoT) this need is not difficult to satisfy. Compared to the other approaches, data driven approaches have the great advantage of not requiring in-depth knowledge specific to the application domain, thus making the contribution of the experts on the final performance of the model less decisive. The contribution of the experts can however be useful to speed up the selection process of the quantities to be used as input, but it has much less weight when compared with the other methods.

In addition, machine learning and data mining techniques may be able to detect relationships between the input parameters and the state of the system that even the experts themselves do not know a priori. Machine-learning-based algorithms can be used to develop predictive scenarios that allow us to extract knowledge about the specific domain

The machine-learning-based approach

Machine-learning-based algorithms automatically extract knowledge from data thanks to the data inputs received, without the need for specific commands from the developer. In these models, the machine can autonomously establish the patterns to be followed to obtain the desired result, this prerogative is typical of artificial intelligence. In the learning process that distinguishes these algorithms, the system receives a set of data necessary for training, estimating the relationships between the input and output data. These relationships represent the parameters of the model estimated by the system.

Problem definition

The choice of a specific model based on machine learning is highly dependent on the goal you want to achieve from the system. In fact, based on the objective, the problem is modeled differently. We can identify two approaches to the problem: diagnostics and prognostics. The purpose of a diagnostic system is to detect and identify a fault when the latter occurs. This, therefore, means monitoring a system, reporting when something is not working as expected, indicating which component is affected by the anomaly, and specifying the type of anomaly.

Prognostics, on the other hand, aim to determine whether a fault is close to occurring or to deduce the probability of occurrence. Obviously, prognostics being an a priori analysis, it can provide a greater contribution in terms of reducing the costs of interventions, but it is a more complex goal to achieve. Another option is to simultaneously use diagnostic and prognostic solutions applied to the same system. In fact, their combination provides two valuable advantages: diagnostics makes it possible to intervene to support decisions in cases where prognostics fail.

This scenario is in fact inevitable, as there are failures that do not follow a pattern that can be predicted, and even failures that are predictable with good precision cannot be identified in all their occurrences. The information obtained through diagnostic applications can be used as an additional input to predictive systems, thus allowing the creation of more sophisticated and precise models.

Binary classification

The simplest way is to represent the fault-finding as a binary classification problem, that is, in which every single input representing the state of the system must be labeled with one of two possible mutually exclusive values. In the event of a diagnostic problem, this means deciding whether the machine is functioning correctly or not correctly, making all possible states fall into these two classes. This is supervised learning, in the sense that the input data is provided with a label that represents the output of the model. In these cases, the system learns to recognize the relationships that bind the input data to the labels.

For prognostics, the interpretation becomes that of deciding whether the machine can fail within a set time interval. The difference between the two meanings is simply due to the different interpretations of the labels. This means that the same model can solve both problems. What will be differentiated is the labeling of the dataset used to carry out the model training phase.

Multiclass classification

The multiclass version is a generalization of binary classification, in which the number of possible labels to choose from is increased. However, only one label must be associated with each input.

The diagnostic case extends the previous case in a very intuitive way, that is, deciding whether the machine is functioning correctly or not correctly, and in the second case, which of the possible anomaly states. While the applications of prognostics you can see the problem is deciding in which time interval before the failure the machine is located, where then the possible labels represent different intervals of proximity to a failure.

Regression

Regression can be used to model prognostic problems. This means that the remaining useful life of a component can be estimated in terms of a continuous number of predetermined time units. In this specific case, the training dataset must only contain data relating to components that have been subject to failures, to allow the labeling of the inputs backward starting from the instant of failure.

This approach to the problem also provides a supervised learning paradigm, in this case, the input data will be associated with continuous output values.

Anomaly detection

A further possible representation for diagnostic problems is to consider it as an anomaly detection problem. This means that the model must be able to establish whether the operation of the machine returns to a normal state or if it deviates from it, that is, re-entering a case of anomaly.

The interpretation of the problem is, therefore, very similar to that of the binary classification. However, this methodology differs from the classification since it falls within the cases of semi-supervised learning, as the model only needs to learn from inputs that represent correct operating states and must, following the training phase, recognize unknown anomalous states, that is, of which the model does not know the characteristics.

Data collection

For an accurate diagnosis of faults in complex machines, it is essential to acquire information through the collection of data, analyze this data using advanced signal processing algorithms and, finally, extract the appropriate functionalities for efficient identification and classification of faults.

The data can be collected through different methodologies—through the measurements of all those physical quantities that somehow describe the state of the machine during its operation. They are obtained through special sensors that convert the physical value into an electrical value called sensor data.

> **Important note**
> Examples of these parameters used are noise, vibrations, pressure, temperature, and humidity, whereby the relevance of each of them strongly depends on the system being monitored.

Typically, an industrial system based on modern automation already has all the data necessary for its diagnostics available. If they are not available, the addition of additional sensors is the first step to implementing a correct fault identification strategy. But the data can also be collected by correlating the static operating conditions of the machine or plant to each instant of time, such as the code of the materials used, the production speed of the machine, and the type of piece produced. In this case, statistical data is defined.

Finally, the data can collect the history of relevant events and actions concerning a machine and its components. This data is called log data and may contain, for example, the history of faults found or repairs and replacements.

After having explored the basic concepts of fault diagnosis, it is now time to practically tackle a simulation problem.

Fault diagnosis model for a motor gearbox

The combustion engine delivers power within a narrow rev range. The tractive force is transmitted to the wheels under varying load situations through gear pairs. The more ratios there are, the better the engine work can be adapted to parameters such as acceleration, slope, load, consumption, and noise. To maintain engine operation within its optimal range of use, it is essential to be able to vary the transmission ratio between the engine itself and the drive wheels. In order to satisfy this need, it is necessary to insert a device generally defined as a mechanical gearbox between the motor and the wheels, which allows the motor to be used in an operating range capable of satisfying the needs of the users.

By using a fixed transmission ratio between the engine and the wheels, a curve is obtained that allows the maximum power of the engine to be exploited at a single-speed value. However, even if you have a stepped gearbox with a finite number of ratios, you can only get close to optimal exploitation of the engine since an infinite number of operating conditions cannot in any case be reached.

The gearbox consists of the following mechanical components: input drive shaft, countershaft, and output drive shaft. On these three components are sprockets of different sizes, meshed with each other, which correspond to the different gears, or torque ratios. Through the use of the clutch and the gear lever, it is possible to freely decide which gear to use according to need; in general, a higher gear corresponds to a higher speed.

Since these are mechanical components subjected to contact with each other, the components of a gearbox are subject to wear, which can also lead to breakage. In the case of breakage, the engine obviously stops and requires the replacement of the gearbox components. An automatic fault identification system can ensure that a problem is detected before the gearbox fails.

In this example, we will use data detected by accelerometers that measured the vibrations of a gearbox in correspondence with two operating conditions: healthy, and broken. The sensors were placed in opposite directions in order to detect all possible changes in operation. These data will be used to train different algorithms for the classification of the operating conditions of the gearbox.

The dataset is available on the Kaggle open data repository, which offers numerous projects to use to train machine learning-based algorithms. In its original version, the dataset is available at the following URL:

```
https://www.kaggle.com/datasets/brjapon/gearbox-fault-diagnosis
```

In this section, we have reduced the dataset's size to avoid the unnecessary burden of the calculations. The dataset offers the measurements of the vibrations detected by two sensors and the corresponding classification of the engine operation: *0* = broken, *1* = healthy.

As always, we will analyze the code line by line:

1. We will start by importing the libraries:

   ```
   import pandas as pd
   import seaborn as sns
   from matplotlib import pyplot as plt
   from sklearn.model_selection import train_test_split
   from sklearn.linear_model import LogisticRegression
   from sklearn.ensemble import RandomForestClassifier
   from sklearn.neural_network import MLPClassifier
   from sklearn.neighbors import KNeighborsClassifier
   from sklearn.inspection import DecisionBoundaryDisplay
   ```

   The `pandas` library is an open source BSD-licensed library that contains data structures and operations to manipulate high-performance numeric values for the Python programming language. The `seaborn` library is a Python library that enhances the data visualization tools of the `matplotlib` module. In the `seaborn` module, there are several features we can use to graphically represent our data. There are methods that facilitate the construction of statistical graphs with `matplotlib`. The `matplotlib` library is a Python library for printing high-quality graphics. Then, we imported a series of models that we will analyze in detail and discuss where we will use them.

2. Now, we can upload the data:

   ```
   data = pd.read_excel('fault.dataset.xlsx')
   ```

   To do this, we'll use the `read_excel` module of the `pandas` library. The `read_excel` method reads an Excel table into a pandas DataFrame.

3. Before starting with data analysis, we will conduct an exploratory analysis to understand how the data is distributed and extract preliminary knowledge. To display the first 10 rows of the DataFrame imported, we can use the `head()` function, as follows:

   ```
   print(data.head(10))
   ```

   The first 10 rows are displayed, as follows:

   ```
             a1         a2  state
   0   2.350390   1.454870      0
   1   2.452970   1.400100      0
   2  -0.241284  -0.267390      0
   3   1.130270  -0.890918      0
   4  -1.296140   0.980479      0
   ```

```
5   -1.650290   1.011530      0
6    0.429159  -1.163700      0
7   -0.191893  -2.945480      0
8    1.417660  -3.317650      0
9    1.699620  -2.446150      0
```

As we can see, there are three columns of data in the dataset: the measurements of the vibrations detected by two sensors (a1, and a2) and the corresponding classification of the engine operation.

4. To extract some information, we can invoke the info() function, as follows:

```
print(data.info())
```

This method prints a concise summary of a DataFrame, including the dtype index, dtypes column, non-null values, and memory usage. The following results are returned:

```
<class 'pandas.core.frame.DataFrame'>
RangeIndex: 20000 entries, 0 to 19999
Data columns (total 3 columns):
 #   Column  Non-Null Count  Dtype
---  ------  --------------  -----
 0   a1      20000 non-null  float64
 1   a2      20000 non-null  float64
 2   state   20000 non-null  int64
dtypes: float64(2), int64(1)
memory usage: 468.9 KB
None
```

Useful information is reported—the numbers of the entries (20,000) and data columns (3). Essentially, the list of all features with the number of elements, the possible presence of missing data, and the type is returned. In this way, we can already get an idea of the type of variables we're about to analyze. In fact, analyzing the results obtained, we note that two types have been identified: float64 (2) and int64 (1).

5. To get a preview of the data contained in it, we can calculate a series of basic statistics. To do so, we'll use the describe() function in the following way:

```
DataStat = data.describe()
print(DataStat)
```

The following results are returned:

```
                 a1             a2           state
count  20000.000000   20000.000000   20000.000000
```

mean	0.024046	0.006011	0.500000
std	5.897926	4.231061	0.500013
min	-36.989500	-23.710700	0.000000
25%	-3.107135	-2.360157	0.000000
50%	-0.043941	0.071639	0.500000
75%	3.011813	2.520958	1.000000
max	33.375500	20.906000	1.000000

The `describe()` function generates descriptive statistics that summarize the central tendency, dispersion, and shape of a dataset's distribution, excluding NaN values. It analyzes both numeric and object series, as well as DataFrame column sets of mixed data types.

6. So far we have not had evidence that the third variable (state) represents a dichotomous variable. To do this, we can use the describe function but pass the arguments as data of the type object:

```
DataStatCat = data.astype('object').describe()
print(DataStatCat)
```

The following results are returned:

	a1	a2	state
count	20000.00000	20000.00000	20000
unique	19888.00000	19877.00000	2
top	-1.93396	1.32888	0
freq	2.00000	2.00000	10000

As we can see, the state variable has only two values, and each value has an occurrence of 1,000.

7. To continue in our preventive visual analysis of the data, let us now use the graphs to help us. For example, we can draw boxplots of the distribution of the data detected by the two sensors (a1, a2):

```
fig, axes = plt.subplots(1,2, figsize=(18, 10))
sns.boxplot(ax=axes[0],x='state', y='a1', data=data)
sns.boxplot(ax=axes[1],x='state', y='a2', data=data)
plt.ylim(-40, 40)
plt.show()
```

A boxplot is a graphical representation used to describe the distribution of a sample by simple dispersion and position indexes. To draw a boxplot of a DataFrame, we used the `seaborn` package. The following chart is returned:

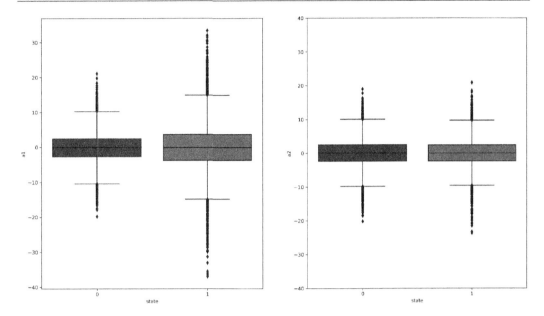

Figure 12.2: A boxplot of sensor measurement data

Through analyzing *Figure 12.2*, we can see that the distributions of the data collected by the two sensors are different. Sensor a1 seems to diversify the vibration values between broken and healthy conditions. This suggests that faults are occurring at the location of this sensor.

8. At this point, it is necessary to divide the data into input (*X*) and target (*Y*), this is necessary to train our algorithm for the classification of the operating conditions:

```
X = data.drop('state', axis = 1)
print('X shape = ',X.shape)
Y = data['state']
print('Y shape = ',Y.shape)
```

9. After separating the target from the input data, it is necessary to divide the data into two groups, the first group will be used to train the classification algorithms, and the second to evaluate its performance.

    To do this, we will use the `train_test_split()` function:

```
X_train, X_test, Y_train, Y_test = train_test_split(X, Y,
test_size = 0.30, random_state = 1)
print('X train shape = ',X_train.shape)
print('X test shape = ', X_test.shape)
print('Y train shape = ', Y_train.shape)
print('Y test shape = ',Y_test.shape)
```

The `train_test_split()` function splits arrays or matrices into random train and test subsets. Four arguments are passed: The first two arguments are X (Input) and Y (target) pandas Dataframe. The last two arguments are as follows:

- `test_size`: This should be between *0.0* and *1.0*, and represents the proportion of the dataset to include in the test split
- `random_state`: This is the seed used by the random number generator

10. Finally, we print the dimensions of the four datasets we obtained on the screen:

```
X train shape =   (14000, 2)
X test shape =    (6000, 2)
Y train shape =   (14000,)
Y test shape =    (6000,)
```

11. After carefully splitting the data, we can train our algorithms. Let's start with a model based on logistic regression:

```
lr_model = LogisticRegression(random_state=0).fit(X_train, Y_train)
lr_model_score = lr_model.score(X_test, Y_test)
print('Logistic Regression Model Score = ', lr_model_score)
```

In a logistic regression model, the response variable is a binary or dichotomous variable. Variables of this type are configured as variables that can only assume two mutually exclusive values, in this case, *0* or *1*. By exploiting statistical methods, it is possible, through the use of logistic regression, to calculate the probability that an observation or input belongs to a class or not. Logistic regression is in fact a predictive analysis that is used to evaluate the relationship between a dependent variable and one or more independent variables, estimating probabilities through a logistic function. The probabilities then turn into binary values in order to make a prediction.

To train a logistic-regression-based model we used the `LogisticRegression ()` function of the `sklearn.linear_model` module. The `sklearn.linear_model` module contains several functions to resolve some problems, such as regression problems and classification problems. The `LogisticRegression ()` function implements regularized logistic regression using the `liblinear` library and the `newton-cg`, `sag`, `saga`, and `lbfgs` solvers. After having trained the model, we tested it with the use of data so far never seen by the algorithm (X_test, Y_test). The following results were obtained:

```
Logistic Regression Model Score =  0.49333333333333335
```

This tells us that only half of the data has been correctly classified. This is a confidence score, a number between *0* and *1* that represents the probability that the forecast model output is correct.

12. So, let's see the results of the classification in visual form. To do this, we will trace a graph showing the two data classes and then trace the contours of the classification domain:

```
ax1 = DecisionBoundaryDisplay.from_estimator(
    lr_model, X_train, response_method="predict",
     alpha=0.5)
ax1.ax_.scatter(X_train.iloc[:,0], X_train.iloc[:,1],
c=Y_train, edgecolor="k")
plt.show()
```

To trace the contours of the classification domain, we used the `DecisionBoundaryDisplay()` function of the `sklearn.inspection` model. The `sklearn.inspection` module helps us identify what influences a model's predictions to better understand them. We can leverage the tools offered by this module to evaluate model assumptions and biases, design a better model, or diagnose problems with model performance. The following diagram is printed:

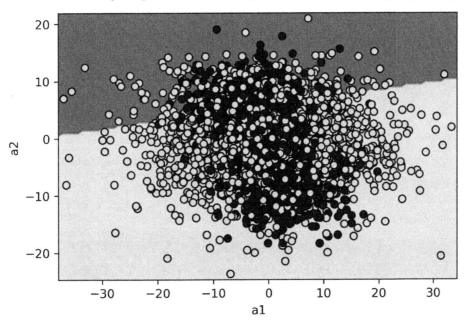

Figure 12.3: The distribution of the two classes of functioning with the decision boundaries returned by the model based on the logistic regression

Figure 12.3 clearly shows that the two classes overlap in the space of the two features (a1, a2). The space is therefore divided by the model into two parts, but this division cannot effectively identify the two classes (score = 0.49).

13. So, let's see if we can improve the performance of the classifier using the random forest algorithm:

    ```
    rm_model = RandomForestClassifier(max_depth=2, random_
    state=0).fit(X_train, Y_train)
    rm_model_score = rm_model.score(X_test, Y_test)
    print('Random Forest Model Score = ', rm_model_score)
    ```

    When you want to get the prediction of a certain event, it is more effective to use a set of predictors rather than the best single predictor, almost always obtaining better predictions. The group of predictors is called an ensemble and this technique is called ensemble learning. An algorithm that takes advantage of this methodology is called the ensemble method. A group of decision trees, each with a different random subset of the training set, that provides a final prediction is called the random forest and is one of the most powerful machine learning algorithms, despite its simplicity. The random forest algorithm is, therefore, a set of decision trees generally trained with the bagging method and can be used for discrete problems, classification, or continuous regression, the latter constituting the present case. This algorithm adds further randomness, in addition to the assignment of subsets of the training set to individual predictors: instead of searching for the best feature as a condition for splitting a node, it finds the best feature in a random subset of the global set of features themselves. This leads to a greater diversity of individual decision trees.

    To train a random-forest-based model, we used the `RandomForestClassifier()` function from the `sklearn.ensemble` module. The following two arguments were passed:

    - `max_depth`: The maximum depth of the tree
    - `random_state`: The randomness of the bootstrapping of the samples used

    In addition to these, the two input variables and the target are passed. Once again, we tested it with the use of data so far never seen by the algorithm (`X_test, Y_test`). The following results were obtained:

    ```
    Random Forest Model Score =  0.5838333333333333
    ```

    We have significantly improved the performance of the classifier, but we are still far from a satisfactory result. Let's now see the visual classification:

    ```
    ax2 = DecisionBoundaryDisplay.from_estimator(
        rm_model, X_train, response_method="predict",
        alpha=0.5)
    ax2.ax_.scatter(X_train.iloc[:,0], X_train.iloc[:,1],
    c=Y_train, edgecolor="k")
    plt.show()
    ```

The following diagram is returned:

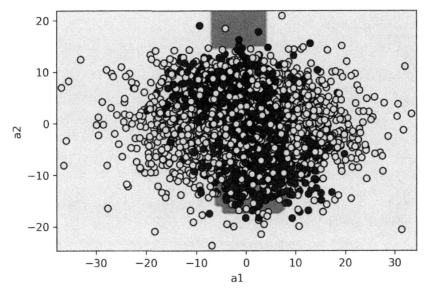

Figure 12.4: The distribution of the two classes that are functioning with the decision boundaries returned by the model based on the random forest classifier

In this case, we can see that the domain has been divided more selectively to demonstrate the performance improvement returned by the classifier.

14. But we are not satisfied; let's see what we get by applying the ANNs that have always behaved well in the classification:

```
mlp_model = MLPClassifier(random_state=1, max_iter=300).
fit(X_train, Y_train)
mlp_model_score = mlp_model.score(X_test, Y_test)
print('Artificial Neural Network Model Score = ', mlp_
model_score)
```

In *Chapter 10*, *Simulating Physical Phenomena Using Neural Networks*, we explored the extreme versatility of neural networks in dealing with both regression problems and classification problems. In this example, we used the MLPClassifier() function from the sklearn.neural_network module. This module includes models based on neural networks. The MLPClassifier() function implements a multi-layer perceptron classifier; two arguments are passed:

- random_state: Random number generation for weights and bias initialization
- max_iter: Maximum number of iterations

In addition to these, the two input variables and the target are passed. Once again, we tested it with the use of data so far never seen by the algorithm (X_test, Y_test). The following results were obtained:

```
Artificial Neural Network Model Score =   0.5955
```

We got a further improvement in the classification; let's see it in a graph:

```
ax3 = DecisionBoundaryDisplay.from_estimator(
    mlp_model, X_train, response_method="predict",
    alpha=0.5)
ax3.ax_.scatter(X_train.iloc[:,0], X_train.iloc[:,1],
c=Y_train, edgecolor="k")
plt.show()
```

The following diagram is printed:

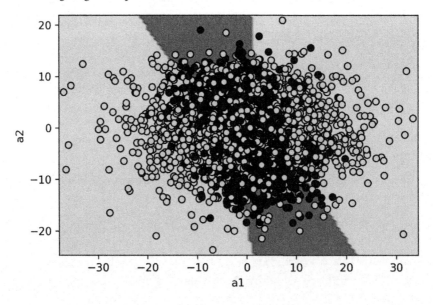

Figure 12.5: Distribution of the two classes that are functioning with the decision boundaries returned by the model based on the ANN classifier

In *Figure 12.5*, we can see a further improvement in the subdivision of the domain confirming the improvement in the performance of the classifier.

15. Finally, let's try to see what happens by delving into a classifier based on the K-nearest neighbors algorithm:

```
kn_model= KNeighborsClassifier(n_neighbors=2).fit(X_
```

```
  train, Y_train)
kn_model_score = kn_model.score(X_test, Y_test)
print('K-nearest neighbors Model score =', kn_model_
score)
```

K-nearest neighbors (KNN) is a supervised machine learning algorithm for classification. It is called a lazy learner, as it does not learn a rule on how to discriminate vector classes, but stores the entire learning dataset. The algorithm itself is quite simple and can be summarized in the following steps:

I.   Choose the number *k* and a metric for the distance

II. Find the *k* elements closest to the sample to be classified

III. Assign the class label with a majority vote

Based on the chosen metric, the KNN algorithm finds the k samples in the learning dataset that are closest (most similar) to the point to be classified. The class of the new point is then determined by a majority vote based on the class to which its k neighbors belong. The main advantage of this memory-based approach is that the classifier adapts immediately as we add learning vectors. On the other hand, the flaw is that the computational complexity for classifying new samples grows at most linearly with the number of learning vectors. Moreover, we cannot a priori ignore any training vector since there is no real learning. Thus, storage space and the number of distances to be calculated can become a nodal problem when working with large datasets.

We used the `K-neighborsClassifier()` function from the `sklearn.neighbors` module. This module contains useful tools for training unsupervised and supervised neighbor-based learning algorithms. Only one argument was passed:

- `n_neighbors`: The number of neighbors to use

The following score was returned:

```
K-nearest neighbors Model score = 0.5531666666666667
```

In this case, we have not obtained an improvement compared to the last two algorithms applied, but in any case, the result is better than the results obtained with the logistic regression.

Let's see what happens in the classification domain:

```
ax4 = DecisionBoundaryDisplay.from_estimator(
    kn_model, X_train, response_method="predict",
      alpha=0.5)
ax4.ax_.scatter(X_train.iloc[:,0], X_train.iloc[:,1],
  c=Y_train, edgecolor="k")
plt.show()
```

The following diagram is printed:

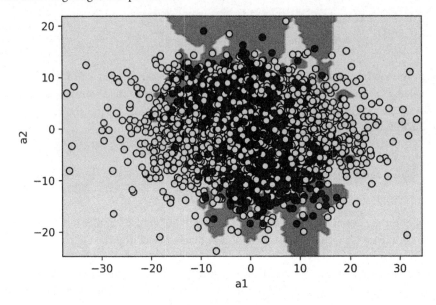

Figure 12.6: The distribution of the two classes that are functioning with the decision boundaries returned by the model based on the KNN algorithm

The boundary decisions seem more localized in correspondence between the two classes, but the difficulty in classifying the data is still evident.

After having seen how to elaborate simulation models of systems for automatic identification of engine gearbox failures, let's now see what happens when we try to simulate a drone failure identification system.

Fault diagnosis system for an unmanned aerial vehicle

Technological development has led to the birth of aircraft with remarkable capabilities capable of autonomously managing flight. This category, known as **unmanned aerial vehicles (UAVs)**, are vehicles that fly unmanned, which brings the advantages of a drastic reduction in operating costs compared to conventional aircraft with a pilot, the ability to operate in environments unsuitable for human presence, and to be used in a timely manner for overhead detection, for example, of natural disasters. Initially used exclusively in the military field – in dull missions (monotonous and long-lasting surveillance and reconnaissance), dirty missions (dangerous for the safety of pilots), and dangerous missions (risky for the lives of pilots) – today they represent the future of modern aeronautics. The enormous potential, the successes in military missions, and the progress made in the field of micro- and nano-technologies are pushing both industries and universities to develop ever more modern, reliable UAV systems capable of being used in a wide spectrum of civilian and military missions.

The wide diffusion of UAVs that is occurring in the world highlights new problems that have not yet been explored. The convenience of using these devices is accompanied by the problems and dangers associated with this technology. One of the problems that have been raised concerns the safety of the flight of such devices – what happens if a UAV crashes to the ground? Furthermore, given the small size of the UAVs, they are hard to identify with radar systems. This feature was immediately exploited by malicious people who used them for fraudulent purposes. The use of drones to transport drugs and smartphones into prisons, up to the use of drones for terrorist attacks, is the news of recent times. In these scenarios, the development of an automatic system for detecting the presence of UAVs in complex urban scenarios becomes crucial.

In this example, we will see how to identify the presence of a UAV based on monitoring the WiFi traffic detected in the vicinity of a sensitive target.

To do this, we will use a dataset available on the UCI Machine Learning Repository site at the following URL:

https://archive.ics.uci.edu/ml/datasets/Unmanned+Aerial+Vehicle+%28UAV%29+Intrusion+Detection

As always, we will analyze the code line by line:

1. We will start by importing the libraries:

    ```
    import pandas as pd
    from sklearn.model_selection import train_test_split
    from sklearn.svm import SVC
    import matplotlib.pyplot as plt
    from sklearn.feature_selection import SelectKBest, chi2
    ```

    The pandas library is an open source BSD-licensed library that contains data structures and operations to manipulate high-performance numeric values for the Python programming language. Then, we imported the `train_test_split()` function from `sklearn.model_selection` module, and the `SVC()` function from `sklearn.svm` module. To trace the diagrams, we first imported the `matplotlib` library, and finally, we imported the `SelectKBest()` and `chi2()` functions from the `sklearn.feature_selection` module. A detailed description of these functions will be proposed and where they will be used.

2. Let's move on to importing the dataset:

    ```
    data = pd.read_excel('UAV_WiFi.xlsx')
    ```

    To do this, we used the `read_excel` module of the pandas library. The `read_excel` method reads an Excel table into a pandas DataFrame.

Before starting with data analysis, we will conduct an exploratory analysis to understand how the data is distributed and extract preliminary knowledge. To display some statistics, we will use the `info()` function as follows:

```
print(data.info())
```

The following data is returned:

```
<class 'pandas.core.frame.DataFrame'>
RangeIndex: 17629 entries, 0 to 17628
Data columns (total 55 columns):
 #   Column                         Non-Null Count  Dtype
---  ------                         --------------  -----
 0   uplink_size_mean               17629 non-null  float64
 1   uplink_size_median             17629 non-null  float64
 2   uplink_size_MAD                17629 non-null  float64
 3   uplink_size_STD                17629 non-null  float64
 4   uplink_size_Skewness           17629 non-null  float64
 5   uplink_size_Kurtosis           17629 non-null  float64
 6   uplink_size_MAX                17629 non-null  float64
 7   uplink_size_MIN                17629 non-null  float64
 8   uplink_size_MeanSquare         17629 non-null  float64
 9   downlink_size_mean             17629 non-null  float64
 10  downlink_size_median           17629 non-null  float64
 11  downlink_size_MAD              17629 non-null  float64
 12  downlink_size_STD              17629 non-null  float64
 13  downlink_size_Skewness         17629 non-null  float64
 14  downlink_size_Kurtosis         17629 non-null  float64
 15  downlink_size_MAX              17629 non-null  int64
 16  downlink_size_MIN              17629 non-null  int64
 17  downlink_size_MeanSquare       17629 non-null  float64
 18  both_links_size_mean           17629 non-null  float64
 19  both_links_size_median         17629 non-null  float64
 20  both_links_size_MAD            17629 non-null  float64
 21  both_links_size_STD            17629 non-null  float64
 22  both_links_size_Skewness       17629 non-null  float64
 23  both_links_size_Kurtosis       17629 non-null  float64
 24  both_links_size_MAX            17629 non-null  float64
 25  both_links_size_MIN            17629 non-null  float64
 26  both_links_size_MeanSquare     17629 non-null  float64
 27  uplink_interval_mean           17629 non-null  float64
 28  uplink_interval_median         17629 non-null  float64
 29  uplink_interval_MAD            17629 non-null  float64
 30  uplink_interval_STD            17629 non-null  float64
 31  uplink_interval_Skewness       17629 non-null  float64
 32  uplink_interval_Kurtosis       17629 non-null  float64
 33  uplink_interval_MAX            17629 non-null  int64
 34  uplink_interval_MIN            17629 non-null  int64
 35  uplink_interval_MeanSquare     17629 non-null  float64
 36  downlink_interval_mean         17629 non-null  float64
 37  downlink_interval_median       17629 non-null  float64
 38  downlink_interval_MAD          17629 non-null  float64
 39  downlink_interval_STD          17629 non-null  float64
 40  downlink_interval_Skewness     17629 non-null  float64
 41  downlink_interval_Kurtosis     17629 non-null  float64
 42  downlink_interval_MAX          17629 non-null  float64
 43  downlink_interval_MIN          17629 non-null  float64
 44  downlink_interval_MeanSquare   17629 non-null  float64
 45  both_links_interval_mean       17629 non-null  float64
 46  both_links_interval_median     17629 non-null  float64
 47  both_links_interval_MAD        17629 non-null  float64
 48  both_links_interval_STD        17629 non-null  float64
 49  both_links_interval_Skewness   17629 non-null  float64
 50  both_links_interval_Kurtosis   17629 non-null  float64
 51  both_links_interval_MAX        17629 non-null  int64
 52  both_links_interval_MIN        17629 non-null  int64
 53  both_links_interval_MeanSquare 17629 non-null  float64
 54  target                         17629 non-null  int64
dtypes: float64(48), int64(7)
memory usage: 7.4 MB
None
```

Figure 12.7: Dataset features list with the number of occurrences and type of data

This list allows us to identify all the features present in the dataset. We can see that there are `17629` data records, `54` features of which `53` represent the input data and the last feature, named target, represents the data classification label (1 = UAV, 0 = No UAV).

Let's extract other statistics from the data:

```
DataStatCat = data.astype('object').describe()
print(DataStatCat)
```

The following data is listed:

```
        uplink_size_mean   ...    target
count       17629.000000   ...     17629
unique      17622.000000   ...         2
top             0.003465   ...         1
freq            2.000000   ...      9760
```

The most interesting thing concerns the target variable, wherein, we have the confirmation that it is binomial data (only two types of occurrences), and that the distribution of the two categories is sufficiently balanced (the frequency of class *1* is equal to 9760, this means that the other class is present with a frequency equal to 17629 - 9760 = 7869).

3. Now, we can separate the input data from the target:

```
X = data.drop('target', axis = 1)
print('X shape = ',X.shape)
Y = data['target']
print('Y shape = ',Y.shape)
```

The following results are displayed:

```
X shape =   (17629, 54)
Y shape =   (17629,)
```

4. To properly train the classifier that will allow us to identify the presence of the UAV near the sensitive target, it is necessary to divide the data into two sets. The first set will be used for training, while the other will be used to verify the correct operation of the classifier:

```
X_train, X_test, Y_train, Y_test = train_test_split(X, Y,
test_size = 0.30, random_state = 1)
print('X train shape = ',X_train.shape)
print('X test shape = ', X_test.shape)
print('Y train shape = ', Y_train.shape)
print('Y test shape = ',Y_test.shape)
```

The train_test_split() function splits arrays or matrices into random train and test subsets. The following results are printed:

```
X train shape =   (12340, 54)
X test shape =   (5289, 54)
Y train shape =   (12340,)
Y test shape =   (5289,)
```

We will, therefore, use 70% of the data for training and the remaining 30% for testing.

5. Now, let's train a support-vector-based classifier:

```
SVC_model = SVC(gamma='scale',
                random_state=0).fit(X_train, Y_train)
SVC_model_score = SVC_model.score(X_test, Y_test)
print('Support Vector Classification Model Score = ',
SVC_model_score)
```

Support vector machines (SVM) are a set of supervised learning methods for regression and pattern classification, developed in the 1990s by Bell AT&T laboratories. They are known as maximum margin classifiers, as they minimize the classification error and maximize the distance margin. SVM can be thought of as an alternative technique for learning polynomial classifiers, as opposed to the classical training techniques of neural networks. Single-layer neural networks have an efficient learning algorithm but are only useful in the case of linearly separable data. Conversely, multilayer neural networks can represent nonlinear functions. But these are difficult to train due to a large number of dimensions of the weight space and because the most popular techniques, such as backpropagation, allow you to obtain the network weights by solving a problem non-convex and unconstrained optimization system, which, consequently, has an indeterminate number of local minima.

The SVM training technique solves both problems. It has an efficient algorithm and is capable of representing complex nonlinear functions. The characteristic parameters of the network are obtained by solving a convex quadratic programming problem with equality or box-type constraints (in which the value of the parameter must be kept within an interval), which provides for a single global minimum.

To train the SVM-based model, we used the `SVC ()` function of the `sklearn.svm` module. The following two parameters were passed:

- `gamma: Kernel coefficient`: We used a scale kernel that fix gamma = 1 / (n_features * X.var())
- Random state: Controls the pseudo-random number generation for the reproducibility of the example

We then tested the model trained on the test data. The following results were obtained:

```
Support Vector Classification Model Score
 =  0.5517110985063339
```

This is the mean accuracy of the model. The accuracy of the classification model returns the percentage of correct predictions on a dataset used for testing that the model has never seen in the training process. This is equivalent to the ratio of the number of correct estimates to the total number of input samples.

Just over 50% of correct classifications do not satisfy us, so let's see how to improve the performance of the classifier. When using the input features to classify a target the different variability of the features can affect the result.

To evaluate the different variability of the data of the input features, we can draw boxplots of the first five features:

```
first_10_columns = X.iloc[:,0:5]
plt.figure(figsize=(10,5))
first_10_columns.boxplot()
```

The following boxplot is displayed:

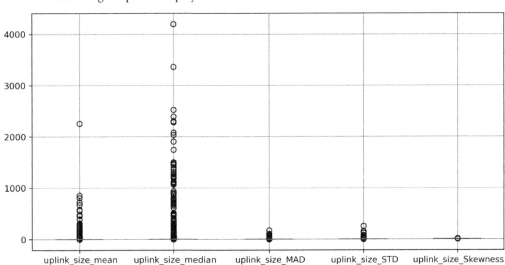

Figure 12.8: A boxplot of the first five features of the input data

We can see that the range of variability is very different already for these first five features; who knows what acacde for the others. In these cases, it is advisable to scale the data. As we saw in *Chapter 10, Simulating Physical Phenomena Using Neural Networks*, we can reduce the values so that they fall within a common range, guaranteeing the same characteristics of variability possessed by the initial dataset. In this way, we will be able to compare variables belonging to different distributions and variables expressed in different units of measurement.

In this example, we will use the min-max method (usually called feature scaling) to get all of the scaled data in the range [0 ~ 1]. The formula to achieve this is as follows:

$$x_{scaled} = \frac{x - x_{min}}{x_{max} - x_{min}}$$

We just have to rewrite this formula in Python:

```
X_scaled = (X-X.min())/(X.max()-X.min())
```

Now, let's check how the data variability has changed by drawing a boxplot of the first five features of the input dataset:

```
first_10_columns = X_scaled.iloc[:,0:5]
plt.figure(figsize=(10,5))
first_10_columns.boxplot()
```

The following boxplot is displayed:

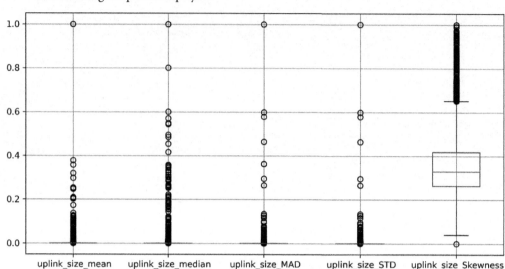

Figure 12.9: A boxplot of the first five features of the input data after the data scaling

After looking at *Figure 12.9* and comparing it with *Figure 12.8*, we can see that now the range of variability of the data has been significantly reduced. Now, all the variables vary in the range [0 – 1].

6. But this may not be enough to improve the performance of the classifier, so let's make a feature selection. Feature selection is based on finding a subset of the original variables, usually iteratively, thus detecting new combinations of variables and comparing prediction errors. The combination of variables that produces the minimum error will be labeled as a selected feature and used as input for the machine learning algorithm. To perform the feature selection, we must set the appropriate criteria in advance. Usually, these selection criteria determine the minimization of a specific predictive error measure for models fit for different subsets. Based on these criteria, the selection algorithms seek a subset of predictors that optimally model the measured responses. Such research is subject to constraints, such as the necessary or excluded characteristics and the size of the subset.

To perform feature selection, we will apply the `SelectKBest()` function as follows:

```
best_input_columns = SelectKBest(chi2, k=10).fit(X_scaled, Y)
sel_index = best_input_columns.get_support()
best_X = X_scaled.loc[: , sel_index]
```

The `SelectKBest()` function selects features according to the k highest scores: this function adopts as a parameter a score function, which is applied to the pair (X_scaled, Y). The score function returns an array of scores. The function then selects only the features that record the highest scores. The `chi2 ()` function was used as the scoring function, which computes chi-squared stats between each non-negative feature and class.

After selecting the best 10 features, we extracted them from the starting dataset and created a new input dataset with only the best 10 features (`best_X`).

Out of curiosity, let's see what they are:

```
feature_selected = best_X.columns.values.tolist()
print("The best 10 feature selected are:", feature_
selected)
```

The following features are selected:

```
The best 10 feature selected are: [' downlink_size_
mean', ' downlink_size_median', ' downlink_size_MAD',
' downlink_size_MAX', ' downlink_size_MIN', ' downlink_
size_MeanSquare', ' uplink_interval_STD', ' uplink_
interval_Kurtosis', ' uplink_interval_MIN', ' both_links_
interval_MIN']
```

Now that we have a new dataset with the best features, let's split the data again (70% for training, 30% for testing):

```
X_train, X_test, Y_train, Y_test = train_test_
split(best_X, Y, test_size = 0.30, random_state = 1)
print('X train shape = ',X_train.shape)
print('X test shape = ', X_test.shape)
print('Y train shape = ', Y_train.shape)
print('Y test shape = ',Y_test.shape)
```

The following subsets are created:

```
X train shape =   (12340, 10)
X test shape =   (5289, 10)
Y train shape =   (12340,)
Y test shape =   (5289,)
```

Now, we can finally re-train the SVM-based classification model:

```
SVC_model = SVC(gamma='auto',random_state=0).fit(X_train,
Y_train)
SVC_model_score = SVC_model.score(X_test, Y_test)
print('Support Vector Classification Model Score = ',
SVC_model_score)
```

The following result is returned:

```
Support Vector Classification Model Score =  1.0
```

Not a bad improvement, we went from a support vector score of 0.55 to 1.0.

Summary

In this chapter, we learned how to approach fault diagnosis using simulation models. We started by exploring the basic concepts of fault diagnosis. Then, we learned how to implement a model for fault diagnosis for a motor gearbox. Finally, we analyzed how to implement a fault diagnosis system for unmanned aerial vehicles.

In the next chapter, we will summarize the simulation modeling processes we looked at in the previous chapters. Then, we will explore the main simulation modeling applications that are used in real life. Finally, we will discover future challenges regarding simulation modeling.

13
What's Next?

In this chapter, we will summarize what has been covered so far in this book and what the next steps are. You will learn how to apply all the skills that you have gained to other projects, as well as the real-life challenges in building and deploying simulation models and other common technologies that data scientists use. By the end of this chapter, you will have a better understanding of the issues associated with building and deploying simulating models and will have broadened your knowledge of additional resources and technologies you can learn about to sharpen your machine learning skills.

In this chapter, we're going to cover the following main topics:

- Summarizing simulation modeling concepts
- Applying simulation models to real life
- The next steps for simulation modeling

Summarizing simulation modeling concepts

Useful in cases where it is not possible to develop a mathematical model capable of effectively representing a phenomenon, simulation models imitate the operations performed over time by a real process. The simulation process involves generating an artificial history of the system to be analyzed; subsequently, the observation of this artificial history is used to trace information regarding the operating characteristics of the system itself and make decisions based on it.

The use of simulation models as a tool to aid decision-making processes has ancient roots and is widespread in various fields. Simulation models are used to study the behavior of a system over time and are built based on a set of assumptions made about the behavior of the system that's expressed using mathematical-logical-symbolic relationships. These relationships are between the various entities that make up the system. The purpose of a model is to simulate changes in the system and predict the effects of these changes on the real system. For example, they can be used in the design phase before the model's actual construction.

> **Important note**
> Simple models are resolved analytically, using mathematical methods. The solution consists of one or more parameters, called **behavior measures**. Complex models are simulated numerically on the computer, where the data is treated as being derived from a real system.

Let's summarize the tools we have available to develop a simulation model.

Generating random numbers

In simulation models, the quality of the final application strictly depends on the possibility of generating good-quality random numbers. In several algorithms, decisions are made based on a randomly chosen value. The definition of random numbers suggests that of random processes through a connection that specifies its characteristics. A random number appears as such because we do not know how it was generated, but once the law within which it was generated is defined, we can reproduce it whenever we want.

Deterministic algorithms do not allow us to generate random number sequences, but simply make pseudo-random sequence generation possible. Pseudo-random sequences differ from random ones in that they are reproducible and, therefore, predictable.

Multiple algorithms are available for generating pseudorandom numbers. In *Chapter 2, Understanding Randomness and Random Numbers*, we analyzed the following in detail:

- **Linear congruential generator** (LCG): This generates a sequence of pseudo-randomized numbers using a piecewise discontinuous linear equation
- **Lagged Fibonacci generator** (LFG): This is based on a generalization of the Fibonacci sequence

More specific methods are added to these to generate uniform distributions of random numbers. The following graph shows a uniform distribution of 1,000 random numbers in the range of 1-10:

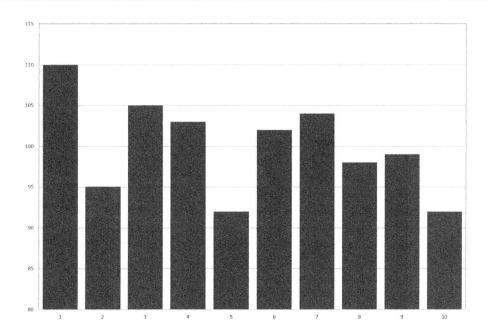

Figure 13.1 – A graph of the distribution of random numbers in the range of 1-10

We analyzed the following methods, both of which we can use to derive a generic distribution, starting from a uniform distribution of random numbers:

- **Inverse transform sampling method**: This method uses inverse cumulative distribution to generate random numbers.
- **Acceptance-rejection method**: This method uses the samples in the region under the graph of its density function.

A pseudo-random sequence returns integers uniformly distributed in each interval, with a very long repetition period and a low correlation level between one element of the sequence and the next.

To self-evaluate the skills that are acquired when generating random numbers, we can try to write some Python code for a bingo card generator. Here, we just limit the numbers from 1 to 90 and make sure that the numbers cannot be repeated and must be equally likely.

Applying Monte Carlo methods

A Monte Carlo simulation is a numerical method based on probabilistic procedures: Vine is widely used in statistics for the resolution of problems that present analytical difficulties that are not otherwise difficult to overcome. This method is based on the possibility of sampling an assigned probability distribution using random numbers. It generates a sequence of events distributed according to the assigned probability. In practice, instead of using a sample of numbers drawn at random, a sequence of numbers that has been obtained with a well-defined iterative process is used. These numbers are called pseudorandom because, although they're not random, they have statistical properties such as those of true random numbers. Many simulation methods can be attributed to the Monte Carlo method, which aims to determine the typical parameters of complex random phenomena.

The following diagram describes the procedure leading from a set of distributions of random numbers to a Monte Carlo simulation:

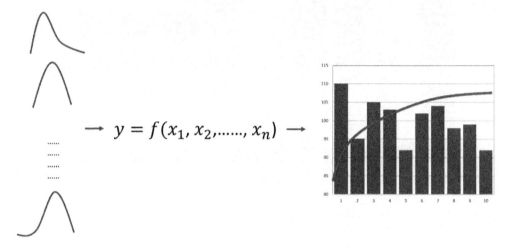

Figure 13.2 – The procedure of a Monte Carlo simulation, starting from a series of distributions of random numbers to one

The Monte Carlo method is essentially a numerical method for calculating the expected value of random variables, that is, an expected value that cannot be easily obtained through direct calculation. To obtain this result, the Monte Carlo method is based on two fundamental theorems of statistics:

- **Law of large numbers**: The simultaneous action of many random factors leads to a substantially deterministic effect.
- **Central limit theorem**: The sum of many independent random variables characterized by the same distribution is approximately normal, regardless of the starting distribution.

A Monte Carlo simulation is used to study the response of a model to randomly generated inputs.

Addressing the Markov decision process

Markov processes are discrete stochastic processes where the transition to the next state depends exclusively on the current state. For this reason, they can be called stochastic processes without memory. The typical elements of a Markovian process are the states in which the system finds itself and the available actions that the decision-maker can carry out in that state. These elements identify two sets: the set of states in which the system can be found, and the set of actions available for each specific state. The action chosen by the decision maker determines a random response from the system, which brings it into a new state. This transition returns a reward that the decision maker can use to evaluate their choice, as shown in the following diagram:

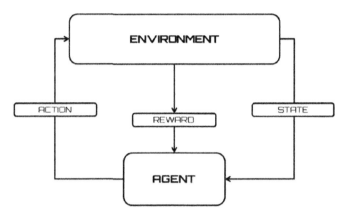

Figure 13.3 – A reward returned from the transition states

Crucial to the system's future choices is the concept of reward, which represents the response of the environment to the action taken. This response is proportional to the weight that the action determines in achieving the objective – it will be positive if it leads to the correct behavior, while it will be negative in the case of a wrong action.

Another fundamental concept in Markovian processes is policy. A policy determines the system's behavior in decision-making. It maps both the states of the environment and the actions to be chosen in those states, representing a set of rules or associations that respond to a stimulus. In a Markov decision-making model, the policy provides a solution that associates a recommended action with each state that can be achieved by the agent. If the policy provides the highest expected utility among the possible actions, it is called an optimal policy. In this way, the system does not have to keep its previous choices in memory. To make a decision, it only needs to execute the policy associated with the current state.

Now, let's consider a practical application of a process that can be treated according to the Markov model. In a small industry, an operating machine works continuously. Occasionally, however, the quality of the products is no longer permissible due to the wear and tear of the spare parts, so the activity must be interrupted, and complex maintenance carried out. It is observed that the deterioration occurs after an operating time T_m of an average of 40 days, while maintenance requires a random time of an average of one day. How is it possible to describe this system with a Markovian model to calculate the probability at a steady state of finding the working machine?

The company cannot bear the downtime of the machine, so it keeps a second one ready to be used as soon as the first one requires maintenance. This second machine is, however, of lower quality, so it breaks after an exponential random work time of five days on average and requires an exponential time of one day on average to start again. As soon as the main machine is reactivated, the use of the secondary is stopped. If the secondary machine breaks before the main are reactivated, the repair team insists only on the main one being used, taking care of the second machine only after restarting the main. How is it possible to describe the system with a Markovian model, calculating the probability at a steady state with both machines stopped?

Think about how you might answer these questions using the knowledge you have gained from this book.

Analyzing resampling methods

In resampling methods, a subset of data is extracted randomly or according to a systematic procedure from an original dataset. The aim is to approximate the characteristics of a sample distribution by reducing the use of system resources.

Resampling methods are methods that repeat simple operations many times, generating random numbers to be assigned to random variables or random samples. These operations require more computational time as the number of repeated operations grows. They are very simple methods to implement and once implemented, they are automatic.

These methods generate dummy datasets from the initial data and evaluate the variability of a statistical property from its variability on all dummy datasets. The methods differ from each other in the way dummy datasets are generated. In the following diagram, you can see some datasets that were generated from an initial random distribution:

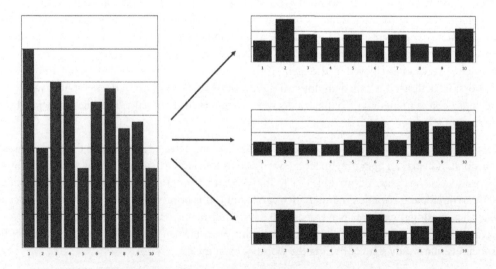

Figure 13.4 – Examples of datasets generated by an initial random distribution

There are many different resampling methods are available. In this book, we analyzed the following methods:

- **Jackknife technique**: Jackknife is based on calculating the statistics of interest for various sub-samples, leaving out one sample observation at a time. The jackknife estimate is consistent for various sample statistics, such as mean, variance, the relation coefficient, the maximum likelihood estimator, and others.
- **Bootstrapping**: The logic of the bootstrap method is to build samples that are not observed, but statistically like those observed. This is achieved by resampling the observed series through an extraction procedure where we reinsert the observations.
- **Permutation test**: Permutation tests are a special case of randomization tests and use a series of random numbers formulated from statistical inferences. The computing power of modern computers has made their widespread application possible. These methods do not require assumptions about data distribution to be met.
- **Cross-validation technique**: Cross-validation is a method used in model selection procedures based on the principle of predictive accuracy. A sample is divided into two subsets, of which the first (the training set) is used for construction and estimation and the second (the validation set) is used to verify the accuracy of the predictions of the estimated model.

Sampling is used if not all of the elements of the population are available. For example, investigations into the past can only be done on available historical data, which is often incomplete.

Exploring numerical optimization techniques

Numerous applications, which are widely used to solve practical problems, make use of optimization methods to drastically reduce the use of resources. Minimizing the cost or maximizing the profit of a choice are techniques that allow us to manage numerous decision-making processes. Mathematical optimization models are an example of optimization methods, in which simple equations and inequalities allow us to express the evaluation and avoid the constraints that characterize the alternative methods.

The goal of any simulation algorithm is to reduce the difference between the values predicted by the model and the actual values returned by the data. This is because a lower error between the actual and expected values indicates that the algorithm has done a good simulation job. Reducing this difference simply means minimizing an objective function that the model being built is based on.

In this book, we have addressed the following optimization methods:

- **Gradient descent**: This method is one of the first methods that was proposed for unconstrained minimization and is based on the use of the search direction in the opposite direction to that of the gradient, or anti-gradient. The interest of the direction opposite to the gradient lies precisely in the fact that, if the gradient is continuous then it constitutes a descent direction that is canceled if and only if the point that's reached is a stationary point.

- **Newton-Raphson**: This method is used for solving numerical optimization problems. In this case, the method takes the form of Newton's method for finding the zeros of a function but is applied to the derivative of the function *f*. This is because determining the minimum point of the function *f* is equivalent to determining the root of the first derivative.
- **Stochastic gradient descent**: This method solves the problem of evaluating the objective function by introducing an approximation of the gradient function. At each step, instead of the sum of the gradients being evaluated in correspondence with each data contained in the dataset, the evaluation of the gradient is used only in a random subset of the dataset.

Using artificial neural networks for simulation

Artificial neural networks (ANNs) are numerical models that have been developed with the aim of reproducing some simple neural activities of the human brain, such as object identification and voice recognition. The structure of an ANN is composed of nodes that, analogous to the neurons present in a human brain, are interconnected with each other through weighted connections, which reproduce the synapses between neurons. The system output is updated until it iteratively converges via the connection weights. The information that's derived from experimental activities is used as input data and the result is processed by the network and returned as output. The input nodes represent the predictive variables that we need to process the dependent variables that represent the output neurons. The following diagram shows the functionality of an artificial neuron:

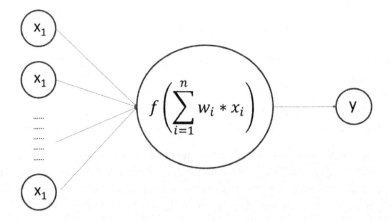

Figure 13.5 – The functionality of an artificial neuron

An ANN's target is the result of calculating the outputs of all the neurons. This means an ANN is a set of mathematical function approximations. A model of this type can simulate the behavior of a real system, such as in pattern recognition. This is the process in which a pattern/signal is assigned to a class. A neural network recognizes patterns by following a training session, in which a set of training patterns are repeatedly presented to the network, with each category that they belong to specified. When a pattern that has never been seen before but belongs to the same category of patterns that it has learned is presented, the network will be able to classify it thanks to the information that was extracted from the training data. Each pattern represents a point in the multidimensional decision space. This space is divided into regions, each of which is associated with a class. The boundaries of these regions are determined by the network through the training process.

Now that we have recapped all the concepts we have learned about throughout this book, let's see how they can be applied to challenges in the real world.

Applying simulation models to real life

The algorithms that we have analyzed in detail throughout this book represent valid tools for simulating real systems. Therefore, they are widely used in real life to carry out research on the possible evolution of a phenomenon, following a possible choice made on it.

Let's look at some specific examples.

Modeling in healthcare

In the healthcare sector, simulation models have a significant weight and are widely used to simulate the behavior of a system to extract knowledge. For example, it is necessary to demonstrate the clinical efficacy of the health intervention under consideration before undertaking an economic analysis. The best available sources are randomized controlled trials. Trials, however, are often designed to leave out the economic aspects, so the key parameters for economic evaluations are generally absent. Therefore, a method is needed to evaluate the effect of disease progression, to limit the bias in the cost-effectiveness analysis. This implies the construction of a mathematical model that describes the natural history of the disease, the impact of the interventions applied on the natural history of the disease, and the results in terms of costs and objectives. The most used techniques are extrapolation, decision analysis, the Markov model, and Monte Carlo simulations:

- In extrapolation, the results of a trial with short follow-up periods are extrapolated beyond the end of the trial itself and various possible scenarios are considered, some more optimistic, in which the benefits associated with an intervention are assumed to be constant over time.

- The Markovian model is frequently used in pharmacoeconomic evaluations, especially following the numerous requests for cost-effectiveness evaluations by government organizations.

- An alternative to calculating the costs and benefits of a therapeutic option is Monte Carlo simulation. As in the Markov model, even in Monte Carlo simulation, precise states of health and the probability of transition are defined. In this case, however, based on the probability of transition and the result of a random number generator, a path is constructed for each patient until the patient themselves reaches the ultimate state of health envisaged by the model. This process is usually repeated for each patient in very large groups (even 10,000 cases), thus providing a distribution of survival times and related costs. The average values of costs and benefits obtained with this model are very similar to those that we would have calculated by applying the Markov model.

- However, Monte Carlo simulation also provides frequency distribution and variance estimates, which allow you to evaluate the level of uncertainty of the results of the model itself. In fact, Monte Carlo simulation is often used to obtain a sensitivity analysis of the results deriving from the application of the Markov model.

Modeling in financial applications

Monte Carlo simulation is normally used to predict the future value of various financial instruments. However, as highlighted previously, it is good to underline that this forecasting method is presented exclusively as an estimate and, therefore, does not provide a precise value as a result. The main financial applications of this method concern pricing options (or derivatives in general) and the evaluation of security portfolios and financial projects. From this, it is immediately evident that they present an element of analogy.

In fact, options, portfolios, and financial projects have a value that's influenced by many sources of uncertainty. The simulation in question does not lend itself to the evaluation of any financial instrument. Securities such as shares and bonds are not normally valued with the method, precisely because their value is subordinated to a lower number of sources of uncertainty.

The options, on the contrary, are derivative securities, the value of which is influenced by the performance of the underlying functions (which may have the most varied content) and by numerous other factors (interest rates, exchange rates, and so on). Monte Carlo simulation allows you to generate pseudorandom values for each of these variables and assign a value to the desired option. It should be noted, however, that the Monte Carlo method is only one of the pricing options available.

Continuing with the category of financial instruments, portfolios are sets of different securities, normally of a varied nature. Portfolios are exposed to a variety of sources of risk. The operational needs of modern financial intermediaries have led to the emergence of calculation methods that aim to monitor the overall risk exposure of their portfolios. The main method in this context is **VaR (value at risk)**, which is often calculated using Monte Carlo simulation.

Ultimately, when a company must evaluate the profitability of a project, it will have to compare the cost of the same with the revenue generated. The initial cost is normally (but not necessarily) certain. The cash flows that are generated, however, are hardly known a priori. The Monte Carlo method allows us to evaluate the profitability of the project by attributing pseudorandom values to the various cash flows.

Modeling physical phenomenon

The simulation of a physical model allows you to experiment with the model by putting it to the test by changing its parameters. The simulation of a model, therefore, allows you to experiment with the various possibilities of the model, as well as its limits, in terms of how the model acts as a framework for the experimentation and organization of our ideas. When the model works, it is possible to remove the scaffolding. In this situation, maybe it turns out that it stands up or something new has been discovered. When constructing a model, reference is made to the ideas and knowledge through which the reality of the phenomenon is formally represented.

Just as there is no univocal way to face and solve problems, there is no univocal way to construct the models that describe the behavior of a given phenomenon. The mathematical description of reality struggles to keep considerations of the infinite, complex, and related aspects that represent a physical phenomenon. If the difficulty is already significant for a physical phenomenon, it will be even greater in the case of a biological phenomenon.

The need to select between relevant and non-relevant variables leads to discrimination between these variables. This choice is made thanks to ideas, knowledge, and the school that those who work on the model come from.

Random phenomena permeate everyday life and characterize various scientific fields, including mathematics and physics. The interpretation of these phenomena experienced a renaissance in the middle of the last century with the formulation of Monte Carlo methods. This milestone was achieved thanks to the intersection of research on the behavior of neurons in a chain reaction and the results achieved in the electronic field with the creation of the first computer. Today, simulation methods based on random number generation are widely used in physics.

One of the key points of quantum mechanics is to determine the energy spectrum of a system. This problem, albeit with some complications, can be resolved analytically in the case of very simple systems. In most cases, however, an analytical solution does not exist. Because of this, there's a need to develop and implement numerical methods capable of solving the differential equations that describe quantum systems. In recent decades, due to technological development and the enormous growth in computing power, we have been able to describe a wide range of phenomena with incredibly high precision.

Modeling fault diagnosis system

To meet the needs of industrial processes in the continuous search for ever higher performance, the control systems are gradually becoming more complex and sophisticated. Consequently, it is necessary to use special supervision, monitoring, and diagnosis devices within the control chain of malfunctions capable of guaranteeing the efficiency, reliability, and safety of the process in question.

Most industrial processes involve the use of diagnostic systems with a high degree of reliability. Examples of types of diagnostic systems are as follows:

- **Diagnostic systems based on hardware redundancy**: These are based on the use of additional/redundant hardware to replicate the signals of the components being monitored. The diagnosis mechanism provides for the analysis of the output signals to the redundant devices. If one of the signals deviates significantly from the others when following this analysis, then the component is classified as malfunctioning.

- **Diagnostic systems based on probability analysis**: This method is based on verifying the likelihood between the output signal generated by a component and the physical law that regulates its operation.

- **Diagnostic systems based on the analysis of signals**: This methodology assumes that when starting from the analysis of certain output signals of a process, it is possible to obtain information regarding any malfunctions.
- **Systems based on machine learning**: These methods provide for the identification and selection of functions and the classification of faults, allowing a systematic approach to fault diagnosis. They can be used in automated and unattended environments.

Modeling public transportation

In recent years, the analysis of issues related to vehicular traffic has taken on an increasingly important role in trying to develop well-functioning transport within cities and on roads in general. Today's transport systems need an optimization process that is coordinated with a development that offers concrete solutions to requests. Through better transportation planning, a process that produces fewer cars in the city and more parking opportunities should lead to a decrease in congestion.

A heavily slowed and congested urban flow can not only inconvenience motorists due to the increase in average travel, but can also make road circulation less safe and increase atmospheric and noise pollution.

Many causes have led to an increase in traffic, but certainly the most important is the strong increase in overall transport demand; this increase is due to factors of a different nature, such as a large diffusion of cars, a decentralization of industrial and city areas, and an often-lacking public transport service.

To try and solve the problem of urban mobility, we need to act with careful infrastructure management and with a traffic network planning program. An important planning tool could be a model of the traffic system, which essentially allows us to evaluate the effects that can be induced on it by interventions in the transport networks, while also allowing us to compare different design solutions. The use of a simulation tool allows us to evaluate some decisional or behavioral problems, quickly and economically, where it is not possible to carry out experiments on the real system. These simulation models represent a valid tool available to technicians and decision-makers in the transport sector for evaluating the effects of alternative design choices. These models allow for a detailed analysis of the solutions being planned at the local level.

There are simulation tools available that allow us to represent traffic accurately and specifically and map its evolution instant by instant, all while taking the geometric aspects of the infrastructure and the real behavior of the drivers into consideration, both of which are linked to the characteristics of the vehicle and the driver. Simulation models allow these to be represented on a small scale and, therefore, at a relatively low cost, as well as the effects and consequences related to the development of a new project. Micro-simulations provide a dynamic vision of the phenomenon since the characteristics of the motion of the individual vehicles (flow, density, and speed) are no longer considered, but real and variable, instant by instant during the simulation.

Modeling human behavior

The study of human behavior in the case of a fire, or cases of a general emergency, presents difficulties that cannot be easily overcome since many of the situations whose data it would be important to know about cannot be simulated in laboratory settings. Furthermore, the reliability of the data drawn from exercises in which there are no surprises or anxiety effects such as stress, as well as the possibility of panic that can occur in real situations, can be considered relative. Above all, the complexity of human behavior makes it difficult to predict the data that would be useful for fire safety purposes.

Studies conducted by scientists have shown that the behaviors of people during situations of danger and emergency are very different. In fact, research has shown that during an evacuation, people will often do things that are not related to escaping from the fire, and these things can constitute up to two-thirds of the time it takes an individual to leave the building. People often want to know what's happening before evacuating, as the alarm does not necessarily convey much information about the situation.

Having a simulation model capable of reproducing a dangerous situation is extremely useful for analyzing the reactions of people in such situations. In general terms, models that simulate evacuations address this problem in three different ways: optimization, simulation, and risk assessment.

The underlying principles of each of these approaches influence the characteristics of each model. Numerous models assume that occupants evacuate the building as efficiently as possible, ignoring secondary activities and those not strictly related to evacuation. The escape routes chosen during the evacuation are considered optimal, as are the characteristics of the flow of people toward the exits. The models that consider many people and that treat the occupants as a homogeneous whole, therefore without giving weight to the specific behavior of the individual, tend toward these aspects.

Next steps for simulation modeling

For most of human history, it has been reasonable to expect that when you die, the world will not be significantly different from when you were born. Over the past 300 years, this assumption has become increasingly outdated. This is because technological progress is continuously accelerating. Technological evolution translates into a next-generation product better than the previous one. This product is therefore a more efficient and effective way of developing the next stage of evolutionary progress. It is a positive feedback circuit. In other words, we are using more powerful and faster tools to design and build more powerful and faster tools. Consequently, the rate of progress of an evolutionary process increases exponentially over time, and the benefits such as speed, economy, and overall power also increase exponentially over time. As an evolutionary process becomes more effective and/or efficient, more resources are then used to encourage the progress of this process. This translates into the second level of exponential growth; that is, the exponential growth rate itself grows exponentially.

This technological evolution also affects the field of numerical simulation, which must be compared with the users' need to have more performant and simpler to make models. The development of a simulation model requires significant skills in model building, experimentation, and analysis. If we want to progress, we need to make significant improvements in the model-building process to meet the demands that come from the decision-making environment.

Increasing the computational power

Numerical simulation is performed by computers, so higher computational powers make the simulation process more effective. The evolution of computational power was governed by **Moore's law**, named after the Intel founder who predicted one of the most important laws regarding the evolution of computational power: every 18 months, the power generated by a chip doubles in computing capacity and halves in price.

When it comes to numerical simulation, computing power is everything. Today's hardware architectures are not so different from those of a few years ago. The only thing that has changed is the power of processing information. In the numerical simulation, information is processed: the more complex a situation becomes, the more the variables involved increase.

The growing processing capacity required for software execution and the increase in the amount of input data has always been satisfied by the evolution of **central processing units** (**CPUs**), according to Moore's law. However, lately, the growth of the computational capacity of CPUs has slowed down and the development of programming platforms has posed new performance requirements that create a strong discontinuity with respect to the hegemony of these CPUs with new hardware architectures in strong diffusion both on the server and on the device's side. In addition, the growing distribution of intelligent applications requires the development of specific architectures and hardware components for the various computing platforms.

Graphical processing units (**GPUs**) were created to perform heavy and complex calculations. They consist of a parallel architecture made up of thousands of small and efficient cores, designed for the simultaneous management of multiple operations.

Field programming gateway array (**FPGA**) architectures are integrated circuits designed to be configured after production based on specific customer requirements. FPGAs contain a series of programmable logic blocks and a hierarchy of reconfigurable interconnections that allow the blocks to be "wired together."

The advancement of hardware affects not only computing power but also storage capacity. We cannot send information at 1 GBps without having a physical place to contain it. We cannot train a simulation architecture without storing a dataset of several terabytes in size. Innovating means seeing opportunities that were not there previously by making use of components that constantly become more efficient. To innovate means to see what can be done by combining a more performant version of the three accelerators.

Machine-learning-based models

Machine learning is a field of computer science that allows computers to learn to perform a task without having to be explicitly programmed for its execution. Evolved from the studies on pattern recognition and theoretical computational learning in the field of artificial intelligence, machine learning explores the study and construction of algorithms that allow computers to learn information from available data and predict new information in light of what has been learned. These algorithms overcome the classic paradigm of strictly static instructions by building a model that automatically learns to predict new data from observations, finding its main use in computing problems where the design and implementation of ad hoc algorithms are not practicable or convenient.

Machine learning has deep links to the field of numerical simulation, which provides methods, theories, and domains of application. In fact, many machine learning problems are formulated as problems regarding minimizing a certain loss function against a specific set of examples (the training set). This function expresses the discrepancy between the values predicted by the model during training and the expected values for each example instance. The goal is to develop a model capable of correctly predicting the expected values in a set of instances never seen before, thus minimizing the loss function. This leads to a greater generalization of prediction skills.

The different machine learning tasks are typically classified into three broad categories, characterized by the type of feedback that the learning system is based on:

- **Supervised learning**: The sample inputs and the desired outputs are presented to the computer, with the aim of learning a general rule capable of mapping the inputs to the outputs.
- **Unsupervised learning**: The computer only provides input data, without any expected output, with the aim of learning some structure in the input data. Unsupervised learning can represent a goal or aim to extrapolate salient features of the data that are useful for executing another machine learning task.
- **Reinforcement learning**: The computer interacts with a dynamic environment in which it must achieve a certain goal. As the computer explores the domain of the problem, it is given feedback in terms of rewards or punishments in order to direct it toward the best solution.

The following diagram shows the different types of machine learning algorithms:

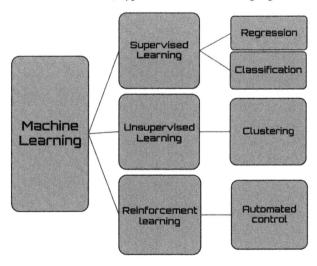

Figure 13.6 – The different types of machine learning algorithms

Automated generation of simulation models

Automated machine learning (**AutoML**) defines applications that can automate the end-to-end process of applying machine learning. Usually, technical experts must process data through a series of preliminary procedures before submitting it to machine learning algorithms. The steps necessary to perform a correct analysis of the data through these algorithms require specific skills that not everyone has. Although it is easy to create a model based on deep neural networks using different libraries, knowledge of the dynamics of these algorithms is required. In some cases, these skills are beyond those possessed by analysts, who must seek the support of industry experts to solve the problem.

AutoML has been developed to create an application that automates the entire machine learning process so that the user can take advantage of these services. Typically, machine learning experts should perform the following activities:

- Preparing the data
- Selecting features
- Selecting an appropriate model class
- Choosing and optimizing model hyperparameters
- Post-processing machine learning models
- Analyzing the results obtained

AutoML automates all these operations. It offers the advantage of producing simpler and faster-to-create solutions that often outperform hand-designed models. There are several AutoML frameworks available, each of which has characteristics that indicate its preferential use.

Summary

In this chapter, we summarized the technologies that we have exposed throughout this book. We have seen how to generate random numbers and have listed the most frequently used algorithms used to generate pseudo-random numbers. Then, we saw how to apply the Monte Carlo methods for numerical simulation based on the assumptions of two fundamental laws: the law of large numbers and the central limit theorem. We then went on to summarize the concepts that Markovian models are based on and then analyzed the various available resampling methods. After that, we explored the most used numerical optimization techniques and learned how to use artificial neural networks for numerical simulation.

Subsequently, we mentioned a series of fields in which numerical simulation is widely used and then looked at the next steps that will allow simulation models to evolve.

In this book, we studied various computational statistical simulations using Python. We started with the basics to understand various methods and techniques to deepen our knowledge of complex topics. At this point, the developers working with a simulation model would be able to put their knowledge to work, adopting a practical approach to the required implementation and associated methodologies so that they're operational and productive in no time. I hope that I have provided detailed explanations of some essential concepts using practical examples and self-assessment questions, exploring numerical simulation algorithms, and providing an overview of the relevant applications to help you make the best models for your needs.

Index

A

acceptance-rejection method 53, 54, 407
accuracy 7
activation functions 319-321
   hyperbolic tangent 320
   Rectified Linear Unit (ReLU) 320
   sigmoid 320
   unit step 320
activity 15
adjacency matrix 345
adjacent edges 344
agent 146, 167
   characteristics 146, 147, 167
agent-based simulation (ABM) 144, 145
   agents 144
   applications 144
   characteristics 145
   environment 144
   rules 144
Amazon stock price trend 290-295
American Standard Code for Information Interchange (ASCII) code 61
anomaly detection 384
Application programming interfaces (APIs) 210
a priori probability 71, 72
artificial neural networks (ANNs) 257, 314, 316, 412
   activation functions 319-321
   architecture 316, 317, 412
   biases 319
   layers 317, 318
   weights 318
artificial neural networks (ANNs), for simulating airfoil self-noise 324, 325
   data, scaling with sklearn 328, 329
   data, splitting 333-335
   data, splitting with pandas 325-328
   data, viewing with Matplotlib 330-333
   multilayer perceptron regressor model 337-339
   multiple linear regression 335-337
Automated machine learning (AutoML) 420

B

Bayes' theorem 73-76
behavior measures 406
bell curve 86
Bellman equation 165, 166
   principle of optimality 166
Bernoulli distribution 28
Bernoulli process 28

binary classification 383
binary cross-entropy
    as loss function 140, 141
binomial distribution 83-85
biological neural networks 314-316
Black-Scholes (BS) model 297, 298
bootstrap definition problem 188
bootstrap method
    versus Jackknife method 192
bootstrapping 187, 364, 411
bootstrapping regression
    applying 192-201
bootstrap resampling
    with Python 188-191
boxplot 330, 388
Brownian motion 27, 286
Broyden-Fletcher-Goldfarb-Shanno
    (LBFGS) algorithm 337

C

calibration 7
cannon pattern 275
cell 274, 275
cellular automata (CA) model
    exploring 273, 274
    game-of-life 275
    Wolfram code 276-280
cellular state 275
central limit theorem 115-119, 408
central processing units (CPUs) 418
certain events 70
chain 27
chi-squared test 43-46
chromosome 259, 260
coefficient of variation (CV)
    estimating 181, 182

compact space 216
complementary events 72
compound probability 74, 75
concrete quality prediction model 346-352
Conjugate Gradient (CG) 253
connected graphs 345
Constrained Optimization By Linear
    Approximation (COBYLA) 253
continuous distributions 76
continuous models 9
continuous stochastic process 26
convolutional neural network (CNN) 341
    convolutional layers 341
    fully connected layers 342
    loss layers 342
    pooling layers 341
    ReLU layers 342
cross-entropy 136, 137
    calculating, in Python 138-140
    iteration phases 137
crossover operation 261
cross-validation 411
    approaching 207
    k-fold cross-validation (k-fold CV) 209
    leave-one-out cross-validation
        (LOOCV) 208, 209
    validation set approach 208
    with Python 210-212
cryptoanalysis 63
cryptography 62
    randomness 63, 64
curse of dimensionality 165
cybersecurity 59

D

daily return 298

data augmentation 93
  brightness and contrast 94
  cropping 94
  distortion 94
  examples 98
  flipping 93
  performing 94-97
  rotation 94
  translation 94
data-based approach
  discovering 382
data collection 384
decision-making workflow 4, 5
deep learning 340
deep neural networks 340, 341
def clause 125
definite integral
  problem, analyzing 122
derivative 218
descent methods 217
deterministic models 8
digraph 344
direct edge 344
discounted cumulative reward 151, 152
discrete distributions 76
Discrete Event Simulation (DES) 14, 15
  activity 15
  entities 15
  events 15
  logical relationships 16
  resources 16
  simulation time 16
  state variables 15
discrete event system 14, 17
discrete models 9
discrete-time stochastic process 27
dog-leg trust-region (dogleg) 253
dynamic models 8

dynamic programming (DP) 165, 358
dynamic systems modeling 17

E

Elman network 342
encrypted/decrypted message
    generator 64-68
entities 15
environment 146
E-step 233
estimates, types
  more likely 370
  optimistic 370
  pessimistic 370
events 7, 15, 70
  certain events 70
  impossible events 70
  random event 70
evolutionary algorithms (EAs) 269
evolutionary computation 257
evolutionary strategies 258
Expectation-Maximization (EM) algorithm
  approaching 232, 234
  for Gaussian mixture 234-240
expected values 43
exploitation 152, 153
exploration 152, 153

F

fault diagnosis 380
  data-based approach 382
  knowledge-based approach 381
  model-based approach 381
fault diagnosis model
  for motor gearbox 385-396
  for unmanned aerial vehicle (UAV) 396-404

feature scaling 328
feedforward neural networks 321
feedforward propagation 321
field programming gateway
     array (FPGA) 418
Finite Mixture Models (FMMs) 234
finite-state machine (FSM) 16
frequency density 81
functions, generating real-
     value distributions
  betavariate (alpha, beta) 58
  expovariate (lambd) 58
  gammavariate (alpha, beta) 58
  gauss (mu, sigma) 58
  lognormvariate (mu, sigma) 59
  normalvariate (mu, sigma) 59
  paretovariate (alpha) 59
  vonmisesvariate (mu, kappa) 59
  weibullvariate (alpha, beta) 59
fuzzy logic (FL) 257

G

Gauss curve 86
Gaussian distribution 33, 86, 234, 235
Gaussian mixture model (GMM) 235
gearbox fault diagnosis
  reference link 385
general-purpose simulation
     system (GPSS) 11, 12
generation 274
genetic algorithm (GA) 258
  applying, for search and
     optimization 263-268
  basics 259, 260
  Bin Packing 258
  combinatorial optimization 258
  design 259
  image processing 259
  machine learning 259
genetic operators 260-263
genetic programming 258
genome 259
genotype 259
global optimality 216
global optimal solutions 215
global optimum value 215
gradient 217
gradient descent 411
  implementing 221-224
gradient descent algorithm 217-219
graphical processing units (GPUs) 418
graph neural networks (GNNs) 343, 345
graphs 343
graph theory 344

H

hash function algorithm 63
Hidden Markov Models (HMMs) 234
hyperbolic tangent 320
hypothesis test 98
  type I error 98
  type II error 98

I

impossible events 70
intelligent agents 146
inverse transform sampling
     method 52, 53, 407
iterative improvement algorithms 240, 241
iterative procedure 359

J

Jackknife method 179-181, 411
  coefficient of variation (CV), estimating 181, 182
  resampling, applying with Python 182-187
  versus bootstrap method 192

K

Keras 93
  data, generating with 93
k-fold cross-validation (k-fold CV) 209
k-fold validation 207
K-nearest neighbors (KNN) 395
knowledge-based approach
  describing 381

L

Lagged Fibonacci generator (LFG) 39-42, 406
lambda 125
latent variable models 234
law of large numbers 115, 408
lazy learner 395
Learmonth-Lewis generator 38
learning rate 220
leave-one-out cross-validation (LOOCV) 208, 209
likelihood 76
Limited-memory BFGS (L-BFGS) 253
linear congruential generator (LCG) 35-37, 406
linear regression analysis 192-201
local optimality 216
logical relationships 16
long short-term memory networks
  analyzing 343

M

machine learning 382, 418
  algorithms 419
  anomaly detection 384
  binary classification 383
  data collection 384
  multiclass classification 383
  problem definition 382
  regression 383
marginal likelihood 76
Markov chain applications 155
  one-dimensional random walk 156-159
  random walks 155
  weather forecasting model 160-165
Markov chains 153
  transition diagram 154, 155
  transition matrix 154
Markov decision processes (MDPs) 148-150
  addressing 408, 409
  agent-environment interaction 149, 150
  elements 357
  goal 150
  history of stochastic process (path) 149
  policy 150
  reward function 150
  state space 149
  state-value function 151
  summarizing 357, 358
Markov processes 146, 357
  agent-environment interface 146-148
  discounted cumulative reward 151, 152
mathematical model 268
matplotlib 221, 330
matyas function 249

Index

Maximum Likelihood Estimation (MLE) 232
MDPtoolbox 359, 360
  used, for addressing tiny forest management problems 363-366
Mealy Machine 16
mean 78
mean absolute error (MAE) 7
mean absolute percentage error (MAPE) 7
mean squared error (MSE) 7, 207, 336
Mersenne Twister algorithm 54
middle-square 34
min-max detection 126, 127
mixed models 234
mixture model 234
model-based approach 381
modeling 5
modern cryptography
  authentication 62
  confidentiality 63
  hash function algorithm 63
  integrity 63
  public key (asymmetric) algorithm 63
  secret key (symmetric) algorithm 63
modulo function 35
Monte Carlo method 115
  application 109, 110
  applying, for Pi estimation 110-114
  components 108
  for stock price prediction 290
Monte Carlo model 8
Monte Carlo simulation 31, 108, 407, 408
  applying 119, 298-302
  numerical optimization 120, 121
  probability distributions, generating 120
  project management 121
  used, for performing numerical integration 122
  used, for scheduling project time 369

Moore's law 418
Moore's Machine 16
motor gearbox
  fault diagnosis model, using 385-396
M-step 233
multi-agent simulation 167, 168
multi-agent system (MAS) 167
  advantages 168
multiclass classification 383
multi-factor analysis
  performing 356
multigraph 344
multilayer perceptron regressor model 337-339
multiple linear regression 335-337
multiplicative 36
multivariate optimization methods 247
mutation operator 262

N

National Association of Securities Dealers Quotation (NASDAQ) 305
  VaR, estimating for assets 305-311
neighbors 274
Nelder-Mead method 248-251
neural networks
  basics 314
  training 322, 323
Newton-Broyden-Fletcher-Goldfarb-Shanno (BFGS) 253
Newton-CG 253
Newton-Raphson method 225, 412
  applying 227-230
  approaching, for numerical optimization 226, 227
  secant method 231
  using, for root finding 225, 226

normal distribution 86-90
numerical integration, with Monte Carlo
    min-max detection 126, 127
    Monte Carlo method, applying 127-129
    numerical solution 124-126
    problem, defining 122-124
    visual representation 129-131
numerical optimization 120, 121
    techniques 214, 411
numpy 36, 221, 236

O

one-dimensional random walk (1D) 156
    simulating 157-159
optimal policy 151, 166, 358, 409
optimization 214
    methodologies 253
optimization problem 214, 215
optimization process 358, 359
oscillator pattern 275

P

parameters 7
partial derivative 218
password 59
password-based authentication
        systems 59, 60
pattern 262
percentage point function (ppf) 310
permutation tests 201, 202, 411
    performing 202-207
Pi estimation
    Monte Carlo method, applying for 110-114
Poisson process 30
policy 150, 409
policy evaluation 364

policy improvement 165, 358, 364
possible interactions
    cooperation 168
    coordination 168
    negotiation 168
posterior probability 76
Powell's conjugate direction
        algorithm 251-253
power analysis 99
    metrics 100
    performing 100-103
    simulation 98
    statistical test, power of 98, 99
    uses 99
predator-prey model 21, 22
principle of optimality 166
prior probability 76
probability
    a priori probability 71, 72
    calculating 71
    complementary events 72
    concepts 70
    definition, with example 71
    events 70
    relative frequency 72, 73
probability density functions
        (PDFs) 77, 78, 108
probability distributions 31, 32, 76
    binomial distribution 83-86
    continuous distributions 76
    discrete distributions 76
    Gaussian distribution 33
    generating 120
    mean 78
    normal distribution 86-90
    probability density function (PDF) 77, 78
    uniform distribution 32, 79-83
    variance 79

probability function 77
probability sampling 179
program structure
　software levels 22
project 369
project management 121, 356, 369
　what-if analysis 356
project manager 369
project time, with Monte Carlo simulation
　scheduling 369
　scheduling grid, defining 369, 370
　task's time, estimating 370
　triangular distribution, exploring 374-377
pseudorandom 31
pseudorandom number generator (PRNG) 34
pseudo-random sequences 406, 407
public key (asymmetric) algorithm 63
Python Imaging Library 95
Python Package Index (pip) 360
Python Schelling model 169-174

R

random.choice() function 57, 58
random distributions, generic methods 51
　acceptance-rejection method 53, 54
　inverse transform sampling method 52, 53
random event 70
random module 54
random number generation
　with Python 54
random number generation algorithms 34
random number generator
　cons 34
　pros 34

random numbers
　generating 406
　properties 33
　with uniform distribution 37-39
random number simulation 31
random one-dimensional walking 29
random password generator 60-62
random.randint() function 56, 57
random.random() function 54, 55
random.sample() function 58
random.seed() function 55, 56
random.uniform() function 56
random walk 28, 29, 155, 156
　Wiener process, addressing as 287, 288
rational agent 167
real system
　simulation 23
real-value distributions
　generating 58
Rectified Linear Unit (ReLU) 320
recurrent neural networks
　examining 342
regression 383
reinforcement learning 419
resampling methods 176, 177
　analyzing 410
resources 16
reward function 150
risk models
　for portfolio management 302
root finding
　Newton-Raphson algorithm,
　　using for 225, 226

S

sampling
  concepts overview 177
  disadvantages 178
  pros 178
  reasoning 178
  working 179
sampling theory 177
SC 256
Schelling's model 168
  in Python 169-174
Schelling's model of segregation 169
SciPy 247, 292
  optimize module 248
secant method 231
secret key (symmetric) algorithm 63
security
  randomness requirements 59
seed 35
sensitivity 7
sensitivity analysis
  direct methods 133
  exploring 131, 132
  global approach 133
  local approach 132
  variance-based sensitivity analysis 133
  with slopes 133
  working 133-135
Shannon entropy 136
sigmoid function 320
simple harmonic oscillator 19-21
Simulated Annealing (SA) 240
  implementing 241-247
simulation 3-5, 13, 23
  end phase 23
  loop phase 22
  start phase 22

simulation-based problem, approaching 9
  data collection 10
  problem analysis 10
  simulation and analysis, of results 13
  simulation model, setting up 10, 11
  simulation model, validating 13
  simulation software selection 11, 12
  software solution, verifying 12, 13
simulation languages 11
simulation modeling
  accuracy 7
  calibration 7
  cons 6
  event 7
  parameters 7
  pros 5, 6
  sensitivity 7
  state variables 6
  system 6
  validation 7
  with neural network techniques 346
simulation modeling, next steps 417
  automated generation 420
  computational power, increasing 418
  machine-learning-based models 418, 419
simulation models 3, 405
  classifying 8
  continuous models 9
  deterministic models 8
  discrete models 9
  dynamic models 8
  static models 8
  stochastic models 9
simulation models, examples
  fault diagnosis system 415, 416
  financial applications 414
  healthcare sector 413, 414
  human behavior 417

physical phenomenon 414, 415
public transportation 416
simulation time 16
 progress, defining ways 16
simulators 11
space of states 26
spaceship pattern 275
standard Brownian motion
 defining 286, 287
 implementing 288, 289
standardized normal distribution 90
state transition graph (STG) 17
state transition table (STT) 17
state-value function 151
state variables 6, 15
static models 8
statistical sampling 177
statistical test 187
 factors, affecting power 99
 power 98, 99
still life pattern 275
stochastic gradient descent (SGD) 231, 232, 338, 412
stochastic models 9
stochastic process 26
 examples 27
stochastic sequence 27
stock price trend
 handling, as time series 295-297
supervised learning 419
support vector machines (SVMs) 328
symbolic regression (SR)
 performing 268-273
synthetic data
 generating 90
 generation methods 91
 real data, versus artificial data 90, 91

semi-structured data, generating 92
structured data, generating 92
system 5, 6, 8

T

time series
 stock price trend, handling as 295-297
tiny forest environment
 states 360
tiny forest management example
 defining 360-363
 probability, changing of fire starting 366-368
tiny forest management problems
 addressing, with MDPtoolbox 363-366
 managing 357
transition diagram 154, 155
transition matrix 154
trial and error method 220, 221
triangular distribution 374-377
truncated Newton's method 253

U

undirected edges 344
uniform distribution 32, 79-82
 testing 42
uniformity test 46-51
unit step activation function 320
unmanned aerial vehicle (UAV) 396
 fault diagnosis system, using 396-404
unsupervised learning 419

V

validation 7
validation set approach 208

Value at Risk (VaR) metric 303, 304, 414
   characteristics 304
   estimating, for NASDAQ assets 305-311
variance 79
   using, as risk measure 302
variance-based sensitivity analysis 133
vertex 345

W

weather forecast
   simulating 160-165
what-if analysis 356
Wiener process 286
   addressing, as random walk 287, 288
Wolfram code
   for CA 276-280
workshop machinery
   managing 17-19

Packt.com

Subscribe to our online digital library for full access to over 7,000 books and videos, as well as industry leading tools to help you plan your personal development and advance your career. For more information, please visit our website.

Why subscribe?

- Spend less time learning and more time coding with practical eBooks and Videos from over 4,000 industry professionals
- Improve your learning with Skill Plans built especially for you
- Get a free eBook or video every month
- Fully searchable for easy access to vital information
- Copy and paste, print, and bookmark content

Did you know that Packt offers eBook versions of every book published, with PDF and ePub files available? You can upgrade to the eBook version at packt.com and as a print book customer, you are entitled to a discount on the eBook copy. Get in touch with us at customercare@packtpub.com for more details.

At www.packt.com, you can also read a collection of free technical articles, sign up for a range of free newsletters, and receive exclusive discounts and offers on Packt books and eBooks.

Other Books You May Enjoy

If you enjoyed this book, you may be interested in these other books by Packt:

Production-Ready Applied Deep Learning

Tomasz Palczewski, Jaejun (Brandon) Lee, Lenin Mookiah

ISBN: 9781803243665

- Understand how to develop a deep learning model using PyTorch and TensorFlow
- Convert a proof-of-concept model into a production-ready application
- Discover how to set up a deep learning pipeline in an efficient way using AWS
- Explore different ways to compress a model for various deployment requirements
- Develop Android and iOS applications that run deep learning on mobile devices
- Monitor a system with a deep learning model in production
- Choose the right system architecture for developing and deploying a model

Data Cleaning and Exploration with Machine Learning

Michael Walker

ISBN: 9781803241678

- Explore essential data cleaning and exploration techniques to be used before running the most popular machine learning algorithms
- Understand how to perform preprocessing and feature selection, and how to set up the data for testing and validation
- Model continuous targets with supervised learning algorithms
- Model binary and multiclass targets with supervised learning algorithms
- Execute clustering and dimension reduction with unsupervised learning algorithms
- Understand how to use regression trees to model a continuous target

Packt is searching for authors like you

If you're interested in becoming an author for Packt, please visit `authors.packtpub.com` and apply today. We have worked with thousands of developers and tech professionals, just like you, to help them share their insight with the global tech community. You can make a general application, apply for a specific hot topic that we are recruiting an author for, or submit your own idea.

Share Your Thoughts

Now you've finished *Hands-On Simulation Modeling with Python, Second Edition*, we'd love to hear your thoughts! Scan the QR code below to go straight to the Amazon review page for this book and share your feedback or leave a review on the site that you purchased it from.

`https://packt.link/r/1-804-61688-5`

Your review is important to us and the tech community and will help us make sure we're delivering excellent quality content.

Download a free PDF copy of this book

Thanks for purchasing this book!

Do you like to read on the go but are unable to carry your print books everywhere? Is your eBook purchase not compatible with the device of your choice?

Don't worry, now with every Packt book you get a DRM-free PDF version of that book at no cost.

Read anywhere, any place, on any device. Search, copy, and paste code from your favorite technical books directly into your application.

The perks don't stop there, you can get exclusive access to discounts, newsletters, and great free content in your inbox daily

Follow these simple steps to get the benefits:

1. Scan the QR code or visit the link below

https://packt.link/free-ebook/9781804616888

2. Submit your proof of purchase
3. That's it! We'll send your free PDF and other benefits to your email directly

Printed in Great Britain
by Amazon